PLANE AND SPHERICAL TRIGONOMETRY

"AND FOUR-PLACE TABLES OF LOGARITHMS"

By

WILLIAM ANTHONY GRANVILLE [PH. D.]

SHEFFIELD SCIENTIFIC SCHOOL, YALE UNIVERSITY

First Publication:

ILLUSTRATED &

PUBLISHED BY

E-KİTAP PROJESİ & CHEAPEST BOOKS

www.cheapestboooks.com

Copyright, 2015 by e-Kitap Projesi

Istanbul

ISBN:

978-1-505-9452-25

PREFACE

It has been the author's aim to treat the subject according to the latest and most approved methods. The book is designed for the use of colleges, technical schools, normal schools, secondary schools, and for those who take up the subject without the aid of a teacher. Special attention has been paid to the requirements of the College Entrance Board. The book contains more material than is required for some first courses in Trigonometry, but the matter has been so arranged that the teacher can make such omissions as will suit his particular needs.

The trigonometric functions are defined as ratios; first for acute angles in right triangles, and then these definitions are extended to angles in general by means of coordinates. The student is first taught to use the natural functions of acute angles in the solution of simple problems involving right triangles. Attention is called to the methods shown in §§ 23–29 for the reduction of functions of angles outside of the first quadrant. In general, the first examples given under each topic are worked out, making use of the natural functions. A large number of carefully graded exercises are given, and the processes involved are summarized into working rules wherever practicable. Illustrative examples are worked out in detail under each topic.

Logarithms are introduced as a separate topic, and attention is called to the fact that they serve to minimize the labor of computation. Granville's *Four-Place Tables of Logarithms* is used. While no radical changes in the usual arrangement of logarithmic tables have been made, several improvements have been effected which greatly facilitate logarithmic computations. Particularly important is the fact that the degree of accuracy which may be expected in a result found by the aid of these tables is clearly indicated. Under each case in the solution of triangles are given two complete sets of examples, — one in which the angles are expressed in degrees and minutes, and another in which the angles are expressed in degrees and the decimal part of a degree. This arrangement, which is characteristic of this book, should be of great advantage to those secondary schools in which college preparation involving both systems is necessary.

To facilitate the drawing of figures and the graphical checking of results a combined ruler and protractor of celluloid is furnished with each copy of the book, and will be found on the inside of the back cover.

In Spherical Trigonometry some simplifications have been introduced in the application of Napier's rule of circular parts to the solution of right spherical triangles. The treatment of oblique spherical triangles is unique. By making use of the Principle of Duality nearly one half of the work usually required in deriving the standard formulas is done away with, and the usual six cases in the solution of oblique spherical triangles have been reduced to three. An attempt has been made to treat the most important applications of Spherical Trigonometry to Geodesy, Astronomy, and Navigation with more clearness and simplicity than has been the rule in elementary treatises.

The author's acknowledgments are due to Professor John C. Tracy for many valuable suggestions in the treatment of Spherical Trigonometry, to Messrs. L. E. Armstrong and C. C. Perkins for verifying the answers to the problems, and to Mr. S. J. Berard for drawing the figures.

W. A. GRANVILLE

SHEFFIELD SCIENTIFIC SCHOOL
    YALE UNIVERSITY

CONTENTS

PLANE TRIGONOMETRY

CHAPTER I

TRIGONOMETRIC FUNCTIONS OF ACUTE ANGLES. SOLUTION OF RIGHT TRIANGLES

CHAPTER II

TRIGONOMETRIC FUNCTIONS OF ANY ANGLE

CHAPTER III

RELATIONS BETWEEN THE TRIGONOMETRIC FUNCTIONS

CHAPTER IV

TRIGONOMETRIC ANALYSIS

CHAPTER V

GENERAL VALUES OF ANGLES. INVERSE TRIGONOMETRIC FUNCTIONS. TRIGONOMETRIC EQUATIONS

CHAPTER VI

GRAPHICAL REPRESENTATION OF TRIGONOMETRIC FUNCTIONS

CHAPTER VII

SOLUTION OF OBLIQUE TRIANGLES

CHAPTER VIII

THEORY AND USE OF LOGARITHMS

CHAPTER IX

ACUTE ANGLES NEAR 0° OR 90°

CHAPTER X

RECAPITULATION OF FORMULAS

SPHERICAL TRIGONOMETRY

CHAPTER I

RIGHT SPHERICAL TRIANGLES

CHAPTER II

OBLIQUE SPHERICAL TRIANGLES

CHAPTER III

APPLICATIONS OF SPHERICAL TRIGONOMETRY TO THE CELESTIAL AND TERRESTRIAL SPHERES

CHAPTER IV

RECAPITULATION OF FORMULAS

PLANE TRIGONOMETRY

CHAPTER I

TRIGONOMETRIC FUNCTIONS OF ACUTE ANGLES
SOLUTION OF RIGHT TRIANGLES

1. Trigonometric functions of an acute angle defined. We shall assume that the student is familiar with the notion of the *angle between two lines* as presented in elementary Plane Geometry. For the present we will confine ourselves to the consideration of acute angles.

Let EAD be an angle less than 90°, that is, an acute angle. From B, any point in one of the sides of the angle, draw a perpendicular to the other side, thus forming a right triangle, as ABC. Let the capital letters A, B, C denote the angles and the small letters a, b, c the lengths of the corresponding opposite sides in the right triangle.* We know in a general way from Geometry that the sides and angles of this triangle are mutually dependent. Trigonometry begins by showing the exact nature of this dependence, and for this purpose employs the *ratios of the sides*. These ratios are called *trigonometric functions*. The six trigonometric functions of any acute angle, as A, are denoted as follows:

$$\begin{array}{lll}
\sin A, & \text{read} & \text{'sine of } A\text{''}; \\
\cos A, & \text{read} & \text{'cosine of } A\text{''}; \\
\tan A, & \text{read} & \text{'tangent of } A\text{''}; \\
\csc A, & \text{read} & \text{'cosecant of } A\text{''}; \\
\sec A, & \text{read} & \text{''secant of } A\text{''}; \\
\cot A, & \text{read} & \text{'cotangent of } A\text{.''}
\end{array}$$

---

* Unless otherwise stated the hypotenuse of a right triangle will always be denoted by c and the right angle by C.

These trigonometric functions (ratios) are defined as follows (see figure):

$$(1)\ \sin A = \frac{\text{opposite side}}{\text{hypotenuse}} \left(= \frac{a}{c}\right); \qquad (4)\ \csc A = \frac{\text{hypotenuse}}{\text{opposite side}} \left(= \frac{c}{a}\right);$$

$$(2)\ \cos A = \frac{\text{adjacent side}}{\text{hypotenuse}} \left(= \frac{b}{c}\right); \qquad (5)\ \sec A = \frac{\text{hypotenuse}}{\text{adjacent side}} \left(= \frac{c}{b}\right);$$

$$(3)\ \tan A = \frac{\text{opposite side}}{\text{adjacent side}} \left(= \frac{a}{b}\right); \qquad (6)\ \cot A = \frac{\text{adjacent side}}{\text{opposite side}} \left(= \frac{b}{a}\right).$$

The essential fact that the numerical value of any one of these functions depends upon the *magnitude* only of the angle A, that is, is independent of the point B from which the perpendicular upon the other side is let fall, is easily established.*

These functions (ratios) are of fundamental importance in the study of Trigonometry. In fact, no progress in the subject is possible without a thorough knowledge of the above six definitions. They are easy to memorize if the student will notice that the three in the first column are reciprocals respectively of those directly opposite in the second column. For,

$$\sin A = \frac{a}{c} = \frac{1}{\frac{c}{a}} = \frac{1}{\csc A}; \qquad \csc A = \frac{c}{a} = \frac{1}{\frac{a}{c}} = \frac{1}{\sin A};$$

$$\cos A = \frac{b}{c} = \frac{1}{\frac{c}{b}} = \frac{1}{\sec A}; \qquad \sec A = \frac{c}{b} = \frac{1}{\frac{b}{c}} = \frac{1}{\cos A};$$

$$\tan A = \frac{a}{b} = \frac{1}{\frac{b}{a}} = \frac{1}{\cot A}; \qquad \cot A = \frac{b}{a} = \frac{1}{\frac{a}{b}} = \frac{1}{\tan A}.$$

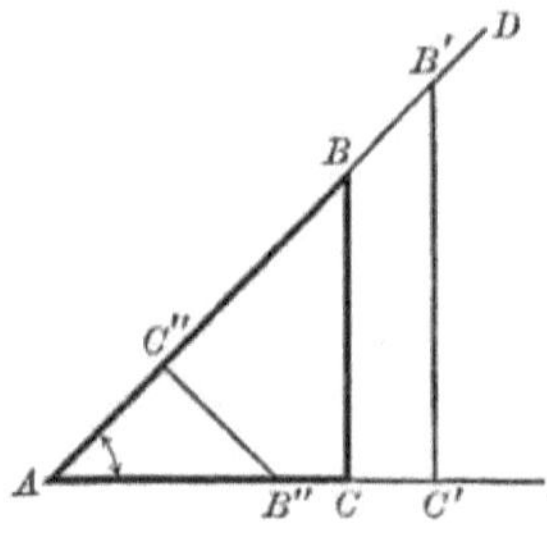

* For, let B' be any other point in AD, and B'' any point in AE. Draw the perpendiculars $B'C'$ and $B''C''$ to AE and AD respectively. The three triangles ABC, $AB'C'$, $AB''C''$, are mutually equiangular since they are right-angled and have a common angle at A. Therefore they are similar, and we have

$$\frac{BC}{AB} = \frac{B'C'}{AB'} = \frac{B''C''}{AB''}.$$

But each of these ratios defines the *sine of* A. In the same manner we may prove this property for each of the other functions. This shows that the size of the right triangle we choose is immaterial; it is only the relative and not the actual lengths of the sides of the triangle that are of importance.

The student should also note that every one of these six ratios will change in value when the angle A changes in size.

If we apply the definitions (1) to (6) inclusive to the acute angle B, there results

$$\sin B = \frac{b}{c}; \qquad\qquad \csc B = \frac{c}{b};$$

$$\cos B = \frac{a}{c}; \qquad\qquad \sec B = \frac{c}{a};$$

$$\tan B = \frac{b}{a}; \qquad\qquad \cot B = \frac{a}{b}.$$

Comparing these with the functions of the angle A, we see that

$$\sin A = \cos B; \qquad\qquad \csc A = \sec B;$$

$$\cos A = \sin B; \qquad\qquad \sec A = \csc B;$$

$$\tan A = \cot B; \qquad\qquad \cot A = \tan B.$$

Since $A + B = 90°$ (i.e. A and B are complementary) the above results may be stated in compact form as follows:

Theorem. *A function of an acute angle is equal to the co-function* * *of its complementary acute angle.*

Ex. 1. Calculate the functions of the angle A in the right triangle where $a = 3$, $b = 4$.

Solution. $c = \sqrt{a^2 + b^2} = \sqrt{9 + 16} = \sqrt{25} = 5.$

Applying (1) to (6) inclusive (p. 2),

$$\sin A = \tfrac{3}{5}; \qquad \csc A = \tfrac{5}{3};$$

$$\cos A = \tfrac{4}{5}; \qquad \sec A = \tfrac{5}{4};$$

$$\tan A = \tfrac{3}{4}; \qquad \cot A = \tfrac{4}{3}.$$

Also find all functions of the angle B, and compare results.

Ex. 2. Calculate the functions of the angle B in the right triangle where $a = 3$, $c = 4$.

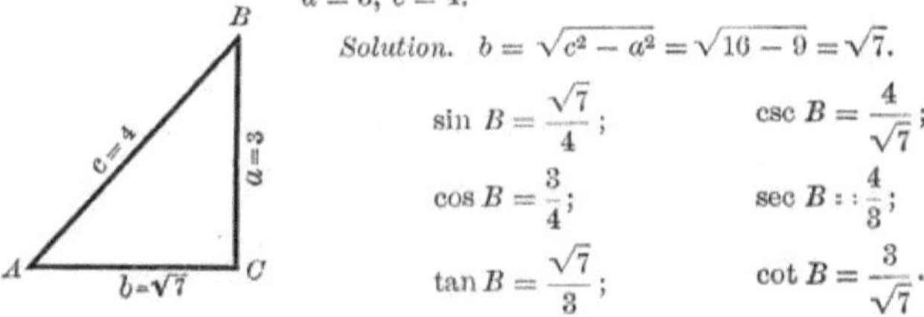

Solution. $b = \sqrt{c^2 - a^2} = \sqrt{16 - 9} = \sqrt{7}.$

$$\sin B = \frac{\sqrt{7}}{4}; \qquad\qquad \csc B = \frac{4}{\sqrt{7}};$$

$$\cos B = \frac{3}{4}; \qquad\qquad \sec B = \frac{4}{3};$$

$$\tan B = \frac{\sqrt{7}}{3}; \qquad\qquad \cot B = \frac{3}{\sqrt{7}}.$$

Also find all functions of the angle A, and compare results.

---

* Sine and cosine are called co-functions of each other. Similarly tangent and cotangent, also secant and cosecant, are co-functions.

Ex. 3. Calculate the functions of the angle A in the right triangle where $a = 2\,mn$, $b = m^2 - n^2$.

Solution.

$$c = \sqrt{a^2 + b^2} = \sqrt{4\,m^2n^2 + m^4 - 2\,m^2n^2 + n^4}$$
$$= \sqrt{m^4 + 2\,m^2n^2 + n^4} = m^2 + n^2.$$

$$\sin A = \frac{2\,mn}{m^2 + n^2}; \qquad \csc A = \frac{m^2 + n^2}{2\,mn};$$

$$\cos A = \frac{m^2 - n^2}{m^2 + n^2}; \qquad \sec A = \frac{m^2 + n^2}{m^2 - n^2};$$

$$\tan A = \frac{2\,mn}{m^2 - n^2}; \qquad \cot A = \frac{m^2 - n^2}{2\,mn}.$$

Ex. 4. In a right triangle we have given $\sin A = \tfrac{4}{5}$ and $a = 80$; find c.

Solution. From (1), p. 2, we have the formula

$$\sin A = \frac{a}{c}.$$

Substituting the values of $\sin A$ and a that are given, there results

$$\frac{4}{5} = \frac{80}{c};$$

and solving, $\qquad\qquad\qquad c = 100.$ *Ans.*

2. Functions of 45°, 30°, 60°. These angles occur very frequently in problems that are usually solved by trigonometric methods. It is therefore important to find the values of the trigonometric functions of these angles and to memorize the results.

(a) To find the functions of 45°.

Draw an isosceles right triangle, as ABC. This makes

$$\text{angle } A = \text{angle } B = 45°.$$

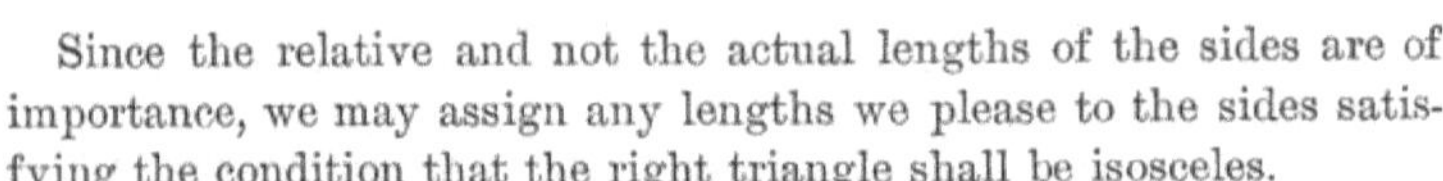

Since the relative and not the actual lengths of the sides are of importance, we may assign any lengths we please to the sides satisfying the condition that the right triangle shall be isosceles.

Let us choose the lengths of the short sides as unity, i.e. let $a = 1$ and $b = 1$.

Then $c = \sqrt{a^2 + b^2} = \sqrt{2}$, and we get

$$\sin 45° = \frac{1}{\sqrt{2}}; \qquad\qquad \csc 45° = \sqrt{2};$$

$$\cos 45° = \frac{1}{\sqrt{2}}; \qquad\qquad \sec 45° = \sqrt{2};$$

$$\tan 45° = 1; \qquad\qquad \cot 45° = 1.$$

(b) *To find the functions of 30° and 60°.*

Draw an equilateral triangle, as ABD. Drop the perpendicular BC from B to AD, and consider the triangle ABC, where

$$\text{angle } A = 60° \text{ and angle } ABC = 30°.$$

Again take the smallest side as unity, i.e. let $b = 1$. This makes

$$c = AB = AD = 2\,AC = 2\,b = 2,$$

and $a = \sqrt{c^2 - b^2} = \sqrt{4 - 1} = \sqrt{3}.$ Therefore

$$\sin 60° = \frac{\sqrt{3}}{2}; \qquad \csc 60° = \frac{2}{\sqrt{3}};$$

$$\cos 60° = \frac{1}{2}; \qquad \sec 60° = 2;$$

$$\tan 60° = \sqrt{3}; \qquad \cot 60° = \frac{1}{\sqrt{3}}.$$

Similarly, from the same triangle,

$$\sin 30° = \frac{1}{2}; \qquad \csc 30° = 2;$$

$$\cos 30° = \frac{\sqrt{3}}{2}; \qquad \sec 30° = \frac{2}{\sqrt{3}};$$

$$\tan 30° = \frac{1}{\sqrt{3}}; \qquad \cot 30° = \sqrt{3}.$$

Writing the more important of these results in tabulated form,* we have

ANGLE	30°	45°	60°
sin	$\frac{1}{2} = .50$	$\frac{1}{\sqrt{2}} = .71+$	$\frac{\sqrt{3}}{2} = .86+$
cos	$\frac{\sqrt{3}}{2} = .86+$	$\frac{1}{\sqrt{2}} = .71+$	$\frac{1}{2} = .50$
tan	$\frac{1}{\sqrt{3}} = .57+$	1	$\sqrt{3} = 1.73+$

The cosecant, secant, and cotangent are easily remembered as being the reciprocals of the sine, cosine, and tangent respectively.

---

* To aid the memory we observe that the numbers in the first (or sine) row are respectively $\sqrt{1}, \sqrt{2}, \sqrt{3}$; each divided by 2.

The second (or cosine) row is formed by reversing the order in the first row.

The last (or tangent) row is formed by dividing the numbers in the first row by the respective numbers in the second row.

The student should become very familiar with the 45° right triangle and the 30°, 60° right triangle. Instead of memorizing the above table we may then get the values of the functions directly from a mental picture of these right triangles.

Ex. 5. Given a right triangle where $A = 60°$, $a = 100$; find c.

Solution. Since we know A (and therefore also any function of A), and the sine of A involves a, which is known, and c, which is wanted, we can find c by using the formula

$$\sin A = \frac{a}{c}. \qquad \text{by (1), p. 2}$$

Substituting $a = 100$, and $\sin A = \sin 60° = \dfrac{\sqrt{3}}{2}$ from the above table, we have

$$\frac{\sqrt{3}}{2} = \frac{100}{c}.$$

Clearing of fractions and solving for c, we get

$$c = \frac{200}{\sqrt{3}} = \frac{200}{1.7+} = 117.6+. \quad Ans.$$

What is the value of B? Following the method illustrated above, show that $b = 58.8+$.

EXAMPLES

Only right triangles are referred to in the following examples.

1. Calculate all the functions of the angle A, having given $a = 8$, $b = 15$.

$$Ans. \quad \sin A = \tfrac{8}{17}, \ \cos A = \tfrac{15}{17}, \ \tan A = \tfrac{8}{15}, \text{ etc.}$$

2. Calculate the functions of the angle B, having given $a = 5$, $c = 7$.

$$Ans. \quad \sin B = \frac{\sqrt{24}}{7}, \ \cos B = \frac{5}{7}, \ \tan B = \frac{\sqrt{24}}{5}, \text{ etc.}$$

3. Calculate the functions of the angle A, having given $b = 2$, $c = \sqrt{11}$.

$$Ans. \quad \sqrt{\frac{7}{11}}, \ \frac{2}{\sqrt{11}}, \ \frac{\sqrt{7}}{2}, \text{ etc.}$$

4. Calculate the functions of the angle B, having given $a = 40$, $c = 41$.

$$Ans. \quad \tfrac{9}{41}, \ \tfrac{40}{41}, \ \tfrac{9}{40}, \text{ etc.}$$

5. Calculate the functions of the angle A, having given $a = p$, $b = q$.

$$Ans. \quad \frac{p}{\sqrt{p^2 + q^2}}, \ \frac{q}{\sqrt{p^2 + q^2}}, \ \frac{p}{q}, \text{ etc.}$$

6. Calculate the functions of the angle A, having given $a = \sqrt{m^2 + mn}$, $c = m + n$.

$$Ans. \quad \frac{\sqrt{m^2 + mn}}{m + n}, \ \frac{\sqrt{mn + n^2}}{m + n}, \ \sqrt{\frac{m}{n}}, \text{ etc.}$$

7. Calculate the functions of the angle B, having given $a = \sqrt{m^2 + n^2}$, $c = m + n$.

$$Ans. \quad \frac{\sqrt{2mn}}{m + n}, \ \frac{\sqrt{m^2 + n^2}}{m + n}, \ \sqrt{\frac{2mn}{m^2 + n^2}}, \text{ etc.}$$

8. Given $\sin A = \frac{3}{5}$, $c = 200.5$; calculate a.   *Ans.* 120.3.

9. Given $\cos A = .44$, $c = 30.5$; calculate b.   *Ans.* 13.42.

10. Given $\tan A = \frac{11}{3}$, $b - \frac{27}{11}$; calculate c.   *Ans.* $\frac{9}{11}\sqrt{130}$.

11. Given $A - 30°$, $a = 25$; calculate c. Also find B and b.
     Ans. $c = 50$, $B = 60°$, $b = 25\sqrt{3}$.

12. Given $B = 30°$, $c = 48$; calculate b. Also find A and a.
     Ans. $b = 24$, $A = 60°$, $a = 24\sqrt{3}$.

13. Given $B = 45°$, $b = 20$; calculate c. Also find A and a.
     Ans. $c = 20\sqrt{2}$, $A = 45°$, $a = 20$.

3. Solution of right triangles. A triangle is composed of six parts, three sides and three angles. To solve a triangle is to find the parts not given. A triangle can be solved if three parts, at least one of which is a side, are given.* A right triangle has one angle, the right angle, always given; hence a right triangle can be solved if two sides, or one side and an acute angle, are given. One of the most important applications of Trigonometry † is the solution of triangles, and we shall now take up the *solution of right triangles.*

The student may have noticed that Examples 11, 12, 13, of the last section were really problems on solving right triangles.

When beginning the study of Trigonometry it is important that the student should draw the figures connected with the problems as accurately as possible. This not only leads to a better understanding of the problems themselves, but also gives a clearer insight into the meaning of the trigonometric functions and makes it possible to test roughly the accuracy of the results obtained. For this purpose the only instruments necessary are a graduated ruler and a protractor. A protractor is an instrument for measuring angles. On the inside of the back cover of this book will be found a Granville's Transparent Combined Ruler and Protractor, with directions for use. The ruler is graduated to inches and centimeters and the protractor to degrees. The student is advised to make free use of this instrument.

4. General directions for solving right triangles.

First step. *Draw a figure as accurately as possible representing the triangle in question.*

Second step. *When one acute angle is known, subtract it from 90° to get the other acute angle.*

---

* It is assumed that the given conditions are consistent, that is, that it is possible to construct the triangle from the given parts.

† The name Trigonometry is derived from two Greek words which taken together mean "I measure a triangle."

Third step. *To find an unknown part, select from (1) to (6), p. 2, a formula involving the unknown part and two known parts, and then solve for the unknown part.*

Fourth step. *Check the values found by noting whether they satisfy relations different from those already employed in the third step. A convenient numerical check is the relation,*

$$a^2 = c^2 - b^2 = (c + b)(c - b).$$

Large errors may be detected by measurement.

Since the two perpendicular sides of a right triangle may be taken as base and altitude, we have at once

$$\textbf{Area of a right triangle} = \frac{ab}{2}.$$

In the last section the functions 30°, 45°, 60°, were found. In more advanced treatises it is shown how to calculate the functions of angles in general.

We will anticipate some of these results by making use of the following table where the values[*] of the trigonometric functions for each degree from 0° to 90° inclusive are correctly given to four or five significant figures.

In looking up the function of an angle between 0° and 45° inclusive, we look for the angle in the extreme left-hand vertical column. The required value of the function will be found on the same horizontal line with the angle, and in the vertical column having that function for a caption at the top. Thus,

$$\sin 15° = .2588,$$
$$\cot 41° = 1.1504, \text{ etc.}$$

Similarly, when looking up the function of an angle between 45° and 90° inclusive we look in the extreme right-hand vertical column. The required value of the function will be found on the same horizontal line with the angle as before, but in the vertical column having that function for a caption at the bottom. Thus,

$$\cos 64° = .4384,$$
$$\sec 85° = 11.474, \text{ etc.}$$

When we have given the numerical value of the function of an angle, and wish to find the angle itself, we look for the given number in the columns having the given function as a caption at the top

---

[*] Also called the *natural values* of the trigonometric functions in contradistinction to their logarithms (see Tables II and III of Granville's *Four-Place Tables of Logarithms*).

TABLE A

NATURAL VALUES OF THE TRIGONOMETRIC FUNCTIONS

Angle	sin	cos	tan	cot	sec	csc	
0°	.0000	1.0000	.0000	∞	1.0000	∞	90°
1°	.0175	.9998	.0175	57.290	1.0002	57.299	89°
2°	.0349	.9994	.0349	28.636	1.0006	28.654	88°
3°	.0523	.9986	.0524	19.081	1.0014	19.107	87°
4°	.0698	.9976	.0699	14.301	1.0024	14.336	86°
5°	.0872	.9962	.0875	11.430	1.0038	11.474	85°
6°	.1045	.9945	.1051	9.5144	1.0055	9.5668	84°
7°	.1219	.9925	.1228	8.1443	1.0075	8.2055	83°
8°	.1392	.9903	.1405	7.1154	1.0098	7.1853	82°
9°	.1564	.9877	.1584	6.3138	1.0125	6.3925	81°
10°	.1736	.9848	.1763	5.6713	1.0154	5.7588	80°
11°	.1908	.9816	.1944	5.1446	1.0187	5.2408	79°
12°	.2079	.9781	.2126	4.7046	1.0223	4.8097	78°
13°	.2250	.9744	.2309	4.3315	1.0263	4.4454	77°
14°	.2419	.9703	.2493	4.0108	1.0306	4.1336	76°
15°	.2588	.9659	.2679	3.7321	1.0353	3.8637	75°
16°	.2756	.9613	.2867	3.4874	1.0403	3.6280	74°
17°	.2924	.9563	.3057	3.2709	1.0457	3.4203	73°
18°	.3090	.9511	.3249	3.0777	1.0515	3.2361	72°
19°	.3256	.9455	.3443	2.9042	1.0576	3.0716	71°
20°	.3420	.9397	.3640	2.7475	1.0642	2.9238	70°
21°	.3584	.9336	.3839	2.6051	1.0711	2.7904	69°
22°	.3746	.9272	.4040	2.4751	1.0785	2.6695	68°
23°	.3907	.9205	.4245	2.3559	1.0864	2.5593	67°
24°	.4067	.9135	.4452	2.2460	1.0946	2.4586	66°
25°	.4226	.9063	.4663	2.1445	1.1034	2.3662	65°
26°	.4384	.8988	.4877	2.0503	1.1126	2.2812	64°
27°	.4540	.8910	.5095	1.9626	1.1223	2.2027	63°
28°	.4695	.8829	.5317	1.8807	1.1326	2.1301	62°
29°	.4848	.8746	.5543	1.8040	1.1434	2.0627	61°
30°	.5000	.8660	.5774	1.7321	1.1547	2.0000	60°
31°	.5150	.8572	.6009	1.6643	1.1666	1.9416	59°
32°	.5299	.8480	.6249	1.6003	1.1792	1.8871	58°
33°	.5446	.8387	.6494	1.5399	1.1924	1.8361	57°
34°	.5592	.8290	.6745	1.4826	1.2062	1.7883	56°
35°	.5736	.8192	.7002	1.4281	1.2208	1.7434	55°
36°	.5878	.8090	.7265	1.3764	1.2361	1.7013	54°
37°	.6018	.7986	.7536	1.3270	1.2521	1.6616	53°
38°	.6157	.7880	.7813	1.2799	1.2690	1.6243	52°
39°	.6293	.7771	.8098	1.2349	1.2868	1.5890	51°
40°	.6428	.7660	.8391	1.1918	1.3054	1.5557	50°
41°	.6561	.7547	.8693	1.1504	1.3250	1.5243	49°
42°	.6691	.7431	.9004	1.1106	1.3456	1.4945	48°
43°	.6820	.7314	.9325	1.0724	1.3673	1.4663	47°
44°	.6947	.7193	.9657	1.0355	1.3902	1.4396	46°
45°	.7071	.7071	1.0000	1.0000	1.4142	1.4142	45°
	cos	sin	cot	tan	csc	sec	Angle

or bottom. If we find it in the column having the given function as a top caption, the required angle will be found on the same horizontal line and in the extreme left-hand column. If the given function is a bottom caption, the required angle will be found in the extreme right-hand column.

Thus, let us find the angle x, having given $\tan x = .7536$.

In the column with *tan* as top caption we find .7536. On the same horizontal line with it, and in the extreme left-hand column, we find the angle $x = 37°$.

Again, let us find the angle x, having given

$$\sin x = .9816.$$

In the column with *sin* as bottom caption we find .9816. On the same horizontal line with it, and in the extreme right-hand column, we find the angle $x = 79°$.

The following examples will further illustrate the use of the table.

Ex. 1. Given $A = 35°$, $c = 267$; solve the right triangle. Also find its area.

Solution. First step. Draw a figure of the triangle indicating the known and unknown parts.

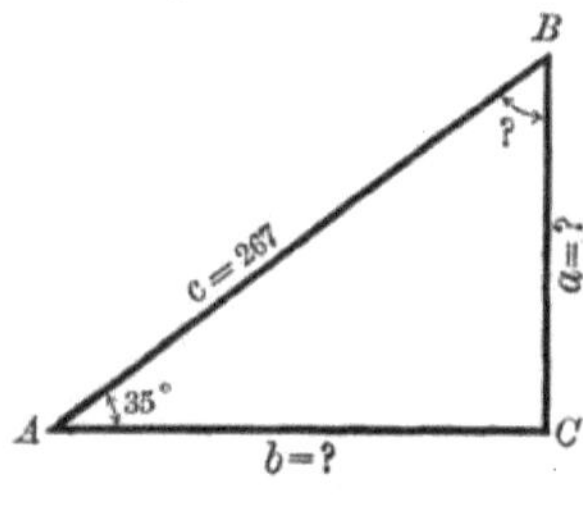

Second step. $B = 90° - A = 90° - 35° = 55°$.

Third step. To find a use formula (1), p. 2, namely,

$$\sin A = \frac{a}{c}.$$

Substituting the value of $\sin A = \sin 35° = .5736$ (found from the table) and $c = 267$, we have

$$.5736 = \frac{a}{267}.$$

Solving for a, we get $a = 153.1.$*

* Multiplying,

$$
\begin{array}{r}
\sin 35° = .5736 \\
267 \\
\hline
40152 \\
34416 \\
11472 \\
\hline
a = 153.1512
\end{array}
$$

Since our table gives not more than the first four significant figures of the sine of an angle, it follows, in general, that all but the first four significant figures of the product are doubtful. The last three figures of the above product should therefore be omitted, for the result will not be more accurate if they are retained. To illustrate this in the above example, suppose we take the sine of 35° from a five-place table, that is, a table which gives the first five significant figures of the sine. Then

$$
\begin{array}{r}
\sin 35° = .57358 \\
267 \\
\hline
401506 \\
344148 \\
114716 \\
\hline
a = 153.14586
\end{array}
$$

Comparing, we see that the two values of a agree in the first four significant figures only. Hence we take $a = 153.1$.

To find b use formula (**2**), p. 2, namely,

$$\cos A = \frac{b}{c}.$$

Substituting as before, we have

$$.8192 = \frac{b}{267},$$

since from the table $\cos A = \cos 35° = .8192$. Hence

$$b = 218.7.$$

Fourth step. By measurements we now check the results to see that there are no large errors. As a numerical check we find that the values of a, b, c satisfy the condition $c^2 = a^2 + b^2$.

To find the area of the triangle we have

$$\text{Area} = \frac{ab}{2} = \frac{153.1 \times 218.7}{2} = 16{,}741.$$

Ex. 2. A ladder 30 ft. long leans against the side of a building, its foot being 15 ft. from the building. What angle does the ladder make with the ground ?

Solution. Our figure shows a right triangle with hypotenuse and side adjacent to the required angle ($= x$) given. Hence

$$\cos x = \tfrac{15}{30} = \tfrac{1}{2} = .5 = .5000.$$

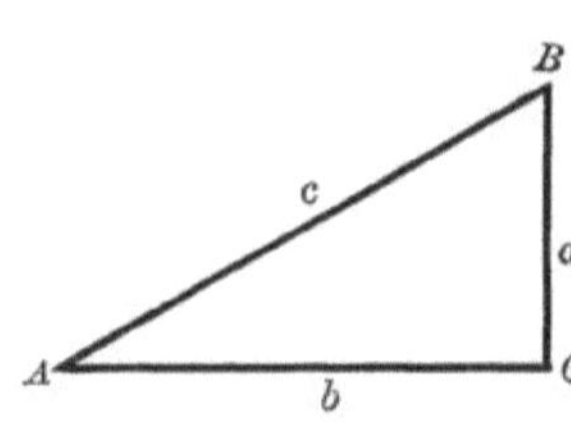

This number is found in the column having *cos* at the bottom and opposite 60°. Hence $x = 60°$. *Ans.*

We shall now derive three formulas by means of which the work of solving right triangles may be simplified. From (**1**), (**2**), (**3**), p. 2,

$$\sin A = \frac{a}{c}, \text{ or,}$$
$$a = c \sin A ;$$
$$\cos A = \frac{b}{c}, \text{ or,}$$
$$b = c \cos A ;$$
$$\tan A = \frac{a}{b}, \text{ or,}$$
$$a = b \tan A.$$

These results may be stated as follows :

(**7**) **Side opposite an acute angle = hypotenuse × sine of the angle.**

(**8**) **Side adjacent an acute angle = hypotenuse × cosine of the angle.**

(**9**) **Side opposite an acute angle = adjacent side × tangent of the angle.**

EXAMPLES

Solve the following right triangles ($C = 90°$).

No.	Given Parts		Required Parts			Area
1	$A = 60°$	$b = 4$	$B = 30°$	$c = 8$	$a = 6.928$	13.856
2	$A = 30°$	$a = 3$	$B = 60°$	$c = 6$	$b = 5.196$	7.794
3	$a = 6$	$c = 12$	$A = 30°$	$B = 60°$	$b = 10.39$	31.18
4	$a = 4$	$b = 4$	$A = 45°$	$B = 45°$	$c = 5.657$	8
5	$a = 2$	$c = 2.8284$	$A = 45°$	$B = 45°$	$b = 2$	2
6	$a = 51.303$	$c = 150$	$A = 20°$	$B = 70°$	$b = 140.95$	3615
7	$B = 51°$	$c = 250$	$A = 39°$	$a = 157.3$	$b = 194.3$	15282
8	$A = 36°$	$c = 1$	$B = 54°$	$a = .5878$	$b = .809$	2378
9	$c = 43$	$a = 38.313$	$A = 63°$	$B = 27°$	$b = 19.52$	373.9
10	$b = 9.696$	$c = 20$	$A = 61°$	$B = 29°$	$a = 17.492$	84.8
11	$a = 137.664$	$c = 240$	$A = 35°$	$B = 55°$	$b = 196.6$	13532
12	$A = 75°$	$a = 80$	$B = 15°$	$b = 21.44$	$c = 82.82$	858
13	$A = 25°$	$a = 30$	$B = 65°$	$b = 64.336$	$c = 70.98$	965
14	$B = 55°$	$b = 10$	$A = 35°$	$a = 7.002$	$c = 12.206$	35
15	$B = 15°$	$b = 20$	$A = 75°$	$a = 74.64$	$c = 77.28$	746
16	$a = 36.4$	$b = 100$	$A = 20°$	$B = 70°$	$c = 106.4$	1820
17	$a = 23.315$	$b = 50$	$A = 25°$	$B = 65°$	$c = 55.17$	582
18	$a = 17.1$	$c = 50$	$A = 20°$	$B = 70°$	$b = 46.985$	402
19	$A = 10°$	$b = 30$	$B = 80°$	$a = 5.289$	$c = 30.46$	79
20	$A = 20°$	$c = 80$	$B = 70°$	$a = 27.36$	$b = 75.176$	1028
21	$B = 86°$	$b = .08$	$A = 4°$	$a = .00559$	$c = .0802$	.0002
22	$B = 32°$	$c = 1760$	$A = 58°$	$b = 932.62$	$a = 1492.5$	695968
23	$a = 30.21$	$c = 33.33$	$A = 65°$	$B = 25°$	$b = 14.086$	213
24	$a = 13.395$	$b = 50$	$A = 15°$	$B = 75°$	$c = 51.77$	335
25	$b = 93.97$	$c = 100$	$A = 20°$	$B = 70°$	$a = 34.2$	1607

26. A tree is broken by the wind so that its two parts form with the ground a right-angled triangle. The upper part makes an angle of 35° with the ground, and the distance on the ground from the trunk to the top of the tree is 50 ft. Find the length of the tree. *Ans.* 96.05 ft.

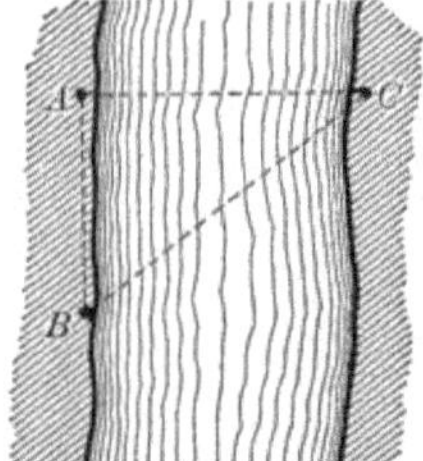

27. In order to find the breadth of a river, a distance AB was measured along the bank, the point A being directly opposite a tree C on the other side. If the angle ABC was observed to be 55° and AB 100 ft., find the breadth of the river. *Ans.* 142.8 ft.

28. Two forts defending a harbor are 2 mi. apart. From one a hostile battleship is observed due south and from the other 15° east of south. How far is the battleship from the nearest fort? *Ans.* 7.464 mi.

29. A vessel whose masts are known to reach 100 ft. above her water line subtends in a vertical plane an angle of 5° to an observer in a rowboat. How far is the boat from the vessel? *Ans.* 1143 ft.

30. The vertical central pole of a circular tent is 20 ft. high, and its top is fastened by ropes 40 ft. long to stakes set in the ground. How far are the stakes from the foot of the pole, and what is the inclination of the ropes to the ground ? *Ans.* 34.6 ft.; 30°.

31. A wedge measures 10 in. along the side and the angle at the vertex is 20°. Find the width of the base. *Ans.* 3.47 in.

32. At two points A, B, 400 yd. apart on a straight horizontal road, the summit of a hill is observed ; at A it is due north with an elevation of 40°, and at B it is due west with an elevation of 27°. Find the height of the hill.

Ans. 522.6 ft.

5. Solution of isosceles triangles. An isosceles triangle is divided by the perpendicular from the vertex to the base into two equal right triangles; hence the solution of an isosceles triangle can be made to depend on the solution of one of these right triangles. The following examples will illustrate the method.

Ex. 1. The equal sides of an isosceles triangle are each 40 in. long, and the equal angles at the base are each 25°. Solve the triangle and find its area.

Solution. $B = 180° - (A + C) = 180° - 50° = 130°$. Drop the perpendicular BD to AC.

$$AD = AB \cos A$$
$$= 40 \cos 25° \quad \text{by (8), p. 11}$$
$$= 40 \times .9063$$
$$= 36.25.$$

Therefore $\qquad AC = 2\, AD = 72.50 \text{ in.}$

To find the area we need in addition the altitude BD.

$$BD = AB \sin A = 40 \sin 25° \qquad \text{by (7), p. 11}$$
$$= 40 \times .4226 = 16.9.$$

Check: $BD = AD \tan 25° = 36.25 \times .4663 = 16.9.$ $\qquad$ by (9), p. 11

Also, $\qquad \text{Area} = \tfrac{1}{2}\, AC \times BD = 603.6 \text{ sq. in.}$

Ex. 2. A barn 60 ft. wide has a gable roof whose rafters are $30\sqrt{2}$ ft. long. What is the pitch of the roof, and how far above the eaves is the ridgepole ?

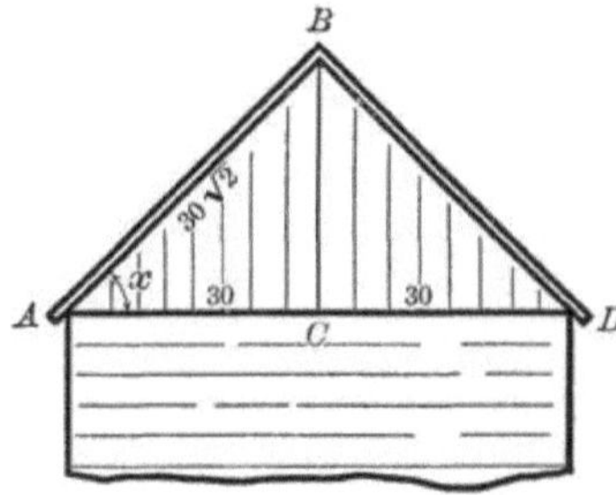

Solution. Drop a perpendicular from B to AD. Then

$$\cos x = \frac{AC}{AB} = \frac{30}{30\sqrt{2}} = \frac{1}{\sqrt{2}}.$$

Hence $x = 45° =$ pitch of the roof.

Also, $BC = AB \sin x \qquad$ by (8), p. 11

$$= 30\sqrt{2} \cdot \frac{1}{\sqrt{2}}$$
$$= 30 \text{ ft.}$$
$$= \text{height of the ridgepole}$$
$$\text{above the eaves.}$$

Check: $AB = \sqrt{\overline{AC}^2 + \overline{BC}^2} = \sqrt{(30)^2 + (30)^2} = \sqrt{1800} = 30\sqrt{2}.$

EXAMPLES

1. The equal sides of an isosceles triangle are each 12 in. long, and the angle at the vertex is 120°. Find the remaining parts and the area.

Ans. Base = 20.78 in.; base angles = 30°; area = 62.35 sq. in.

2. The equal angles of an isosceles triangle are each 35°, and the base is 393.18 in. Find the remaining parts.

Ans. Vertex angle = 110°; equal sides = 240 in.

3. Given the base 300 ft. and altitude 150 ft. of an isosceles triangle; solve the triangle.

Ans. Vertex angle = 90°; equal angles = 45°; equal sides = 212.13 ft.

4. The base of an isosceles triangle is 24 in. long and the vertical angle is 48°; find the remaining parts and the area.

Ans. Equal angles = 66°; equal sides = 53.5 in. ; area = 586.5 sq. in.

5. Each of the equal sides of an isosceles triangle is 50 ft. and each of its equal angles is 40°. Find the base, the altitude, and the area of the triangle.

Ans. Alt. = 32.14 ft.; base = 76.6 ft.; area = 1231 sq. ft.

6. The base of an isosceles triangle is 68.4 ft. and each of its equal sides is 100 ft. Find the angles, the height, and the area.

Ans. 40°, 70°; 93.97 ft.; 3213.8 sq. ft.

7. The base of an isosceles triangle is 100 ft. and its height is 35.01 ft. Find its equal sides and the angles. *Ans.* 61.04 ft.; 35°, 110°.

8. The base of an isosceles triangle is 100 ft. and the equal angles are each 65°. Find the equal sides, the height, and the area.

Ans. 118.3 ft.; 107.23 ft.; 5361.5 sq. ft.

9. The ground plan of a barn measures 40 × 80 ft. and the pitch of the roof is 45°; find the length of the rafters and the area of the whole roof, the horizontal projection of the cornice being 1 ft. *Ans.* 29.7 ft.; 4870 sq. ft.

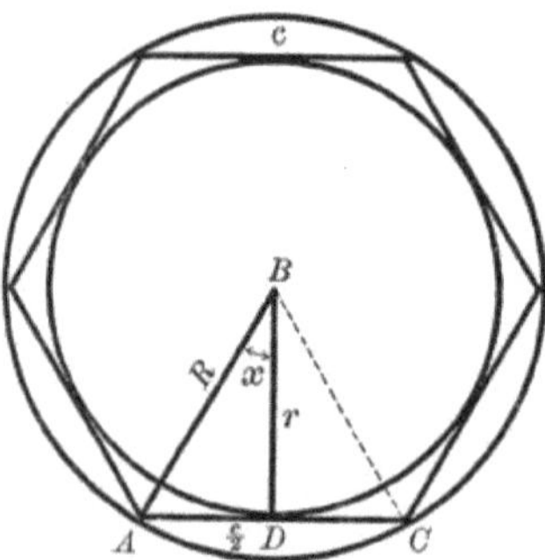

6. Solution of regular polygons. Lines drawn from the center of a regular polygon of n sides to the vertices are the radii of the circumscribed circle and divide the polygon into n equal isosceles triangles. The perpendiculars from the center to the sides of the polygon are the radii of the inscribed circle and divide these n equal isosceles triangles into $2\,n$ equal right triangles. Hence the solution of a regular polygon depends on the solution of one of these right triangles.

From Geometry we know that the central angle $ABC = \dfrac{360°}{n}$; hence in the right triangle ABD the

$$\text{angle } x = \frac{180°}{n}.$$

Also, $\qquad AD = \dfrac{c}{2}$ = half the length of one side,

$$AB = R = \text{radius of circumscribed circle,}$$
$$BD = r = \text{radius of inscribed circle,}$$
$$p = nc = \text{perimeter of polygon,}$$
$$\frac{pr}{2} = \text{area of polygon.}$$

EXAMPLES

1. One side of a regular decagon is 10 in.; find radii of inscribed and circumscribed circles and area of polygon.

Solution. Since $n = 10$, in this example we have

$$x = \frac{180°}{n} = \frac{180°}{10} : 18°.$$

Then $\qquad R = \dfrac{5}{\sin 18°} = \dfrac{5}{.3090} = 16.18$ in.,

and $\qquad r = \dfrac{5}{\tan 18°} = \dfrac{5}{.3249} = 15.39$ in.

Check: $r = R \cos 18° = 16.18 \times .9511$
$$= 15.39.$$

Also, $\quad p = 10 \times 10 = 100$ in.
$$= \text{perimeter of polygon ;}$$

hence $\qquad \dfrac{pr}{2} = \dfrac{100 \times 15.39}{2} = 769.5$ sq. in.
$$= \text{area.}$$

2. The side of a regular pentagon is 24 ft.; find R, r, and area.
$$\textit{Ans.} \quad 20.42 \text{ ft.}; \ 16.52 \text{ ft.}; \ 991.2 \text{ sq. ft.}$$

3. Find the remaining parts of a regular polygon, having given

(a) $n = 9$, $\quad c = 12$.   *Ans.* $R = 17.54$; $r = 16.48$; area $= 889.9$.
(b) $n = 18$, $R = 10$. $r = 9.848$; $c = 3.472$; area $= 307.8$.
(c) $n = 20$, $R = 20$. $r = 19.75$; $c = 6.256$; area $= 1236$.
(d) $n = 12$, $\ r = 8$. $R = 8.28$; $\ c = 4.29$; area $= 206$.

4. The side of a regular hexagon is 24 ft. Find the radii of the inscribed and circumscribed circles; also find the difference between the areas of the hexagon and the inscribed circle, and the difference between the areas of the hexagon and the circumscribed circle. *Ans.* $R = 24$ ft.; $r = 20.8$ ft.; 139.1 sq. ft.; 311 sq. ft.

5. If c be the side of a regular polygon of n sides, show that

$$R = \frac{1}{2} c \csc \frac{180°}{n} \text{ and } r = \frac{1}{2} c \cot \frac{180°}{n}.$$

6. If r be the radius of a circle, show that the side of the regular inscribed polygon of n sides is $2\,r\sin\dfrac{180°}{n}$, and that the side of the regular circumscribed polygon is $2\,r\tan\dfrac{180°}{n}$.

7. Interpolation. In the examples given so far we have needed the functions of such angles only as were explicitly given in our table; that is, the number of degrees in the angle involved was given by a whole number. It is evident that such will not always be the case. In general, our problems will involve angles expressed in degrees and parts of a degree, as $28.4°$, $5.63°$, $10°\,13'$, $72°\,27.4'$, $42°\,51'\,16''$, etc.

In order to find from the table the numerical value of the function of such an angle not given in the table, or to find the angle corresponding to a given numerical value of some function not found in the table, we use a process called *interpolation*. This is based on the assumption that a change in the angle causes a proportional change in the value of each function, and conversely, *provided these changes are small.** To illustrate; from the table we have

$$\sin 38° = .6157$$
$$\sin 37° = .6018$$

Subtracting, $\qquad\qquad\qquad\qquad .0139 =$ difference for one degree; that is, at $37°$ a change of one degree in the angle causes a change in the value of the sine of $.0139$. If, then, x is any other small change in the angle from $37°$, and d the corresponding change in the value of the sine, we must have, near $37°$,

$$1° : x :: .0139 : d,$$
$$\therefore\ d = .0139\,x,$$

if x is expressed in the *decimal parts of a degree.*

For example, let us tabulate the values of the sines of all angles from $37°$ to $38°$ at intervals of 0.1 of a degree.

x	d	
$0.1°$	$.0014$	$\therefore\ \sin 37.1° = .6018 + .0014 = .6032$
$0.2°$	$.0028$	$\therefore\ \sin 37.2° = .6018 + .0028 = .6046$
$0.3°$	$.0042$	$\therefore\ \sin 37.3° = .6018 + .0042 = .6060$
$0.4°$	$.0056$	$\therefore\ \sin 37.4° = .6018 + .0056 = .6074$
$0.5°$	$.0070$	$\therefore\ \sin 37.5° = .6018 + .0070 = .6088$
$0.6°$	$.0083$	$\therefore\ \sin 37.6° = .6018 + .0083 = .6101$
$0.7°$	$.0097$	$\therefore\ \sin 37.7° = .6018 + .0097 = .6115$
$0.8°$	$.0111$	$\therefore\ \sin 37.8° = .6018 + .0111 = .6129$
$0.9°$	$.0125$	$\therefore\ \sin 37.9° = .6018 + .0125 = .6143$

---

* This condition is most important. The change in value of the cotangent for one degree is very large when the angle is very small. In this case the table would therefore lead to very inaccurate results if interpolation was used for cotangents of small angles (see Chapter IX, p. 178).

The following examples will further illustrate the process of interpolating.

(a) *To find the function of a given angle when the angle is not found in the table.*

Ex. 1. Find $\sin 32.8°$.

Solution. The sine of $32.8°$ must lie between $\sin 32°$ and $\sin 33°$. From the table on p. 9,

$$\sin 33° = .5446$$
$$\sin 32° = .5299$$
$$.0147 = \text{difference in the sine (called the}$$

tabular difference) corresponding to a difference of $1°$ in the angle. Now in order to find $\sin 32.8°$, we must find the difference in the sine corresponding to $.8°$ and add it to $\sin 32°$, for the sine will be increased by just so much when the angle is increased from $32°$ to $32.8°$. Denoting by d the difference corresponding to $.8°$, we have

$$1° : .8° :: .0147 : d,$$

or, $$d = .0118.$$

Hence $$\sin 32° = .5299$$
$$\underline{d = .0118} = \text{difference for } .8°$$
$$\therefore \; \sin 32.8° = .5417. \quad Ans.$$

Ex. 2. Find $\tan 47° 25'$.

Solution. The tangent of $47° 25'$ must lie between $\tan 47°$ and $\tan 48°$. From the table,

$$\tan 48° = 1.1106$$
$$\tan 47° = 1.0724$$
$$.0382 = \text{tabular difference correspond-}$$

ing to a difference of $60' (= 1°)$ in the angle. Denoting by d the difference corresponding to $25'$, we have

$$60' : 25' :: .0382 : d,$$

or, $$d = .0159.$$

Hence $$\tan 47° = 1.0724$$
$$\underline{d = \quad .0159} = \text{difference for } 25'.$$
$$\therefore \; \tan 47° 25' = 1.0883. \quad Ans.$$

Ex. 3. Find $\cos 68.57°$.

Solution. The cosine of $68.57°$ must lie between $\cos 68°$ and $\cos 69°$. From the table,

$$\cos 68° = .3746$$
$$\cos 69° = .3584$$
$$.0162 = \text{tabular difference correspond-}$$

ing to a difference of $1°$ in the angle. Denoting by d the difference corresponding to $.57°$, we have

$$1° : .57° :: .0162 : d,$$

or, $$d = .0092.$$

Since the cosine decreases as the angle increases, this difference must be subtracted * from $\cos 68°$ in order to get $\cos 68.57°$.

Hence
$$\cos 68° = .3746$$
$$d = .0092 = \text{difference for } .57°.$$
$$\therefore \ \cos 68.57° = .3654. \quad Ans.$$

(*b*) *To find an angle when the given numerical value of a function of the angle is not found in the table.*

Ex. 4. Find the angle whose tangent is .4320.

Solution. This problem may also be stated : Having given $\tan x = .4320$, to find the angle x. We first look up and down the columns with *tan* at top or bottom, until we find two numbers between which .4320 lies. These are found to be .4245 and .4452, the former being $\tan 23°$ and the latter $\tan 24°$. We then know that the required angle x must lie between $23°$ and $24°$. To find how far ($= y$) beyond $23°$ the angle x lies, we first find the difference between $\tan 23°$ and $\tan x$; thus,

$$\tan x = .4320$$
$$\tan 23° = .4245$$
$$.0075 = \text{difference in the tangent corre-}$$
sponding to the excess of the angle x over $23°$; denote this excess by y. Also,

$$\tan 24° = .4452$$
$$\tan 23° = .4245$$
$$.0207 = \text{tabular difference correspond-}$$
ing to a difference of $1°$ in the angle. Then, as before,

$$1° : y :: .0207 : .0075,$$
or,
$$y = .36°.$$
Hence
$$x = 23° + y = 23.36°. \quad Ans.$$

In case we want the angle expressed in degrees and minutes, we can either multiply $.36°$ by 60, giving $21.6'$ so that the required angle is $23° \, 21.6'$, or else we can find y in minutes at once by using instead the proportion

$$60' : y :: .0207 : .0075,$$
or,
$$y = 21.6'.$$
Hence
$$x = 23° + y = 23° \, 21.6'. \quad Ans.$$

EXAMPLES

1. Verify the following :

(a) $\sin 51.6° = .7836.$ (f) $\csc 80.3° = 1.0145.$ (k) $\sec 25° \, 2.5' = 1.1038.$

(b) $\tan 27.42° = .5188.$ (g) $\sin 43° \, 18' = .6858.$ (l) $\csc 72° \, 54' = 1.0463.$

(c) $\cos 79.9° = .1753.$ (h) $\cos 84° \, 42' = .0924.$ (m) $\sin 58° \, 36.2' = .8536.$

(d) $\cot 65.62° = .4532.$ (i) $\tan 31° \, 7.8' = .6041.$

(e) $\sec 12.37° = 1.0238.$ (j) $\cot 11° \, 43.4' = 4.8278.$

---

* In the case of the sine, tangent, and secant this difference is always added, because these functions increase when the angle increases (the angle being acute). In the case of the cosine, cotangent, and cosecant, however, this difference is always subtracted, because these functions decrease when the angle increases. It is always the function of the *smaller of the two angles* that this difference is added to or subtracted from.

2. Find the angle x, having given

(a) $\sin x = .5280.$		*Ans.*	$x = 31.87°.$
(b) $\tan x = .6344.$			$x = 32.39°.$
(c) $\sec x = 1.2122.$			$x = 34.41°.$
(d) $\cos x = .9850.$			$x = 9.93°.$
(e) $\cot x = 3.5249.$			$x = 15.85°.$
(f) $\csc x = 1.7500.$			$x = 34.85°.$
(g) $\sin x = .9425.$			$x = 70° 28.8'.$
(h) $\cos x = .2118.$			$x = 77.77°.$
(i) $\tan x = 1.1652.$			$x = 49° 21.6'.$
(j) $\cot x = .0803.$			$x = 85.41°.$
(k) $\sec x = 4.6325.$			$x = 77.51°.$
(l) $\csc x = 1.2420.$			$x = 53.63°.$
(m) $\sin x = .7100.$			$x = 45° 14.4'.$
(n) $\cos x = .9999.$			$x = 0° 30'.$
(o) $\tan x = .9845.$			$x = 44° 33'.$
(p) $\cot x = 8.6892.$			$x = 6° 36'.$

8. Terms occurring in trigonometric problems. *The vertical line* at a point is the line which coincides with the plumb line through that point.

A horizontal line at a point is a line which is perpendicular to the vertical line through that point.

A vertical plane at a point is a plane which contains the vertical line through that point.

The horizontal plane at a point is the plane which is perpendicular to the vertical line through that point.

A vertical angle is one lying in a vertical plane.

A horizontal angle is one lying in a horizontal plane.

The angle of elevation of an object above the horizontal plane of the observer is the vertical angle between the line drawn from the observer's eye to the object, and a horizontal line through the eye.

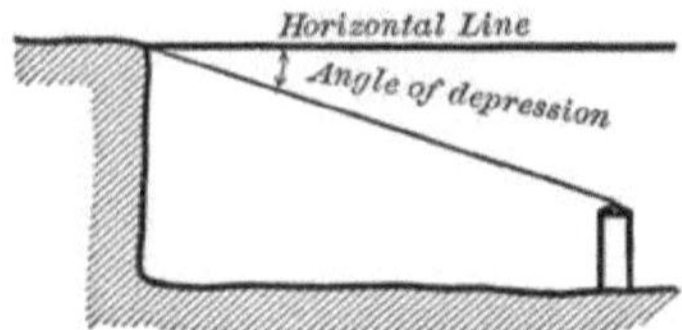

The angle of depression of an object below the horizontal plane of the observer is the vertical angle between the line drawn from the observer's eye to the object, and a horizontal line through the eye.

The horizontal distance between two points is the distance from one of the two points to the vertical line drawn through the other.

The vertical distance between two points is the distance from one of the two points to the horizontal plane through the other.

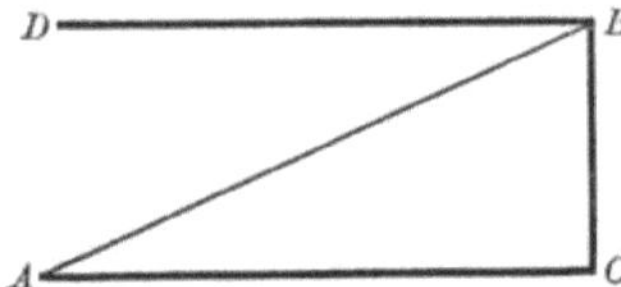

Thus, let BC be the vertical line at B, and let the horizontal plane at A cut this vertical line in C; then AC is called the horizontal distance between A and B and BC the vertical distance.

The Mariner's Compass is divided into 32 equal parts; hence each part $= 360° \div 32 = 11\frac{1}{4}°$. The following figure shows how the different divisions are designated. North, south, east, and west are called the *cardinal points*, and on paper these directions are usually taken as upward, downward, to the right, and to the left respectively. The direction of an object from an observer at C may be given in several ways. Thus, A in the figure is said to bear N.E. by E. from C, or from C the bearing of A is N.E. by E. In the same way the bearing of C from A is S.W. by W. The point A is 3 points north of east and 5 points east of north. Also, E. $33\frac{3}{4}°$ N. means the same as N.E. by E.

In order to illustrate the application of the trigonometric functions (ratios) to the solution of practical examples, we shall now give a variety of problems on finding heights, distances, angles, areas, etc. In solving these problems it is best to follow some definite plan. In general we may proceed as follows:

(*a*) Construct a drawing to some convenient scale which will show the relations between the given and the required lines and angles.

(*b*) If necessary draw any auxiliary lines that will aid in the solution, and decide on the simplest steps that will solve the problem.

(*c*) Write down the formulas needed, make the calculations, and check the results.

EXAMPLES

Solve the following right triangles ($C = 90°$).

No.	Given Parts		Required Parts		
1	$a = 60$	$c = 100$	$A = 36° 52'$	$B = 53° 8'$	$b = 80$
2	$a = 16.98$	$c = 18.7$	$A = 65° 14'$	$B = 24° 46'$	$b = 7.834$
3	$a = 147$	$c = 184$	$A = 53° 2'$	$B = 36° 58'$	$b = 110.67$
4	$A = 34° 15'$	$a = 843.2$	$B = 55° 45'$	$c = 1498.5$	$b = 1238$
5	$A = 31° 14.2'$	$c = 2.934$	$B = 58° 45.8'$	$a = 1.522$	$b = 2.509$
6	$B = 47.26°$	$c = 4.614$	$A = 42.74°$	$a = 3.131$	$b = 3.387$
7	$A = 23.5°$	$c = 627$	$B = 66.5°$	$a = 250$	$b = 575$
8	$A = 28° 5'$	$c = 2280$	$B = 61° 55'$	$a = 1073$	$b = 2011$
9	$B = 43.8°$	$b = 50.94$	$A = 46.2°$	$a = 53.18$	$c = 73.6$
10	$B = 6° 12.3'$	$c = 3721$	$A = 83° 47.7'$	$a = 3699$	$b = 402.2$
11	$a = .624$	$c = .91$	$A = 43° 18'$	$B = 46° 42'$	$b = .6622$
12	$a = 5$	$b = 2$	$A = 68° 12'$	$B = 21° 48'$	$c = 5.385$
13	$a = 101$	$b = 116$	$A = 41° 3'$	$B = 48° 57'$	$c = 153.8$
14	$A = 43.5°$	$c = 11.2$	$B = 46.5°$	$a = 7.71$	$b = 8.124$
15	$B = 68° 50'$	$a = 729.3$	$A = 21° 10'$	$b = 1884$	$c = 2020$
16	$A = 58.65°$	$c = 35.73$	$B = 31.35°$	$a = 30.51$	$b = 18.59$
17	$B = 10.85°$	$c = .7264$	$A = 79.15°$	$a = .7133$	$b = .1367$
18	$a = 24.67$	$b = 33.02$	$A = 36° 46'$	$B = 53° 14'$	$c = 41.22$
19	$B = 21° 33' 51''$	$a = .821$	$A = 68° 26' 9''$	$b = .3244$	$c = .8827$
20	$A = 74° 0' 18''$	$c = 275.62$	$B = 15° 59' 42''$	$a = 264.9$	$b = 75.93$
21	$A = 64° 1.3'$	$b = 200.05$	$B = 25° 58.7'$	$a = 410.5$	$c = 456.7$
22	$b = .02497$	$c = .04792$	$A = 58° 36'$	$B = 31° 24'$	$a = .0409$
23	$b = 1.4367$	$c = 3.4653$	$A = 65° 30'$	$B = 24° 30'$	$a = 3.153$

24. The length of a kite string is 250 yd., and the angle of elevation of the kite is 40°. Find the height of the kite, supposing the line of the kite string to be straight. *Ans.* 160.7 yd.

25. At a point 200 ft. in a horizontal line from the foot of a tower the angle of elevation of the top of the tower is observed to be 60°. Find the height of the tower. *Ans.* 346 ft.

26. A stick 10 ft. in length stands vertically on a horizontal plane, and the length of its shadow is 8.391 ft. Find the angle of elevation of the sun. *Ans.* 50°.

27. From the top of a rock that rises vertically 80 ft. out of the water the angle of depression of a boat is found to be 30°; find the distance of the boat from the foot of the rock. *Ans.* 138.57 ft.

28. Two ships leave the same dock at the same time in directions S.W. by S. and S.E. by E. at rates of 9 and 9.5 mi. per hour respectively. Find their distance apart after 1 hr. *Ans.* 13.1 mi.

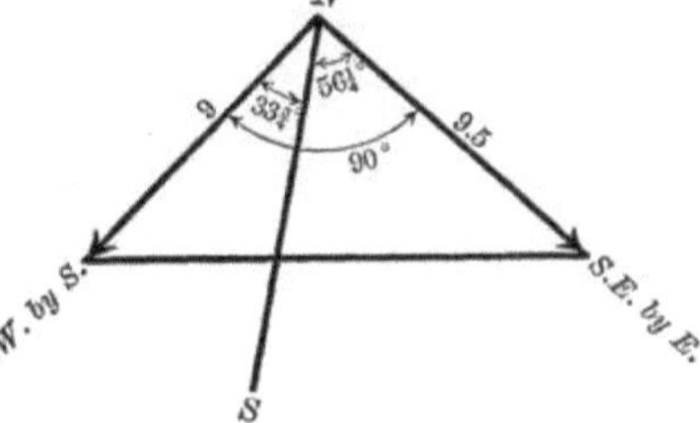

29. From the top of a tower 120 ft. high the angle of depression of an object on a level with the base of the tower is 27° 43′. What is the distance of the object from the top and bottom of the tower? *Ans.* 258 ft., 228 ft.

30. A ship is sailing due east at the rate of 7.8 mi. an hour. A headland is observed to bear due north at 10.37 A.M. and 33° west of north at 12.43 P.M. Find the distance of the headland from each point of observation.

Ans. 25.22 mi., 30.07 mi.

31. A ship is sailing due east at a uniform rate of speed. At 7 A.M. a lighthouse is observed bearing due north, 10.32 mi. distant, and at 7.30 A.M. it bears 18° 13′ west of north. Find the rate of sailing of the ship and the bearing of the lighthouse at 10 A.M. *Ans.* 6.79 mi. per hour, 63° 14′ W. of N.

32. From the top of a tower the angle of depression of the extremity of a horizontal base line 1000 ft. in length, measured from the foot of the tower, is observed to be 21° 16′ 37″. Find the height of the tower. *Ans.* 389.5 ft.

33. The length of the side of a regular octagon is 12 in. Find the radii of the inscribed and circumscribed circles. *Ans.* 14.49 in., 15.68 in.

34. What is the angle of elevation of an inclined plane if it rises 1 ft. in a horizontal distance of 40 ft.? *Ans.* 1° 26′.

35. A ship is sailing due N.E. at the rate of 10 mi. an hour. Find the rate at which she is moving due north. *Ans.* 7.07 mi. per hour.

36. A ladder 40 ft. long may be so placed that it will reach a window 33 ft. high on one side of the street, and by turning it over without moving its foot it will reach a window 21 ft. high on the other side. Find the breadth of the street. *Ans.* 56.68 ft.

37. At a point midway between two towers on a horizontal plane the angles of elevation of their tops are 30° and 60° respectively. Show that one tower is three times as high as the other.

38. A man in a balloon observes that the bases of two towers, which are a mile apart on a horizontal plane, subtend an angle of 70°. If he is exactly above the middle point between the towers, find the height of the balloon.

Ans. 3770 ft.

39. In an isosceles triangle each of the equal angles is 27° 8′ and each of the equal sides 3.088. Solve the triangle. *Ans.* Base = 5.496.

40. What is the angle of elevation of a mountain slope which rises 238 ft. in a horizontal distance of one eighth of a mile? *Ans.* 19° 50′.

41. If a chord of 41.36 ft. subtends an arc of 145° 37′, what is the radius of the circle? *Ans.* 21.65 ft.

42. If the diameter of a circle is 3268 ft., find the angle at the center subtended by an arc whose chord is 1027 ft. *Ans.* 36° 49′.

43. From each of two stations east and west of each other the angle of elevation of a balloon is observed to be 45°, and its bearings N.W. and N.E. respectively. If the stations are 1 mi. apart, find the height of the balloon.

Ans. 3733 ft.

44. In approaching a fort situated on a plain, a reconnoitering party finds at one place that the fort subtends an angle of 10°, and at a place 200 ft. nearer the fort that it subtends an angle of 15°. How high is the fort and what is the distance to it from the second place of observation?

Hint. Denoting the height by y and the distance by x, we have

$$y = x \tan 15°,$$ by (9), p. 11

also, $$y = (x + 200) \tan 10°.$$ by (9), p. 11

Solve these two simultaneous equations for x and y, substituting the values of $\tan 15°$ and $\tan 10°$ from the table on p. 9. *Ans.* $x = 385$ ft, $y = 103$ ft.

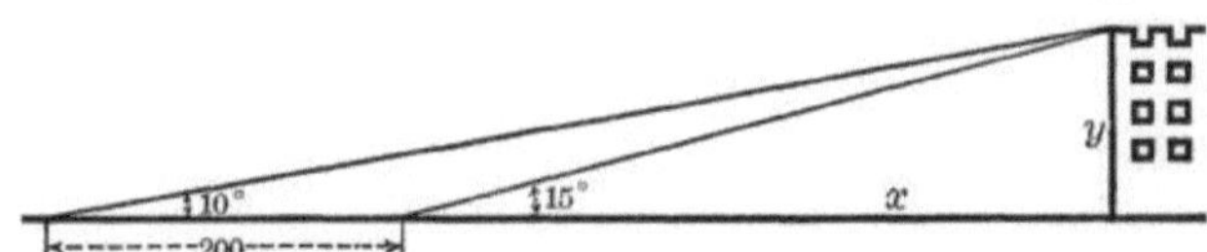

45. A cord is stretched around two wheels with radii of 7 ft. and 1 ft. respectively, and with their centers 12 ft. apart. Prove that the length of the cord is $12\sqrt{3} + 10\pi$ ft.

46. A flagstaff 25 ft. high stands on the top of a house. From a point on the plain on which the house stands, the angles of elevation of the top and the bottom of the flagstaff are observed to be 60° and 45° respectively. Find the height of the house. *Ans.* 33.94 ft.

47. A man walking on a straight road observes at one milestone a house in a direction making an angle of 30° with the road, and at the next milestone the angle is 60°. How far is the house from the road? *Ans.* 1524 yd.

48. Find the number of square feet of pavement required for the shaded portion of the streets shown in the figure, all the streets being 50 ft. wide.

$$Ans. \quad \frac{28750}{\sqrt{3}} + 7500 = 24094.$$

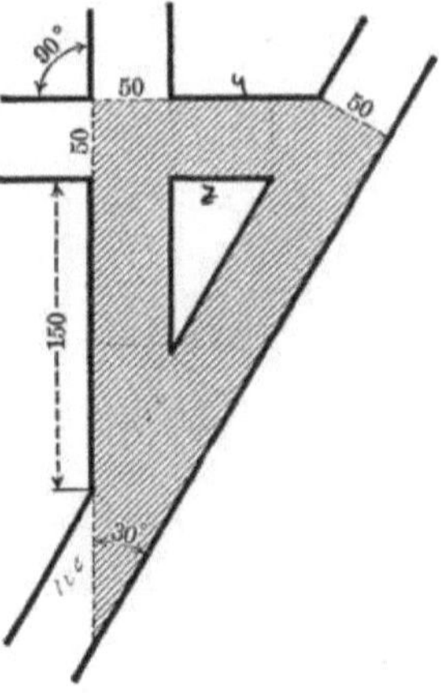

TRIGONOMETRIC FUNCTIONS OF ANY ANGLE

9. Generation of angles. The notion of an angle, as usually presented in Elementary Geometry, is not general enough for the purposes of Trigonometry. We shall have to deal with positive and negative angles of any magnitude. Such a conception of angles may be formed as follows:

An angle may be considered as generated by a line which first coincides with one side of the angle, then revolves about the vertex, and finally coincides with the other side.

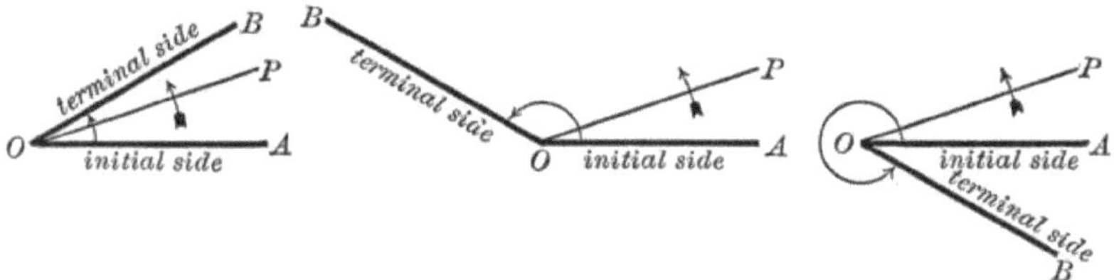

This line is called the *generating line* of the angle. In its first position it is said to coincide with the *initial side* of the angle, and in its final position with the *terminal side* of the angle.

Thus, the angle *A OB* is generated by the line *OP* revolving about *O* in the direction indicated from the initial side *OA* to the terminal side *OB*.

10. Positive and negative angles. In the above figures the angles were generated by revolving the generating line *counter-clockwise ;* mathematicians have agreed to call such angles *positive.* Below are

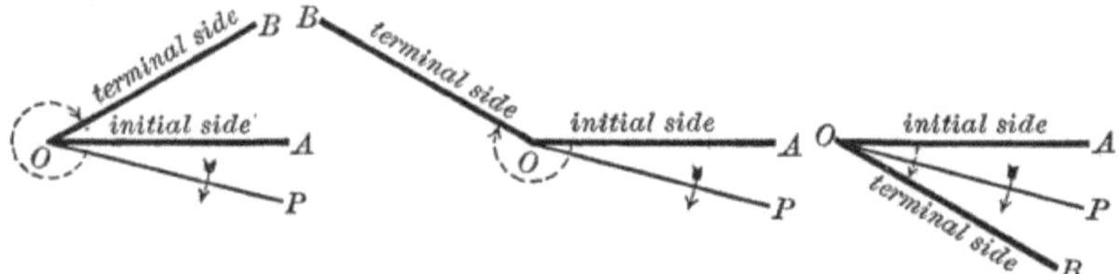

angles having the same initial and terminal sides as those above, but the *angles are different* since they have been generated by revolving the generating line *clockwise ;* such angles are said to be *negative.**

---

* The arcs with arrowheads will be drawn full when indicating a positive angle, and dotted when indicating a negative angle.

11. Angles of any magnitude. Even if angles have the same initial and terminal sides, and have been generated by rotation in the same direction, they may be different. Thus, to generate one right angle, the generating line rotates into the position OB as shown in Fig. *a*. If, however, the generating line stops in the position OB after making one complete revolution, as shown in Fig. *b*, then we have generated an angle of magnitude five right angles ; or, if two complete revolutions were first made, as shown in Fig. *c*, then we have

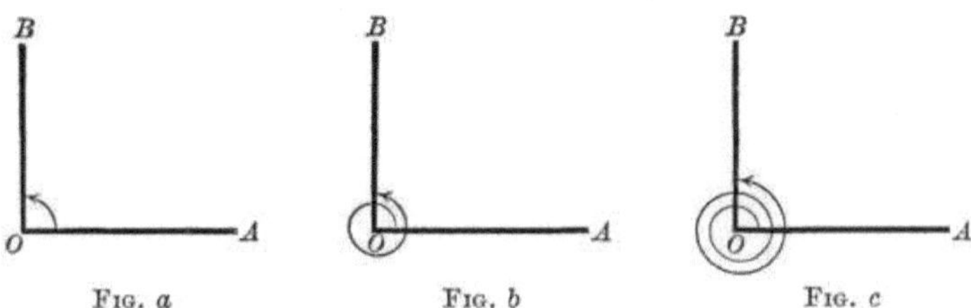

Fɪɢ. a   Fɪɢ. b   Fɪɢ. c

generated an angle of magnitude nine right angles ; and so on indefinitely. This also shows that positive angles may have any magnitude whatever. Similarly, by making complete revolutions clockwise, it is seen that negative angles may have any magnitude.*

12. The four quadrants. It is customary to divide the plane about the vertex of an angle into four parts called *quadrants*, by passing two mutually perpendicular lines through the vertex. Thus, if O is the vertex, the different quadrants are named as indicated in the figure below, **the initial side being horizontal and drawn to the right.**

An angle is said to be (or lie) in a certain quadrant when its terminal side lies in that quadrant.

In the figures shown on the previous page, only the least positive and negative angles having the given initial and terminal sides are indicated by the arcs. As a matter of fact there are an infinite number of positive and negative angles in each case which have the same initial and terminal sides, all differing in magnitude by multiples of 360°. The following examples will illustrate the preceding discussion.

* Thus, the minute hand of a clock generates -4 rt. ∡ every hour, i.e. -96 rt. ∡ every day.

EXAMPLES

1. Show that 1000° lies in the fourth quadrant.

Solution. $1000° = 720° + 280° = 2 \times 360° + 280°$. Hence we make two complete revolutions in the positive direction and 280° beyond, and the terminal side of 280° lies in the fourth quadrant.

2. Show that $-568°$ lies in the second quadrant.

Solution. $-568° = -360° - 208°$. Hence we make one complete revolution in the negative direction and 208° beyond in the negative direction, and the terminal side of $-208°$ lies in the second quadrant.

3. In what quadrants are the following angles ?

(a) 225°.	(e) 651°.	(i) 540°.	(m) 1500°.
(b) 120°.	(f) $-150°$.	(j) 420°.	(n) 810°.
(c) $-315°$.	(g) $-75°$.	(k) $-910°$.	(o) $-540°$.
(d) $-240°$.	(h) $-1200°$.	(l) $-300°$.	(p) 537°.

13. Rectangular coördinates of a point in a plane. In order to define the functions of angles not acute, it is convenient to introduce the

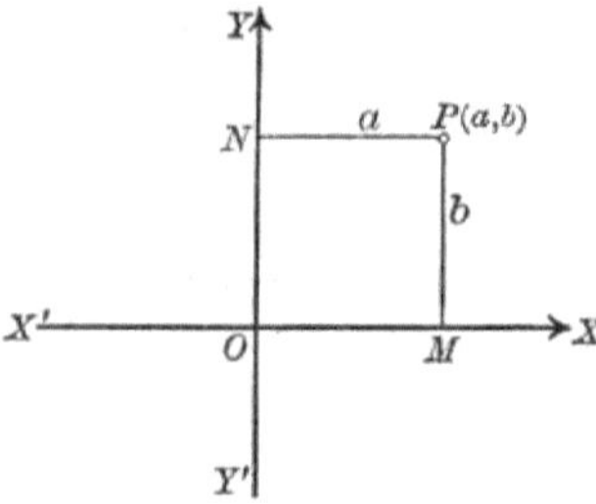

notion of *coördinates.* Let $X'X$ be a horizontal line and $Y'Y$ a line perpendicular to it at the point O. Any point in the plane of these lines (as P) is determined by its *distance* and *direction* from each of the perpendiculars $X'X$ and $Y'Y$. Its distance from $Y'Y$ (as $NP = a$) is called the *abscissa* of the point, and its distance from $X'X$ (as $MP = b$) is called the *ordinate* of the point.

Abscissas measured to the *right* of $Y'Y$ are *positive.*

Abscissas measured to the *left* of $Y'Y$ are *negative.*

Ordinates measured *above* $X'X$ are *positive.*

Ordinates measured *below* $X'X$ are *negative.*

The abscissa and ordinate taken together are called the *coördinates* of the point and are denoted by the symbol (a, b).

The lines $X'X$ and $Y'Y$ are called the *axes of coördinates*, $X'X$ being the *axis of abscissas* or the *axis of X*, and $Y'Y$ the *axis of ordinates* or the *axis of Y;* and the point O is called the *origin of coördinates.*

The axes of coördinates divides the plane into four parts called quadrants (just as in the previous section), the figure indicating the proper signs of the coordinates in the different quadrants.

To *plot a point* is to locate it from its coördinates. The most convenient way to do this is to first count off from O along $X'X$ a number of divisions equal to the abscissa, to the right or left according as the abscissa is positive or negative.

Then from the point so determined count off a number of divisions equal to the ordinate, upward or downward according as the ordinate is positive or negative. The work of plotting points is much simplified by the use of *coordinate* or *plotting paper*, constructed by ruling off the plane into equal squares, the sides being parallel

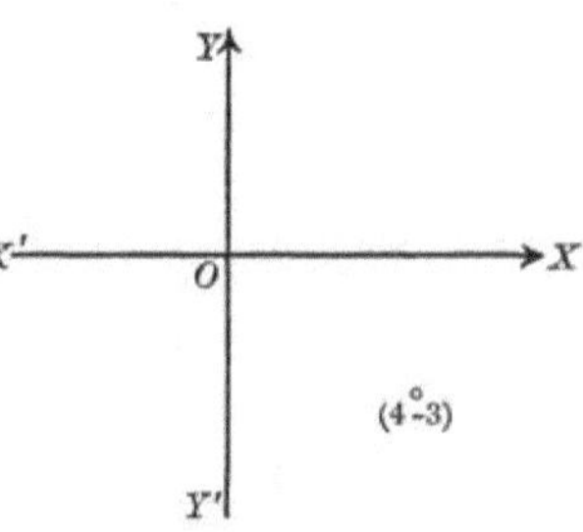

to the axes. Thus, to plot the point $(4, -3)$, count off four divisions from O on the axis of X to the right, and then three divisions downward from the point so determined on a line parallel to the axis of Y. Similarly, the following figures show the plotted points $(-2, 3)$, $(-3, -4)$, $(0, 3)$.

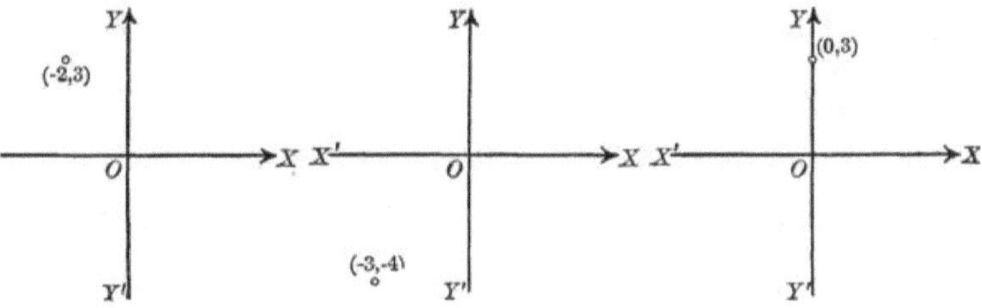

14. Distance of a point from the origin. Represent the abscissa of a point P by a and the ordinate by b, and its distance from the origin by h. Then

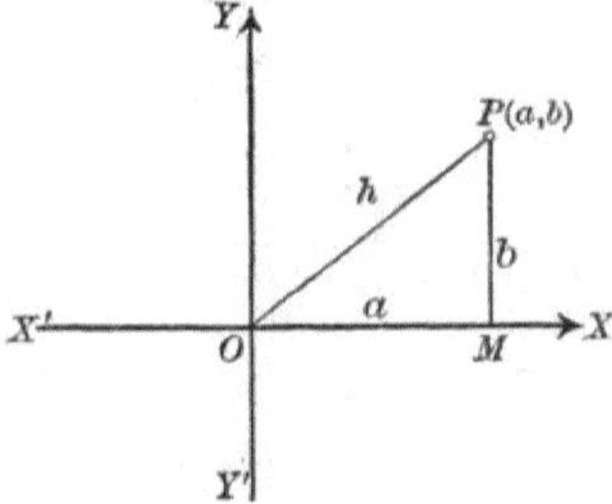

$$h = \sqrt{a^2 + b^2},$$

since h is the hypotenuse of a right triangle whose sides are a and b. Although h may be either positive or negative, it will be sufficient for our purposes to treat it as being always positive.

In order to become familiar with the notion of coordinates, the student should plot a large number of points.

EXAMPLES

1. (a) Plot accurately the points $(5, 4)$, $(-3, 4)$, $(-2, -4)$, $(5, -1)$, $(6, 0)$, $(-5, 0)$, $(0, 4)$, $(0, -3)$.

(b) What is the distance of each point from the origin?

Ans. $\sqrt{41}$, 5, $2\sqrt{5}$, etc.

2. Plot accurately the points $(1, 1)$, $(-1, -1)$, $(-1, 1)$, $(\sqrt{3}, 1)$, $(\sqrt{3}, -1)$, $(-\sqrt{3}, -1)$, and find the distance of each one from the origin.

3. Plot accurately the points $(\sqrt{2}, 0)$, $(-5, -10)$, $(3, -2\sqrt{2})$, $(10, 3)$, $(0, 0)$, $(0, -\sqrt{5})$, $(3, -5)$, $(-4, 5)$.

15. Trigonometric functions of any angle defined. So far the six trigonometric functions have been defined only for acute angles (§ 1, p. 2). Now, however, we shall give a new set of definitions which will apply to any angle whatever, and which agree with the definitions already given for acute angles.

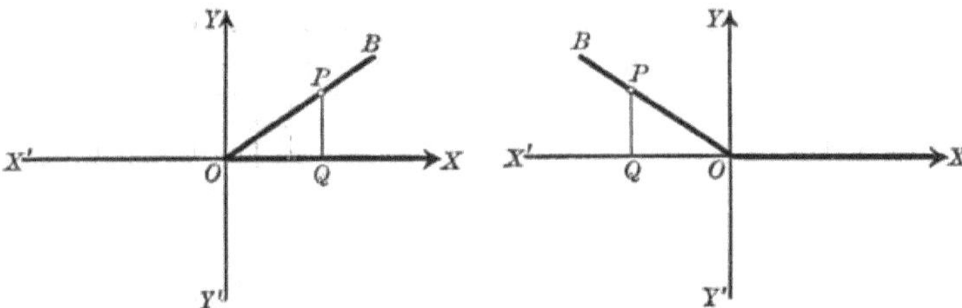

<table>
<tr><td align="center">Angle in first quadrant</td><td align="center">Angle in second quadrant</td></tr>
</table>

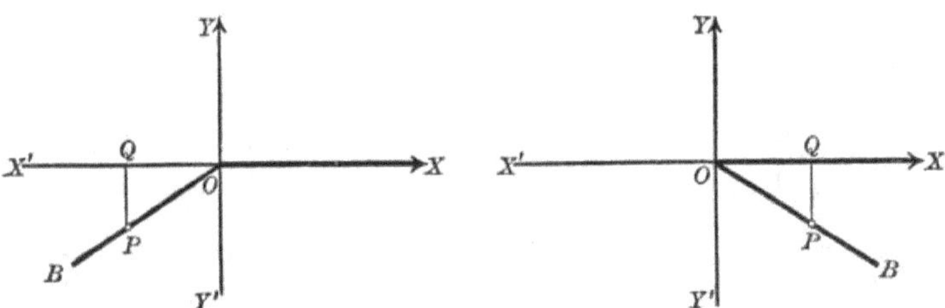

<table>
<tr><td align="center">Angle in third quadrant</td><td align="center">Angle in fourth quadrant</td></tr>
</table>

Take the origin of coördinates at the vertex of the angle and the initial side as the axis of X. Draw an angle XOB in each quadrant.

From any point P on the terminal side OB of the angle draw PQ perpendicular to the initial side, or the initial side produced. In every case OQ is the abscissa and QP the ordinate of the point P.

Denoting by XOB any one of these angles, their functions are defined as the following ratios:

$$(10)\quad \sin XOB = \frac{QP}{OP} = \frac{\text{ordinate}}{\text{hypotenuse}}; \qquad (13)\quad \csc XOB = \frac{OP}{QP} = \frac{\text{hypotenuse}}{\text{ordinate}};$$

$$(11)\quad \cos XOB = \frac{OQ}{OP} = \frac{\text{abscissa}}{\text{hypotenuse}}; \qquad (14)\quad \sec XOB = \frac{OP}{OQ} = \frac{\text{hypotenuse}}{\text{abscissa}};$$

$$(12)\quad \tan XOB = \frac{QP}{OQ} = \frac{\text{ordinate}}{\text{abscissa}}; \qquad (15)\quad \cot XOB = \frac{OQ}{QP} = \frac{\text{abscissa}}{\text{ordinate}}.\,{}_*$$

To the above six functions may be added the *versed sine* (written versin) and *coversed sine* (written coversin), which are defined as follows:

$$\text{versin}\,XOB = 1 - \cos XOB; \quad \text{coversin}\,XOB = 1 - \sin XOB.$$

16. Algebraic signs of the trigonometric functions. Bearing in mind the rule for the algebraic signs of the abscissas and ordinates of points given in § 13, p. 26, and remembering that the hypotenuse OP is always taken as positive (§ 14, p. 27), we have at once, from the definitions of the trigonometric functions given in the last section, that:

*In **I Quadrant**, all the functions are positive.*
*In **II Quadrant**, sin and csc are positive; all the rest are negative.*
*In **III Quadrant**, tan and cot are positive; all the rest are negative.*
*In **IV Quadrant**, sec and cos are positive; all the rest are negative.*

These results are also exhibited in the following

Rule for Signs

All functions not indicated in each quadrant are negative.

This rule for signs is easily memorized if the student remembers that reciprocal functions of the same angle must necessarily have the same sign, i.e.

sin and **csc** *have the same sign,*
cos and **sec** *have the same sign,*
tan and **cot** *have the same sign.*

17. Having given the value of a trigonometric function, to construct geometrically all the angles which satisfy the given value, and to find the values of the other five functions. Here we will make use of the

---

* As in acute angles it is seen that the functions in one column are the reciprocals of the functions in the other.

notion of coördinates, assuming as before that each angle has its vertex at the origin, and its initial side coinciding with the axis of X. It remains, then, only to fix the terminal side of each angle, or, what amounts to the same thing, to determine one point (not the origin) in the terminal side. When one function only is given, it will appear that two terminal sides satisfying the given condition may be constructed. Thus, if we have given $\tan x = \frac{2}{3}$, we may write

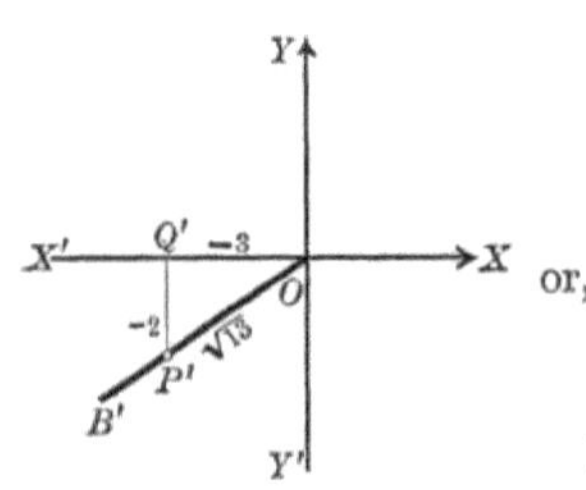

FIG. a

$$\tan x = \frac{2}{3} = \frac{-2}{-3} = \frac{\text{ordinate}}{\text{abscissa}}, \quad \textbf{(12)}, \text{ p. 29}$$

and hence, taking $\tan x = \frac{2}{3}$, one terminal side is determined by the origin and (3, 2), giving the angle XOB (in the first quadrant).

The other terminal side, taking $\tan x = \dfrac{-2}{-3}$, is determined by the origin and $(-3, -2)$, giving the angle XOB' (in the third quadrant). Hence all the angles x * which satisfy the condition

$$\tan x = \tfrac{2}{3}$$

FIG. b

have the initial side OX, and the terminal side OB,   Fig. a

or,   have the initial side OX and the terminal side OB'.   Fig. b

Let us now determine the values of all the functions. From Fig. a,

$$OP = \sqrt{\overline{OQ}^2 + \overline{QP}^2} = \sqrt{9 + 4} = \sqrt{13} \text{ (always positive).}$$

Hence by § 15, p. 29, from Fig. a,

$$\sin XOB = \frac{2}{\sqrt{13}}; \qquad\qquad \csc XOB = \frac{\sqrt{13}}{2},$$

$$\cos XOB = \frac{3}{\sqrt{13}}; \qquad\qquad \sec XOB = \frac{\sqrt{13}}{3};$$

$$\tan XOB = \frac{2}{3}; \qquad\qquad \cot XOB = \frac{3}{2}.$$

---

* It is evident that, corresponding to each figure, there are an infinite number of both positive and negative angles differing by multiples of 360° which satisfy the given condition.

Similarly, from Fig. *b*,

$$\sin XOB' = -\frac{2}{\sqrt{13}}; \qquad\qquad \csc XOB' = -\frac{\sqrt{13}}{2},$$

$$\cos XOB' = -\frac{3}{\sqrt{13}}; \qquad\qquad \sec XOB' = -\frac{\sqrt{13}}{3};$$

$$\tan XOB' = \frac{2}{3}; \qquad\qquad \cot XOB' = \frac{3}{2}.$$

Or, denoting by x any angle which satisfies the given condition, we may write down these results in more compact form as follows:

$$\sin x = \pm\frac{2}{\sqrt{13}}; \qquad\qquad \csc x = \pm\frac{\sqrt{13}}{2},$$

$$\cos x = \pm\frac{3}{\sqrt{13}}; \qquad\qquad \sec x = \pm\frac{\sqrt{13}}{3};$$

$$\tan x = \frac{2}{3}; \qquad\qquad \cot x = \frac{3}{2}.$$

The method is further illustrated in the following examples:

Ex. 1. Having given $\sin x = -\frac{1}{3}$, construct the angle x. Also find the values of the other five functions.

Solution. Here we may write, by **(10)**, p. 29,

$$\sin x = -\frac{1}{3} = \frac{-1}{3} = \frac{\text{ordinate}}{\text{hypotenuse}} \qquad \text{(hypotenuse always positive).}$$

Since abscissa $= \pm\sqrt{(\text{hypot.})^2 - (\text{ord.})^2} = \pm\sqrt{9 - 1} = \pm 2\sqrt{2}$, one terminal side is determined by the origin and $(-2\sqrt{2}, -1)$, giving the angle XOB in the third quadrant. Here

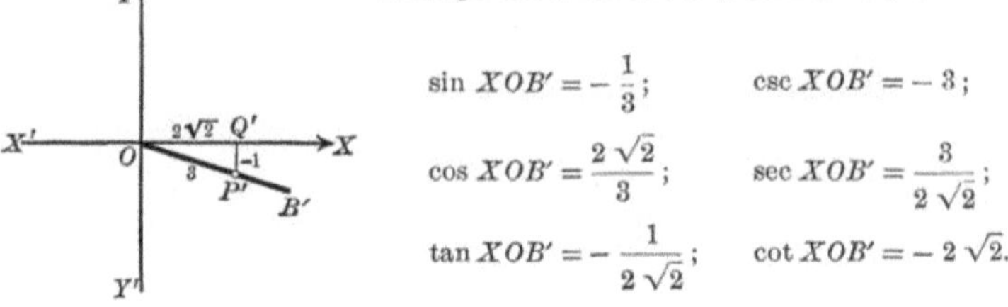

$$\sin XOB = -\frac{1}{3}; \qquad\qquad \csc XOB = -3;$$

$$\cos XOB = -\frac{2\sqrt{2}}{3}; \qquad\qquad \sec XOB = -\frac{3}{2\sqrt{2}};$$

$$\tan XOB = \frac{1}{2\sqrt{2}}; \qquad\qquad \cot XOB = 2\sqrt{2}.$$

The other terminal side is determined by the origin and $(2\sqrt{2}, -1)$ giving the angle XOB' in fourth quadrant. Here

$$\sin XOB' = -\frac{1}{3}; \qquad\qquad \csc XOB' = -3;$$

$$\cos XOB' = \frac{2\sqrt{2}}{3}; \qquad\qquad \sec XOB' = \frac{3}{2\sqrt{2}};$$

$$\tan XOB' = -\frac{1}{2\sqrt{2}}; \qquad\qquad \cot XOB' = -2\sqrt{2}.$$

Or, denoting by x any angle which satisfies the given condition, we have

$$\sin x = -\frac{1}{3}; \qquad\qquad \csc x = -3;$$

$$\cos x = \pm\frac{2\sqrt{2}}{3}; \qquad\qquad \sec x = \pm\frac{3}{2\sqrt{2}};$$

$$\tan x = \pm\frac{1}{2\sqrt{2}}; \qquad\qquad \cot x = \pm 2\sqrt{2}.$$

Ex. 2. Having given $\cot x = \dfrac{m}{n}$, find all the other functions of x.

Solution. Here we may write, by **(15)**, p. 29,

$$\cot x = \frac{m}{n} = \frac{-m}{-n} = \frac{\text{abscissa}}{\text{ordinate}},$$

and hypotenuse $= \sqrt{m^2 + n^2}$.

Hence one terminal side is determined by the origin and (m, n), and the other terminal side by the origin and $(-m, -n)$. Therefore

$$\sin x = \pm\frac{n}{\sqrt{m^2 + n^2}}; \qquad\qquad \csc x = \pm\frac{\sqrt{m^2 + n^2}}{n};$$

$$\cos x = \pm\frac{m}{\sqrt{m^2 + n^2}}; \qquad\qquad \sec x = \pm\frac{\sqrt{m^2 + n^2}}{m};$$

$$\tan x = \frac{n}{m}; \qquad\qquad \cot x = \frac{m}{n}.*$$

EXAMPLES

In each of the following examples construct geometrically the angle x, and compute the values of all the functions of x.

Given.

1. $\sin x = \dfrac{3}{5}.$ 6. $\tan x = \dfrac{a}{b}.$ 11. $\tan x = -\sqrt{7}.$

2. $\cos x = -\dfrac{1}{3}.$ 7. $\sin x = c.$ 12. $\sin x = -\dfrac{2}{3}.$

3. $\cot x = -3.\dagger$ 8. $\cos x = \dfrac{a^2 - b^2}{a^2 + b^2}.$ 13. $\tan x = 2.5.$

4. $\sec x = -\dfrac{5}{3}.$ 9. $\csc x = -\sqrt{3}.$ 14. $\sec x = p.$

5. $\csc x = \dfrac{13}{5}.$ 10. $\cos x = \dfrac{m}{c}.$

---

* When m and n have the same sign, x represents angles in the first and third quadrants. When m and n have opposite signs, x represents angles in the second and fourth quadrants.

$\dagger\ \cot x = -3 = \dfrac{-3}{1} = \dfrac{3}{-1}.$

ANSWERS

	Quadrant	sin	cos	tan	csc	sec	cot
1.	I	$\dfrac{3}{5}$	$\dfrac{4}{5}$	$\dfrac{3}{4}$	$\dfrac{5}{3}$	$\dfrac{5}{4}$	$\dfrac{4}{3}$
	II	$\dfrac{3}{5}$	$-\dfrac{4}{5}$	$-\dfrac{3}{4}$	$\dfrac{5}{3}$	$-\dfrac{5}{4}$	$-\dfrac{4}{3}$
2.	II	$\dfrac{2\sqrt{2}}{3}$	$-\dfrac{1}{3}$	$-2\sqrt{2}$	$\dfrac{3}{2\sqrt{2}}$	-3	$-\dfrac{1}{2\sqrt{2}}$
	III	$-\dfrac{2\sqrt{2}}{3}$	$-\dfrac{1}{3}$	$2\sqrt{2}$	$-\dfrac{3}{2\sqrt{2}}$	-3	$\dfrac{1}{2\sqrt{2}}$
3.	II	$\dfrac{1}{\sqrt{10}}$	$-\dfrac{3}{\sqrt{10}}$	$-\dfrac{1}{3}$	$\sqrt{10}$	$-\dfrac{\sqrt{10}}{3}$	-3
	IV	$-\dfrac{1}{\sqrt{10}}$	$\dfrac{3}{\sqrt{10}}$	$-\dfrac{1}{3}$	$-\sqrt{10}$	$\dfrac{\sqrt{10}}{3}$	-3
4.	II	$\dfrac{4}{5}$	$-\dfrac{3}{5}$	$-\dfrac{4}{3}$	$\dfrac{5}{4}$	$-\dfrac{5}{3}$	$-\dfrac{3}{4}$
	III	$-\dfrac{4}{5}$	$-\dfrac{3}{5}$	$\dfrac{4}{3}$	$-\dfrac{5}{4}$	$-\dfrac{5}{3}$	$\dfrac{3}{4}$
5.	I	$\dfrac{5}{13}$	$\dfrac{12}{13}$	$\dfrac{5}{12}$	$\dfrac{13}{5}$	$\dfrac{13}{12}$	$\dfrac{12}{5}$
	II	$\dfrac{5}{13}$	$-\dfrac{12}{13}$	$-\dfrac{5}{12}$	$\dfrac{13}{5}$	$-\dfrac{13}{12}$	$-\dfrac{12}{5}$
6.		$\pm\dfrac{a}{\sqrt{a^2+b^2}}$	$\pm\dfrac{b}{\sqrt{a^2+b^2}}$	$\dfrac{a}{b}$	$\pm\dfrac{\sqrt{a^2+b^2}}{a}$	$\pm\dfrac{\sqrt{a^2+b^2}}{b}$	$\dfrac{b}{a}$
7.		c	$\pm\sqrt{1-c^2}$	$\pm\dfrac{c}{\sqrt{1-c^2}}$	$\dfrac{1}{c}$	$\pm\dfrac{1}{\sqrt{1-c^2}}$	$\pm\dfrac{\sqrt{1-c^2}}{c}$
8.	I, IV	$\pm\dfrac{2ab}{a^2+b^2}$	$\dfrac{a^2-b^2}{a^2+b^2}$	$\pm\dfrac{2ab}{a^2-b^2}$	$\pm\dfrac{a^2+b^2}{2ab}$	$\dfrac{a^2+b^2}{a^2-b^2}$	$\pm\dfrac{a^2-b^2}{2ab}$
9.	III, IV	$-\dfrac{1}{\sqrt{3}}$	$\mp\sqrt{\dfrac{2}{3}}$	$\pm\dfrac{1}{\sqrt{2}}$	$-\sqrt{3}$	$\mp\sqrt{\dfrac{3}{2}}$	$\pm\sqrt{2}$
10.		$\pm\dfrac{\sqrt{c^2-m^2}}{c}$	$\dfrac{m}{c}$	$\pm\dfrac{\sqrt{c^2-m^2}}{m}$	$\pm\dfrac{c}{\sqrt{c^2-m^2}}$	$\dfrac{c}{m}$	$\pm\dfrac{m}{\sqrt{c^2-m^2}}$
11.	II, IV	$\pm\dfrac{\sqrt{14}}{4}$	$\mp\dfrac{\sqrt{2}}{4}$	$-\sqrt{7}$	$\pm\dfrac{4}{\sqrt{14}}$	$\mp\dfrac{4}{\sqrt{2}}$	$-\dfrac{1}{\sqrt{7}}$
12.	III, IV	$-\dfrac{2}{3}$	$\mp\dfrac{\sqrt{5}}{3}$	$\pm\dfrac{2}{\sqrt{5}}$	$-\dfrac{3}{2}$	$\mp\dfrac{3}{\sqrt{5}}$	$\pm\dfrac{\sqrt{5}}{2}$
13.	I, III	$\pm\dfrac{5}{\sqrt{29}}$	$\pm\dfrac{2}{\sqrt{29}}$	$\dfrac{5}{2}$	$\pm\dfrac{\sqrt{29}}{5}$	$\pm\dfrac{\sqrt{29}}{2}$	$\dfrac{2}{5}$
14.		$\pm\dfrac{\sqrt{p^2-1}}{p}$	$\dfrac{1}{p}$	$\pm\sqrt{p^2-1}$	$\pm\dfrac{p}{\sqrt{p^2-1}}$	p	$\pm\dfrac{1}{\sqrt{p^2-1}}$

18. Five of the trigonometric functions expressed in terms of the sixth. For this purpose it is again convenient to use the definitions of the functions which depend on the notion of coordinates (§ 13, p. 26). The following examples will illustrate the method.

Ex. 1. Express, in terms of $\sin x$, the other five functions of x.

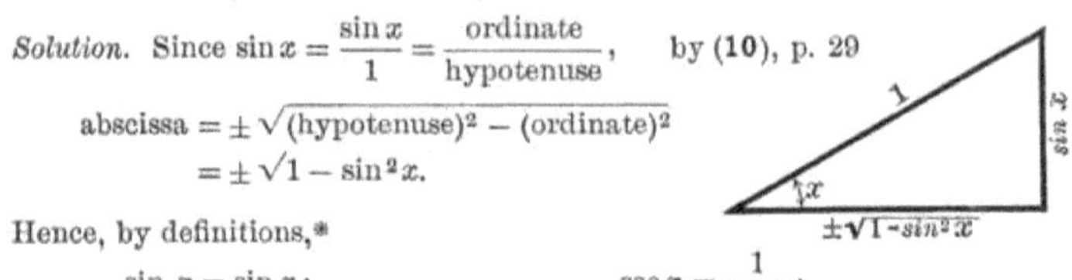

Solution. Since $\sin x = \dfrac{\sin x}{1} = \dfrac{\text{ordinate}}{\text{hypotenuse}}$,   by **(10)**, p. 29

$$\text{abscissa} = \pm \sqrt{(\text{hypotenuse})^2 - (\text{ordinate})^2}$$
$$= \pm \sqrt{1 - \sin^2 x}.$$

Hence, by definitions,*

$$\sin x = \sin x\,;\qquad\qquad \csc x = \frac{1}{\sin x}\,;$$

$$\cos x = \pm \sqrt{1 - \sin^2 x}\,;\qquad\qquad \sec x = \pm \frac{1}{\sqrt{1 - \sin^2 x}}\,;$$

$$\tan x = \pm \frac{\sin x}{\sqrt{1 - \sin^2 x}}\,;\qquad\qquad \cot x = \pm \frac{\sqrt{1 - \sin^2 x}}{\sin x}\,.$$

Ex. 2. Express, in terms of $\tan x$, the other five functions of x.

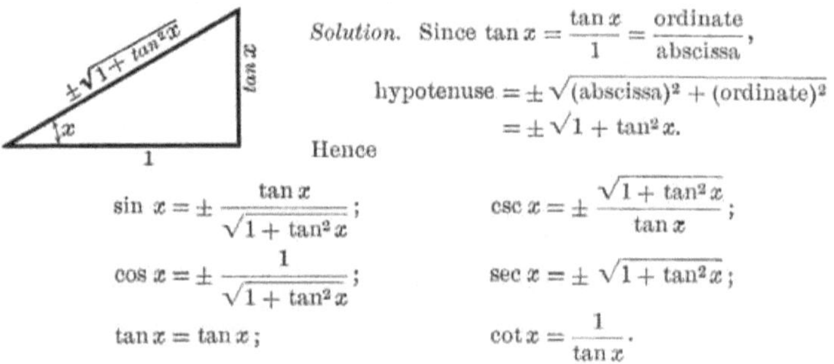

Solution. Since $\tan x = \dfrac{\tan x}{1} = \dfrac{\text{ordinate}}{\text{abscissa}}$,

$$\text{hypotenuse} = \pm \sqrt{(\text{abscissa})^2 + (\text{ordinate})^2}$$
$$= \pm \sqrt{1 + \tan^2 x}.$$

Hence

$$\sin x = \pm \frac{\tan x}{\sqrt{1 + \tan^2 x}}\,;\qquad\qquad \csc x = \pm \frac{\sqrt{1 + \tan^2 x}}{\tan x}\,;$$

$$\cos x = \pm \frac{1}{\sqrt{1 + \tan^2 x}}\,;\qquad\qquad \sec x = \pm \sqrt{1 + \tan^2 x}\,;$$

$$\tan x = \tan x\,;\qquad\qquad \cot x = \frac{1}{\tan x}\,.$$

Ex. 3. Having given $\sec x = \frac{5}{4}$, find the values of the other five functions.

Solution. Since $\sec x = \dfrac{5}{4} = \dfrac{\text{hypotenuse}}{\text{abscissa}}$,   by **(14)**, p. 29

$$\text{ordinate} = \pm \sqrt{(\text{hypotenuse})^2 - (\text{abscissa})^2}$$
$$= \pm \sqrt{25 - 16}$$
$$= \pm 3.$$

Hence

$$\sin x = \pm \tfrac{3}{5}\,;\qquad\qquad \csc x = \pm \tfrac{5}{3}\,;$$

$$\cos x = \tfrac{4}{5}\,;\qquad\qquad \sec x = \tfrac{5}{4}\,;$$

$$\tan x = \pm \tfrac{3}{4}\,;\qquad\qquad \cot x = \pm \tfrac{4}{3}\,.$$

---

* It is convenient to draw a right triangle (as above) to serve as a check on the numerical part (not the algebraic signs) of our work. We then refer to the definitions of the functions of an acute angle (p. 2) where

$$\text{adjacent side corresponds to the abscissa,}$$

and

$$\text{opposite side corresponds to the ordinate.}$$

EXAMPLES

1. Express, in terms of $\cos x$, the other five functions of x.

$$Ans. \quad \sin x = \pm \sqrt{1 - \cos^2 x}; \quad \csc x = \pm \frac{1}{\sqrt{1 - \cos^2 x}};$$

$$\cos x = \cos x; \quad \sec x = \frac{1}{\cos x};$$

$$\tan x = \pm \frac{\sqrt{1 - \cos^2 x}}{\cos x}; \quad \cot x = \pm \frac{\cos x}{\sqrt{1 - \cos^2 x}}.$$

2. Express, in terms of $\cot x$, the other five functions of x.

$$Ans. \quad \sin x = \pm \frac{1}{\sqrt{1 + \cot^2 x}}; \quad \csc x = \pm \sqrt{1 + \cot^2 x};$$

$$\cos x = \pm \frac{\cot x}{\sqrt{1 + \cot^2 x}}; \quad \sec x = \pm \frac{\sqrt{1 + \cot^2 x}}{\cot x};$$

$$\tan x = \frac{1}{\cot x}; \quad \cot x = \cot x.$$

3. Express, in terms of $\sec x$, the other five functions of x.

$$Ans. \quad \sin x = \pm \frac{\sqrt{\sec^2 x - 1}}{\sec x}; \quad \csc x = \pm \frac{\sec x}{\sqrt{\sec^2 x - 1}};$$

$$\cos x = \frac{1}{\sec x}; \quad \sec x = \sec x;$$

$$\tan x = \pm \sqrt{\sec^2 x - 1}; \quad \cot x = \pm \frac{1}{\sqrt{\sec^2 x - 1}}.$$

4. Express, in terms of $\csc x$, the other five functions of x.

$$Ans. \quad \sin x = \frac{1}{\csc x}; \quad \csc x = \csc x;$$

$$\cos x = \pm \frac{\sqrt{\csc^2 x - 1}}{\csc x}; \quad \sec x = \pm \frac{\csc x}{\sqrt{\csc^2 x - 1}};$$

$$\tan x = \pm \frac{1}{\sqrt{\csc^2 x - 1}}; \quad \cot x = \pm \sqrt{\csc^2 x - 1}.$$

5. Having given $\sec x = -\tfrac{17}{8}$, find the values of the other five functions of x.

$$Ans. \quad \sin x = \pm \tfrac{15}{17}; \quad \csc x = \pm \tfrac{17}{15};$$

$$\cos x = -\tfrac{8}{17}; \quad \sec x = -\tfrac{17}{8};$$

$$\tan x = \mp \tfrac{15}{8}; \quad \cot x = \mp \tfrac{8}{15}.$$

6. Having given $\sin x = a$, find the values of the other functions of x.

$$Ans. \quad \sin x = a; \quad \csc x = \frac{1}{a};$$

$$\cos x = \pm \sqrt{1 - a^2}; \quad \sec x = \pm \frac{1}{\sqrt{1 - a^2}};$$

$$\tan x = \pm \frac{a}{\sqrt{1 - a^2}}; \quad \cot x = \pm \frac{\sqrt{1 - a^2}}{a}.$$

7. Having given $\cot x = \sqrt{2}$, find the values of the other functions of x.

$$Ans. \quad \sin x = \pm \frac{1}{\sqrt{3}}; \quad \csc x = \pm \sqrt{3};$$

$$\cos x = \pm \sqrt{\frac{2}{3}}; \quad \sec x = \pm \sqrt{\frac{3}{2}};$$

$$\tan x = \frac{1}{\sqrt{2}}; \quad \cot x = \sqrt{2}.$$

19. Line definitions of the trigonometric functions. The definitions of the trigonometric functions given in § 15, p. 29, are called the *ratio definitions.* From these we shall now show how the functions of any

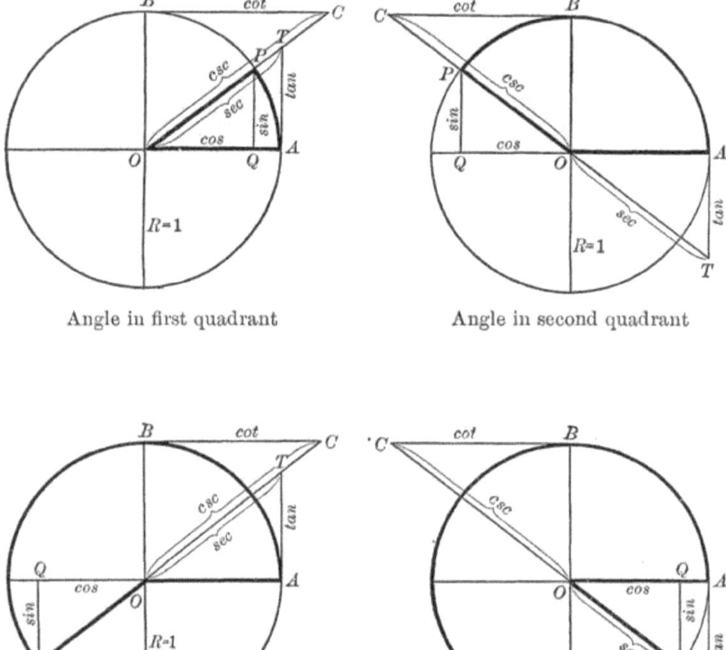

Angle in first quadrant

Angle in second quadrant

Angle in third quadrant

Angle in fourth quadrant

angle may be represented by the numerical measures of the lengths of lines drawn as shown above in connection with a unit circle (i.e. a circle with radius unity).

Applying these ratio definitions, we get

$$\sin AOP = \frac{QP}{OP\,(=1)} = \mathbf{\textit{QP}};$$

$$\cos AOP = \frac{OQ}{OP\,(=1)} = \mathbf{\textit{OQ}};$$

$$\tan AOP = \frac{QP}{OQ} = \frac{AT}{OA\,(=1)}^{\,*} = \mathbf{\textit{AT}};$$

$$\sec AOP = \frac{OP}{OQ} = \frac{OT}{OA\,(=1)}^{\,*} = \mathbf{\textit{OT}};$$

$$\cot AOP = \frac{OQ}{QP} = \frac{BC}{OB\,(=1)}^{\,\dagger} = \mathbf{\textit{BC}};$$

$$\csc AOP = \frac{OP}{QP} = \frac{OC}{OB\,(=1)}^{\,\dagger} = \mathbf{\textit{OC}}.$$

From these results the so-called *line definitions* of the trigonometric functions may be stated as follows:

*The **sin** equals the length of the perpendicular drawn from the extremity of the terminal radius to the horizontal diameter.*

*The **cos** equals the length of the line drawn from the center to the foot of this perpendicular.*

*The **tan** equals the length of a line drawn tangent to the circle from the right-hand extremity of the horizontal diameter and meeting the terminal radius produced.*

*The **sec** equals the distance from the center to the point of intersection of this tangent with the terminal radius produced.*

*The **cot** equals the length of a line drawn tangent to the circle from the upper extremity of the vertical diameter and meeting the terminal radius produced.*

*The **csc** equals the distance from the center to the point of intersection of this cotangent with the terminal radius produced.*

Algebraic signs must, however, be attached to these lengths so as to agree with the rule for the signs of the trigonometric functions on p. 29. We observe that

sin *and* **tan** *are* **positive** *if measured* **upward** *from the horizontal diameter, and* **negative** *if measured* **downward** *;*

cos *and* **cot** *are* **positive** *if measured to the* **right** *of the vertical diameter, and* **negative** *if measured to the* **left** *;*

sec *and* **csc** *are* **positive** *if measured in the* **same direction as the terminal side of the angle**, *and* **negative** *if measured in the* **opposite direction**.

---

* Since triangles OQP and OAT are similar.

† Since triangles OQP and OBC are similar.

20. Changes in the values of the functions as the angle varies.

(*a*) **The sine.** Let x denote the variable angle AOP.

As x decreases, the sine decreases through the values Q_1P_1, Q_2P_2, etc., and as x approaches zero as a limit, the sine approaches zero as a limit. This is written

$$\sin 0° = 0.$$

As x increases from $0°$ and approaches $90°$ as a limit, the sine is positive, and increases from zero through the values Q_3P_3, Q_4P_4, etc., and approaches $OB(=1)$ as a limit. This is written

$$\sin 90° = 1.$$

As x increases from $90°$ and approaches $180°$ as a limit, the sine is positive and decreases from $OB(=1)$ through Q_5P_5, etc., and approaches zero as a limit. This is written

$$\sin 180° = 0.$$

As x increases from $180°$ and approaches $270°$ as a limit, the sine is negative and increases in numerical value from zero through Q_6P_6, etc., and approaches the limit $OB'(=-1)$. This is written

$$\sin 270° = -1.$$

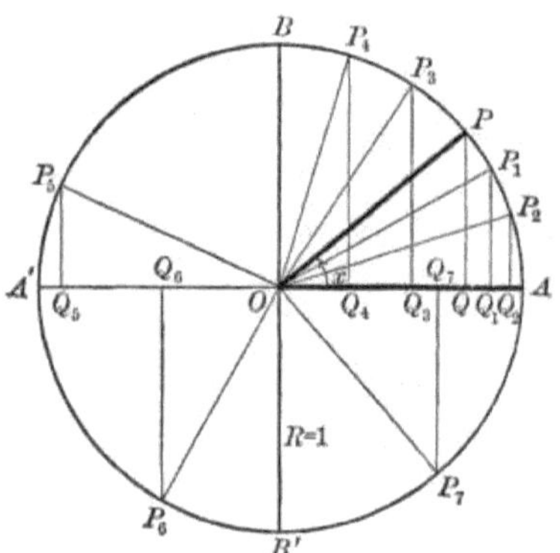

As x increases from $270°$ and approaches $360°$ as a limit, the sine is negative and decreases in numerical value from $OB'(=-1)$ through Q_7P_7, etc., and approaches the limit zero. This is written

$$\sin 360° = 0.$$

(*b*) **The cosine.** Using the last figure, we see that as x decreases, the cosine increases through the values OQ_1, OQ_2, etc., and as x approaches zero as a limit, the cosine approaches the limit $OA(=1)$. This is written

$$\cos 0° = 1.$$

As x increases from $0°$ and approaches $90°$ as a limit, the cosine is positive and decreases from $OA(=1)$ through the values OQ_3, OQ_4, etc., and approaches the limit zero. This is written

$$\cos 90° = 0.$$

As x increases from 90° and approaches 180° as a limit, the cosine is negative and increases in numerical value from zero through OQ_5, etc., and approaches the limit $OA'(=-1)$. This is written

$$\cos 180° = -1.$$

As x increases from 180° and approaches 270° as a limit, the cosine is negative and decreases in numerical value from $OA'(--1)$ through OQ_6, etc., and approaches the limit zero. This is written

$$\cos 270° = 0.$$

As x increases from 270° and approaches 360° as a limit, the cosine is positive and increases from zero through OQ_7, etc., and approaches the limit $OA (=1)$. This is written

$$\cos 360° = 1.$$

(c) **The tangent.** Let x denote the variable angle AOT.

As x decreases, the tangent decreases through the values AT_1, AT_2, etc., and as x approaches zero as a limit, the tangent approaches the limit zero. This is written

$$\tan 0° = 0.$$

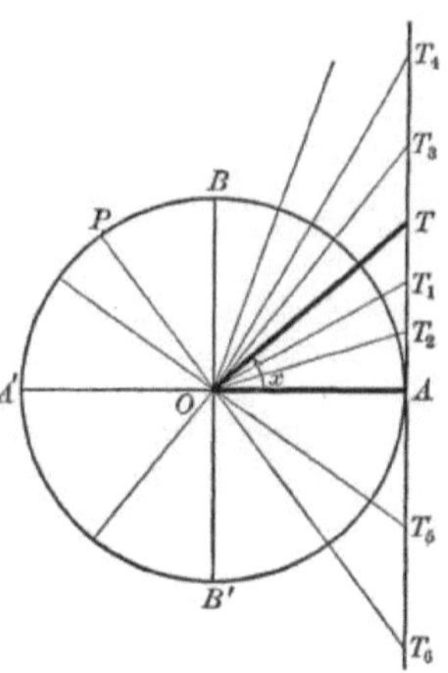

As x increases from 0° and approaches 90° as a limit, the tangent is positive and increases from zero through the values AT_3, AT_4, etc., without limit, i.e. beyond any numerical value. This is written

$$\tan 90° = +\infty.^*$$

Now suppose the angle x to be equal to the angle AOP and let it approach 90° as a limit; then the corresponding tangent AT_6 is negative and increases in numerical value without limit. This is written

$$\tan 90° = -\infty.$$

We see, then, that the limit of the tangent will be $+\infty$ or $-\infty$ according as x is increasing or decreasing as it approaches the limit 90°. As one statement these last two results are written

$$\tan 90° = \infty,$$

when, as in this book, no distinction is made for the manner in which the angle approaches the limit 90°.

---

* $+\infty$ is read *plus infinity*.
 $-\infty$ is read *minus infinity*.
 ∞ is read simply *infinity*.

As x increases from 90° and approaches 180° as a limit, the tangent is negative and decreases in numerical value from $-\infty$ through AT_6, AT_5, etc., and approaches the limit zero. This is written

$$\tan 180° = 0.$$

As x increases from 180° and approaches 270° as a limit, the tangent is positive and increases from zero through AT_3, AT_4, etc., without limit. This is written

$$\tan 270° = \infty.$$

As x increases from 270° and approaches 360° as a limit, the tangent is negative and decreases in numerical value from $-\infty$ through AT_6, AT_5, etc., and approaches the limit zero. This is written

$$\tan 360° = 0.$$

(d) **The secant.** Using the last figure, we see that as x decreases, the secant decreases through the values OT_1, OT_2, etc., and approaches $OA\,(=1)$ as a limit. This is written

$$\sec 0° = 1.$$

As x increases from 0° and approaches 90° as a limit, the secant is positive and increases from $OA\,(=1)$ through OT_3, OT_4, etc., without limit. This is written

$$\sec 90° = \infty.$$

As x increases from 90° and approaches 180° as a limit, the secant is negative and decreases in numerical value from $-\infty$ through OT_6, OT_5, etc., and approaches minus $OA\,(=-1)$ as a limit. This is written

$$\sec 180° = -1.$$

As x increases from 180° and approaches 270° as a limit, the secant is negative and increases in numerical value from minus $OA\,(=-1)$ through OT_3, OT_4, etc., without limit. This is written

$$\sec 270° = \infty.$$

As x increases from 270° and approaches 360° as a limit, the secant is positive and decreases from $+\infty$ through OT_6, OT_5, etc., and approaches the limit $OA\,(=1)$. This is written

$$\sec 360° = 1.$$

(*e*) **The cotangent.** Let x denote the variable angle AOC.

As x decreases, the cotangent increases through the values BC_1, BC_2, etc., and as x approaches 0° as a limit, the cotangent increases without limit. This is written

$$\cot 0° = \infty.$$

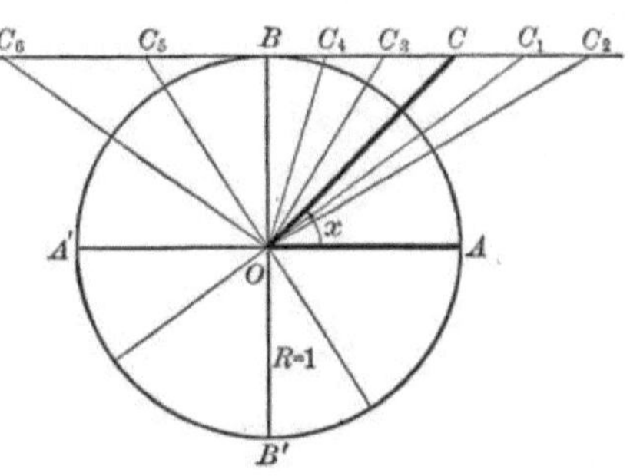

As x increases from 0° and approaches 90° as a limit, the cotangent is positive and decreases from $+\infty$ through the values BC_3, BC_4, etc., and approaches the limit zero. This is written

$$\cot 90° = 0.$$

As x increases from 90° and approaches 180° as a limit, the cotangent is negative and increases in numerical value from zero through BC_5, BC_6, etc., without limit. This is written

$$\cot 180° = \infty.$$

As x increases from 180° and approaches 270° as a limit, the cotangent is positive and decreases from $+\infty$ through BC_3, BC_4, etc., and approaches the limit zero. This is written

$$\cot 270° = 0.$$

As x increases from 270° and approaches 360° as a limit, the cotangent is negative and increases in numerical value from zero through BC_5, BC_6, etc., without limit. This is written

$$\cot 360° = \infty.$$

(*f*) **The cosecant.** Using the last figure, we see that as x decreases, the cosecant increases through the values OC_1, OC_2, etc., and as x approaches 0° as a limit, the cosecant increases without limit. This is written

$$\csc 0° = \infty.$$

As x increases from 0° and approaches 90° as a limit, the cosecant is positive and decreases from $+\infty$ through OC_3, OC_4, etc., and approaches the limit $OB(=1)$. This is written

$$\csc 90° = 1.$$

42 PLANE TRIGONOMETRY

As x increases from 90° and approaches 180° as a limit, the cosecant is positive and increases from $OB(=1)$ through OC_5, OC_6, etc., without limit. This is written

$$\csc 180° = \infty.$$

As x increases from 180° and approaches 270° as a limit, the cosecant is negative and decreases in numerical value from $-\infty$ through OC_3, OC_4, etc., and approaches the limit minus $OB(=-1)$. This is written $\csc 270° = -1.$

As x increases from 270° and approaches 360° as a limit, the cosecant is negative and increases in numerical value from minus $OB(=-1)$ through OC_5, OC_6, etc., without limit. This is written

$$\csc 360° = \infty.$$

These results may be written in tabulated form as follows : *

	0°	90°	180°	270°	360°
sin	0	1	0	1	0
cos	1	0	1	0	1
tan	0	∞	0	∞	0
cot	∞	0	∞	0	∞
sec	1	∞	1	∞	1
csc	∞	1	∞	1	∞

It is of importance to note that as an angle varies its

sine and **cosine** *can only take on values between* -1 *and* $+1$ *inclusive ;*

tangent *and* **cotangent** *can take on any values whatever ;*

secant *and* **cosecant** *can take on any values whatever, except those lying between* -1 *and* $+1$.

EXAMPLES

1. Prove the following:

(a) $\sin 0° + \cos 90° = 0.$

(b) $\sin 180° + \cos 270° = 0.$

(c) $\cos 0° + \tan 0° = 1.$

(d) $\tan 180° + \cot 90° = 0.$

(e) $\sin 270° - \sin 90° = -2.$

(f) $\cos 0° + \sin 90° = 2.$

(g) $\cos 180° + \sin 270° = -2.$

(h) $\sec 0° + \csc 90° = 2.$

(i) $\sec 180° - \sec 0° = -2.$

(j) $\cos 90° - \cos 270° = 0.$

(k) $\sin 90° + \cos 90° + \csc 90° + \cot 90° = 2.$

(l) $\cos 180° + \sec 180° + \sin 180° + \tan 180° = -2.$

(m) $\tan 360° - \sin 270° - \csc 270° + \cos 360° = 3.$

* The above table is easily memorized if the student will notice that the first four columns are composed of squares of four blocks each, in which the numbers on the diagonals are the same ; also the first two columns are identical with the next two if 1 be replaced by -1; also the first and last columns are identical.

2. Compute the values of the following expressions:

(a) $a \sin 0° + b \cos 90° - c \tan 180°$.   *Ans.* 0.

(b) $a \cos 90° - b \tan 180° + c \cot 90°$.   0.

(c) $a \sin 90° - b \cos 360° + (a - b) \cos 180°$.   0.

(d) $(a^2 - b^2) \cos 360° - 4\,ab \sin 270°$. $a^2 + 4\,ab - b^2$.

21. Angular measure. There are two systems in general use for the measurement of angles. For elementary work in mathematics and for engineering purposes the system most employed is

*Degree measure, or the sexagesimal system.** *The unit angle is* **one degree**, *being the angle subtended at the center of a circle by an arc whose length equals* $\frac{1}{360}$ *of the circumference of the circle.* The degree is subdivided into 60 *minutes*, and the minute into 60 *seconds*. Degrees, minutes, and seconds are denoted by symbols. Thus 63 degrees 15 minutes 36 seconds is written 63° 15′ 36″. Reducing the seconds to the decimal part of a minute, the angle may be written 63° 15.6′. Reducing the minutes to the decimal part of a degree, the angle may also be written 63.26°.† It has been assumed that the student is already familiar with this system of measuring angles, and the only reason for referring to it here is to compare it with the following newer system.

22. Circular measure. *The unit angle is* **one radian,** *being the angle subtended at the center of a circle by an arc whose length equals the length of the radius of the circle.*

Thus, in the figure, if the length of the arc AB equals the radius of the circle, then

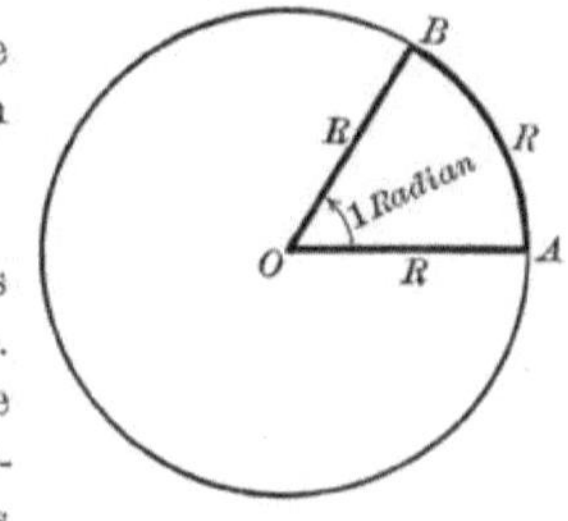

angle $AOB = 1$ radian.

The *circular measure* of an angle is its magnitude expressed in terms of radians. This system was introduced early in the last century. It is now used to a certain extent in practical work, and is universally used in the higher branches of mathematics.

Both of the above systems will be used in what follows in this book.‡

---

* Invented by the early Babylonians, whose tables of weights and measures were based on a scale of 60. This was probably due to the fact that they reckoned the year at 360 days. This led to the division of the circumference of a circle into 360 degrees. A radius laid off as a chord would then cut off 60 degrees.

† To reduce seconds to the decimal part of a minute we divide the number of seconds by 60. Similarly, we reduce minutes to the decimal part of a degree. See Conversion Tables on p. 17 of Granville's *Four-Place Tables of Logarithms.*

‡ A third system is the *Centesimal* or *French System.* The unit is *one grade*, being $\frac{1}{100}$ of a right angle. Each grade is divided into 100 *minutes* and each minute into 100 *seconds.* This system has not come into general use.

Now let us find the relation between the old and new units. From Geometry we know that the circumference of a circle equals $2\pi R$; and this means that the radius may be measured off on the circumference 2π times.* But by the above definition each radius measured off on the circumference subtends an angle of *one radian* at the center, and we also know that the angles about O equal 360°. Therefore

$$2\pi \text{ radians} = 360°,$$
$$\pi \text{ radians} = 180°,$$
$$1 \text{ radian} = \frac{180°}{\pi} = \frac{180°}{3.1416}, \text{ or,}$$

(16) $\qquad\qquad$ **1 radian = 57.2957° +.**

It therefore follows at once that:

To reduce radians to degrees, *multiply the number of radians by* $57.2957 \left(=\dfrac{180}{\pi}\right).$

To reduce degrees to radians, *divide the number of degrees by* $57.2957 \left(=\dfrac{180}{\pi}\right).$

Since 360 degrees $= 2\pi$ radians,

$$1 \text{ degree} = \frac{\pi}{180} \text{ radian} = \frac{3.1416}{180} \text{ radian, or,}$$

(17) $\qquad\qquad$ **1 degree = .01745 radian.**

Hence the above rules may also be stated as follows:

To reduce radians to degrees, *divide the number of radians by* $01745 \left(=\dfrac{\pi}{180}\right).$

To reduce degrees to radians, *multiply the number of degrees by* $.01745 \left(=\dfrac{\pi}{180}\right).$

The student should now become accustomed to expressing angles in circular measure, thus:

$$360° = 2\pi \text{ radians,} \qquad\qquad 60° = \frac{\pi}{3} \text{ radians,}$$

$$180° = \pi \text{ radians,} \qquad\qquad 30° = \frac{\pi}{6} \text{ radians,}$$

$$90° = \frac{\pi}{2} \text{ radians,} \qquad\qquad 45° = \frac{\pi}{4} \text{ radians,}$$

$$270° = \frac{3\pi}{2} \text{ radians,} \qquad\qquad 15° = \frac{\pi}{12} \text{ radians, etc.}$$

* The student should carefully observe that we do not lay off these radii as chords.

When writing the trigonometric functions of angles expressed in circular measure it is customary to omit the word "radians," thus:

$\sin (\pi$ radians) is written simply $\sin \pi$ and $= \sin 180°$,

$\tan \left(\dfrac{\pi}{2}$ radians$\right)$ is written simply $\tan \dfrac{\pi}{2}$ and $= \tan 90°$,

$\cot \left(\dfrac{3\pi}{4}$ radians$\right)$ is written simply $\cot \dfrac{3\pi}{4}$ and $= \cot 135°$,

$\cos \left(\dfrac{5\pi}{6}$ radians$\right)$ is written simply $\cos \dfrac{5\pi}{6}$ and $= \cos 150°$,

$\csc (1$ radian) is written simply $\csc 1$ and $= \csc 57.29°$,

$\sec (\tfrac{1}{2}$ radian) is written simply $\sec \tfrac{1}{2}$ and $= \sec 28.69°$, etc.

Since the number of times that the radius of a circle can be measured off on an arc of the same circle determines the number of radians in the angle subtended at the center by that arc, we have

(18) **Number of radians in angle** $= \dfrac{\text{length of subtending arc}}{\text{length of radius}}.$

Hence, knowing any two of the three quantities involved, the third may easily be found.

Ex. 1. What is the circular measure of the angle subtended by an arc of length 3.7 in. if the radius of the circle is 2 in.? Also express the angle in degrees.

Solution. Substituting in (**18**), we have

$$\text{Number of radians} = \frac{3.7}{2} = 1.85. \quad Ans.$$

To reduce this angle to degrees, we have, from (**16**),

$$1.85 \times 57.2957° = 105.997°. \quad Ans.$$

Ex. 2. What is the radius of a circle in which an arc of length 64 in. subtends an angle of 2.5 radians?

Solution. Substituting in (**18**), $2.5 = \dfrac{64}{R},$

$$R = 25.6 \text{ in.} \quad Ans.$$

EXAMPLES

1. In what quadrant does an angle lie * if its sine and cosine are both negative? if sine is positive and cosine negative? if sine is negative and cosine positive? if cosine and tangent are both negative? if cosine is positive and tangent negative? if sine and cotangent are both negative? if sine is negative and secant positive?

2. What signs must the functions of the acute angles of a right triangle have? Why?

* That is, in what quadrant will its terminal side lie?

3. What functions of an angle of an oblique triangle may be negative? Why?

4. In what quadrant do each of the following angles lie?

$$\frac{5\pi}{12}; \ -\frac{\pi}{6}; \ -\frac{7\pi}{3}; \ \frac{14\pi}{3}; \ -\frac{11\pi}{4}; \ \frac{15\pi}{16}; \ \frac{\pi+2}{6}; \ -\frac{3\pi+2}{5}; \ 2; \ \frac{1}{4}; \ -1; \ -\frac{5}{2}.$$

5. Determine the signs of the six trigonometrical functions for each one of the angles in the last example.

6. Express the following angles in degrees:

$$1.3; \ \frac{1}{2}; \ \frac{2\pi}{3}; \ -2.5; \ -\frac{3\pi}{8}; \ \frac{\pi+1}{6}; \ -3; \ -2.8; \ \frac{3\pi+2}{5}.$$

$$Ans. \ 74.4844°; \ 28.6478°; \ 120°; \ -143.239°; \ -67.5°;$$
$$39.549°; \ -171.887°; \ -160.4279°; \ 130.92°.$$

7. Express the following angles in circular measure: $22\tfrac{1}{2}°$; $60°$; $135°$; $-720°$; $990°$; $-120°$; $-100.28°$; $45.6°$; $142°\ 43.2'$; $-243.87°$; $125°\ 23'\ 19''$.

$$Ans. \ 0.3926; \ 1.0472; \ 2.3562; \ -12.56; \ 17.2788; \ -2.0944;$$
$$-1.75; \ .7958; \ 2.4909; \ -4.2563; \ 2.1884.$$

8. Express in degrees and in radians:

    (a) Seven tenths of four right angles.
    (b) Five fourths of two right angles.
    (c) Two thirds of one right angle.

$$Ans. \ (a)\ 252°,\ \frac{7\pi}{5}; \ (b)\ 225°,\ \frac{5\pi}{4}; \ (c)\ 60°,\ \frac{\pi}{3}.$$

9. Find the number of radians in an angle at the center of a circle of radius 25 ft., which intercepts an arc of $37\tfrac{1}{2}$ ft.   *Ans.* 1.5.

10. Find the length of the arc subtending an angle of $4\tfrac{1}{2}$ radians at the center of a circle whose radius is 25 ft.   *Ans.* $112\tfrac{1}{2}$ ft.

11. Find the length of the radius of a circle at whose center an angle of 1.2 radians is subtended by an arc whose length is 9.6 ft.   *Ans.* 8 ft.

12. Find the length of an arc of 80° on a circle of 4 ft. radius.   *Ans.* $5\tfrac{3}{6}\tfrac{7}{3}$ ft.

13. Find the number of degrees in an angle at the center of a circle of radius 10 ft. which intercepts an arc of 5π ft.   *Ans.* 90°.

14. Find the number of radians in an angle at the center of a circle of radius $3\tfrac{2}{11}$ inches, which intercepts an arc of 2 ft.   *Ans.* $4\tfrac{4}{5}$.

15. How long does it take the minute hand of a clock to turn through $-1\tfrac{2}{3}$ radians?

$$Ans. \ \frac{50}{\pi} \ \text{min.}$$

16. What angle in circular measure does the hour hand of a clock describe in 39 min. $22\tfrac{1}{2}$ sec.?

$$Ans. \ -\frac{7\pi}{64} \ \text{rad.}$$

17. A wheel makes 10 revolutions per second. How long does it take to turn through 2 radians, taking $\pi = \tfrac{22}{7}$?   *Ans.* $\tfrac{7}{220}$ sec.

18. A railway train is traveling on a curve of half a mile radius at the rate of 20 mi. per hour. Through what angle has it turned in 10 sec.?

$$Ans. \ 6\tfrac{4}{11} \ \text{degrees.}$$

19. The angle subtended by the sun at the eye of an observer is about half a degree. Find approximately the diameter of the sun if its distance from the observer be 90,000,000 mi.   *Ans.* 785,400 mi.

23. Reduction of trigonometric functions to functions of acute angles. The values of the functions of different angles are given in trigonometric tables, such, for instance, as the one on p. 9. These tables, however, give the trigonometric functions of angles between $0°$ and $90°$ only, while in practice we sometimes have to deal with positive angles greater than $90°$ and with negative angles. We shall now show that the trigonometric functions of an angle of any magnitude whatever, positive or negative, can be expressed in terms of the trigonometric functions of a positive angle less than $90°$, that is, of an acute angle. In fact, we shall show, although this is of less importance, that the functions of any angle can be found in terms of the functions of a positive angle less than $45°$.

In the next eighteen sections x and y denote acute angles.

24. Functions of complementary angles. To make our discussion complete we repeat the following from p. 3.

Theorem. *A function of an acute angle is equal to the co-function of its complementary acute angle.*

Ex. Express $\sin 72°$ as the function of a positive angle less than $45°$.

Solution. Since $90° - 72° = 18°$, $72°$ and $18°$ are complementary, and we get

$$\sin 72° = \cos 18°. \quad Ans.$$

EXAMPLES

1. Express the following as functions of the complementary angle :

(a) $\cos 68°$.	(e) $\cot 9.167°$.	(i) $\csc 52° \; 18'$.
(b) $\tan 48.6°$.	(f) $\sin 72° \; 51' \; 43''$.	(j) $\cot \dfrac{2\pi}{5}$.
(c) $\sec 81° \; 16'$.	(g) $\cos \dfrac{\pi}{6}$.	
(d) $\sin \dfrac{\pi}{3}$.	(h) $\sec 19° \; 29.8'$.	(k) $\sin 1.2$.
		(l) $\tan 66° \; 22.3'$.

2. Show that in a right triangle any function of one of the acute angles equals the co-function of the other acute angle.

3. If A, B, C are the angles of any triangle, prove that

$$\sin \tfrac{1}{2} A = \cos \tfrac{1}{2} (B + C).$$

25. Reduction of functions of angles in the second quadrant.

First method. In the unit circle whose center is O (see figure on next page), let AOP' be any angle in the second quadrant. The functions of any such angle are the same as the corresponding functions of the positive angle $AOP' = 180° - P'OQ'$. Let x be the measure of the acute angle $P'OQ'$, and construct $AOP = P'OQ' = x$.

Now draw the lines representing all the functions of the supplemental angles x and $180° - x$. From the figure

$$\text{angle } QOP = \text{angle } P'OQ', \qquad \text{by construction}$$
$$OP = OP'. \qquad \text{equal radii}$$

Therefore the right triangles OPQ and $OP'Q'$ are equal, giving

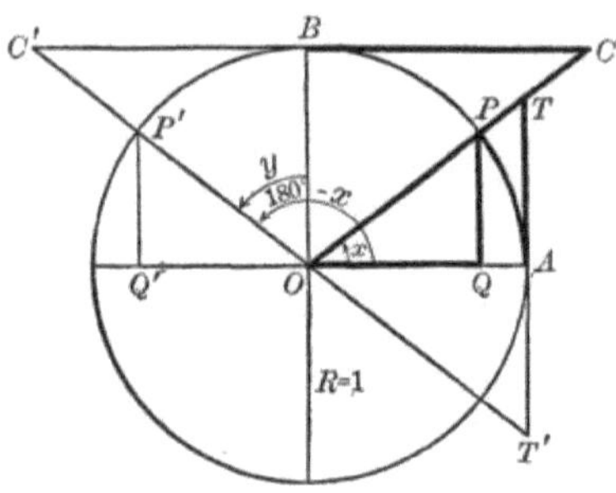

$$OQ' = OQ.$$

But $OQ' = \cos(180° - x)$ and $OQ = \cos x$; hence $\cos(180° - x)$ equals $\cos x$ in numerical value. Since they have opposite signs, however, we get

$$\cos(180° - x) = -\cos x.$$

Also, from the same triangles,

$$Q'P' = QP.$$

But $Q'P' = \sin(180° - x)$ and $QP = \sin x$, and since they have the same sign, we get

$$\sin(180° - x) = \sin x.$$

Similarly, the two right triangles OTA and $OT'A$ may be proven equal, giving

$$AT' = AT \text{ and } OT' = OT,$$

or, $\quad \tan(180° - x) = -\tan x$ and $\sec(180° - x) = -\sec x.$

In the same manner, by proving the right triangles OBC and OBC' equal, we get

$$BC' = BC \text{ and } OC' = OC,$$

or, $\quad \cot(180° - x) = -\cot x$ and $\csc(180° - x) = \csc x.$

Collecting these results, we have

$$\sin(180° - x) = \sin x; \qquad \csc(180° - x) = \csc x;$$
$$\cos(180° - x) = -\cos x; \qquad \sec(180° - x) = -\sec x;$$
$$\tan(180° - x) = -\tan x; \qquad \cot(180° - x) = -\cot x.$$

Hence we have the

Theorem. *The functions of an angle in the second quadrant equal numerically the same-named functions of the acute angle between its terminal side and the terminal side of 180°. The algebraic signs, however, are those for an angle in the second quadrant.*

Ex. 1. Express $\sin 123°$ as the function of an acute angle, and find its value.

Solution. Since $180° - 123° = 57°$,

$$\sin 123° = \sin (180° - 57°) = \sin 57° = .8387 \text{ (p. 9).} \quad \textit{Ans.}$$

Ex. 2. Find the value of $\sec \dfrac{5\,\pi}{6}$.

Solution. $\sec \dfrac{5\,\pi}{6} = \sec 150° = \sec (180° - 30°) = -\sec 30° = -\dfrac{2}{\sqrt{3}}.$ *Ans.*

Ex. 3. Find $\tan 516°$.

Solution. $516°$ is an angle in the second quadrant, for $516° - 360° = 156°$.
Hence $\tan 516° = \tan 156°* = \tan (180° - 24°) = -\tan 24° = -.4452.$ *Ans.*

Second method. The angle AOP' may also be written $90° + y$, where y measures the acute angle BOP'. Since the angles BOP' and $P'OQ'$ are complementary, we have, from theorem on p. 47,

$$\begin{array}{ll}
\sin x = \cos y; & \csc x = \sec y; \\
\cos x = \sin y; & \sec x = \csc y; \\
\tan x = \cot y; & \cot x = \tan y.
\end{array}$$

Since $180° - x = 90° + y$, we get, combining the above results with the results on the previous page,

$$\begin{array}{ll}
\sin (90° + y) = \cos y; & \csc (90° + y) = \sec y; \\
\cos (90° + y) = -\sin y; & \sec (90° + y) = -\csc y; \\
\tan (90° + y) = -\cot y; & \cot (90° + y) = -\tan y.
\end{array}$$

Hence we have the

Theorem. *The functions of an angle in the second quadrant equal numerically the co-named functions of the acute angle between its terminal side and the terminal side of $90°$. The algebraic signs, however, are those for an angle in the second quadrant.*

Ex. 4. Find the value of $\cos 109°$.

Solution. Since $109° = 90° + 19°$,

$$\cos 109° = \cos (90° + 19°) = -\sin 19° = -.3256. \quad \textit{Ans.}$$

Ex. 5. Find the value of $\cos \dfrac{19\,\pi}{4}$.

Solution. $\dfrac{19\,\pi}{4} = 855° = 720° + 135°.$

Therefore

$$\cos \frac{19\,\pi}{4} = \cos 855° = \cos 135° = \cos (90° + 45°) = -\sin 45° = -\frac{1}{\sqrt{2}}. \quad \textit{Ans.}$$

The above two methods teach us how to do the same thing, namely, *how to find the functions of an angle in the second quadrant in terms of the functions of an acute angle.* The first method is generally to be preferred, however, as the name of the function does not change, and hence we are less likely to make a mistake.

* The above theorem was proven for an angle of any magnitude whatever whose terminal side lies in the second quadrant. The generating line of the angle may have made one or more complete revolutions before assuming the position of the terminal side. In that case we should first (if the revolutions have been counter-clockwise, i.e. in the positive direction) subtract such a multiple of $360°$ from the angle that the remainder will be a positive angle less than $360°$.

EXAMPLES

1. Construct a table of sines, cosines, and tangents of all angles from 0° to 180° at intervals of 30°.

Ans.

	0°	30°	60°	90°	120°	150°	180°
sin	0	$\dfrac{1}{2}$	$\dfrac{\sqrt{3}}{2}$	1	$\dfrac{\sqrt{3}}{2}$	$\dfrac{1}{2}$	0
cos	1	$\dfrac{\sqrt{3}}{2}$	$\dfrac{1}{2}$	0	$-\dfrac{1}{2}$	$-\dfrac{\sqrt{3}}{2}$	-1
tan	0	$\dfrac{1}{\sqrt{3}}$	$\sqrt{3}$	∞	$-\sqrt{3}$	$-\dfrac{1}{\sqrt{3}}$	0

2. Construct a table of sines, cosines, and tangents of all angles from 90° to 180° at intervals of 15°, using table on p. 9.

Ans.

	90°	105°	120°	135°	150°	165°	180°
sin	1.0000	.9659	.8660	.7071	.5000	.2588	0.0000
cos	0.0000	$-.2588$	$-.5000$	$-.7071$	$-.8660$	$-.9659$	-1.0000
tan	∞	-3.7321	-1.7321	-1.0000	$-.5774$	$-.2679$	0.0000

3. Construct a table of sines, cosines, and tangents of all angles from 90° to 135° at intervals of 5°.

4. Express the following as functions of an acute angle :

(a) $\sin 138°$.
(b) $\tan 883°$.
(c) $\cos 165°\ 20'$.
(d) $\sec 102°\ 18'$.

(e) $\cot \dfrac{4\pi}{5}$.
(f) $\cot 170.48°$.
(g) $\csc 317°$.

(h) $\sin \dfrac{13\pi}{5}$.
(i) $\cos 2.58$.
(j) $\tan 1.5$.

5. Find values of the following :

(a) $\sin 128° = .788$.
(b) $\cos 160° = -.9397$.
(c) $\tan 135° = -1$.
(d) $\sec \dfrac{2\pi}{3} = -2$.
(e) $\cot \dfrac{11\pi}{4} = -1$.
(f) $\csc 835° = 1.1034$.

(g) $\sin \dfrac{8\pi}{7}$.
(h) $\tan 108°\ 15'$.
(i) $\cos 173°\ 9.4'$.
(j) $\tan \dfrac{5\pi}{6}$.
(k) $\cos 496.7°$.
(l) $\sec 168.42°$.

(m) $\cot 95°\ 14'$.
(n) $\csc 126°\ 42.8'$.
(o) $\sin \dfrac{7\pi}{9}$.
(p) $\cos 500°$.
(q) $\tan 870°$.
(r) $\sec 1.9°$.
(s) $\tan 1$.

6. Express the following as functions of an acute angle less than 45° :

(a) $\sin 106° = \cos 16°$.
(b) $\cos 148.3° = -\cos 31.7°$.
(c) $\tan 862°$.
(d) $\sec 794°\ 52'$.

(e) $\csc \dfrac{11\pi}{12}$.
(f) $\cos \dfrac{23\pi}{9}$.

26. Reduction of functions of angles in the third quadrant.

First method. In the unit circle whose center is O, let AOP' be any angle in the third quadrant. The functions of any such angle are the same as the corresponding functions of the positive angle $AOP' = 180° + Q'OP'$. Let x be the measure of the acute angle $Q'OP'$, and construct $AOP = Q'OP' = x$.

Now drawing the lines representing all the functions of the angles x and $180° + x$, we get, just as in the previous case,

$$\sin(180° + x) = -\sin x; \qquad \csc(180° + x) = -\csc x;$$
$$\cos(180° + x) = -\cos x; \qquad \sec(180° + x) = -\sec x;$$
$$\tan(180° + x) = \tan x; \qquad \cot(180° + x) = \cot x.$$

Hence we have the

Theorem. *The functions of an angle in the third quadrant equal numerically the same-named functions of the acute angle between its terminal side and the terminal side of 180°. The algebraic signs, however, are those for an angle in the third quadrant.*

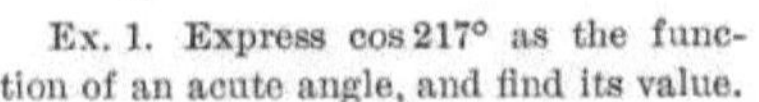

Ex. 1. Express $\cos 217°$ as the function of an acute angle, and find its value.

Solution. Since $217° - 180° = 37°$,

$$\cos 217° = \cos(180° + 37°) = -\cos 37° = -.7986. \ \textit{Ans.}$$

Ex. 2. Find value of $\csc 225°$.

Solution. $\csc 225° = \csc(180° + 45°) = -\csc 45° = -\sqrt{2}. \ \textit{Ans.}$

Ex. 3. Find value of $\sin 600°$.

Solution. $600°$ is an angle in the third quadrant, for $600° - 360° = 240°$.

Hence $\sin 600° = \sin 240° = \sin(180° + 60°) = -\sin 60° = -\dfrac{\sqrt{3}}{2}. \ \textit{Ans.}$

Second method. The angle AOP' may also be written $270° - y$, where y measures the acute angle $P'OB'$. Since the angles $P'OB'$ and $Q'OP'$ $(= AOP)$ are complementary, we have, from theorem on p. 47, combined with the above results, remembering that $180° + x = 270° - y$,

$$\sin(270° - y) = -\cos y; \qquad \csc(270° - y) = -\sec y;$$
$$\cos(270° - y) = -\sin y; \qquad \sec(270° - y) = -\csc y;$$
$$\tan(270° - y) = \cot y; \qquad \cot(270° - y) = \tan y.$$

Hence we have the

Theorem. *The functions of an angle in the third quadrant equal numerically the co-named functions of the acute angle between its terminal side and the terminal side of 270°. The algebraic signs, however, are those of an angle in the third quadrant.*

Ex. 4. Find $\sin 259°$.

Solution. Since $270° - 11° = 259°$,

$$\sin 259° = \sin(270° - 11°) = -\cos 11° = -.9816. \quad Ans.$$

As in the last case, the first method is generally to be preferred.

EXAMPLES

1. Construct a table of sines, cosines, and tangents of all angles from $0°$ to $270°$ at intervals of $45°$.

Ans.

	0°	45°	90°	135°	180°	225°	270°
sin	0	$\dfrac{1}{\sqrt{2}}$	1	$\dfrac{1}{\sqrt{2}}$	0	$-\dfrac{1}{\sqrt{2}}$	-1
cos	1	$\dfrac{1}{\sqrt{2}}$	0	$-\dfrac{1}{\sqrt{2}}$	-1	$-\dfrac{1}{\sqrt{2}}$	0
tan	0	1	∞	-1	0	1	∞

2. Construct a table of sines, cosines, and tangents of all angles from $180°$ to $270°$ at intervals of $15°$, using table on p. 9.

Ans.

	180°	195°	210°	225°	240°	255°	270°
sin	0	$-.2588$	$-.5000$	$-.7071$	$-.8660$	$-.9659$	-1.0000
cos	-1.0000	$-.9659$	$-.8660$	$-.7071$	$-.5000$	$-.2588$	0
tan	0	.2679	.5774	1.0000	1.7321	3.7321	∞

3. Construct a table of sines, cosines, and tangents of all angles from $135°$ to $270°$ at intervals of $5°$.

4. Express the following as functions of an acute angle:

(a) $\tan 200°$.

(b) $\sin 583°$.

(c) $\cos 224° \ 26'$.

(d) $\sec 260° \ 40'$.

(e) $\cot \dfrac{7\pi}{5}$.

(f) $\csc 4.3$.

(g) $\sin 128°$.

(h) $\cos 998.7°$.

(i) $\sin \dfrac{16\pi}{5}$.

(j) $\cos \dfrac{8\pi}{3}$.

5. Find values of the following:

(a) $\tan 235° = 1.4281$.

(b) $\cot 1300° = 1.1918$.

(c) $\sin 212° \ 16'$.

(d) $\cos \dfrac{4\pi}{3} = -\dfrac{1}{2}$.

(e) $\sec \dfrac{7\pi}{6} = -\dfrac{2}{\sqrt{3}}$.

(f) $\sin 609°$.

(g) $\cos \dfrac{13\pi}{12}$.

(h) $\tan 4$.

(i) $\cot \dfrac{29\pi}{9}$.

(j) $\csc \dfrac{21\pi}{4}$.

(k) $\sin 228.4°$.

(l) $\tan 255° \ 27.8'$.

(m) $\cot 185° \ 52'$.

(n) $\cos 587°$.

(o) $\csc \dfrac{8\pi}{7}$.

(p) $\sin 262° \ 10'$.

(q) $\cos 204.86°$.

(r) $\tan \dfrac{9\pi}{8}$.

6. Express the following as functions of an acute angle less than 45°:

(a) $\cos \dfrac{17\,\pi}{12}$.

(b) $\tan 236.5°$.

(c) $\sin 594°$.

(d) $\sec \dfrac{5\,\pi}{4}$.

(e) $\sin \dfrac{23\,\pi}{7}$.

(f) $\cos 260° \, 53.4'$.

27. Reduction of functions of angles in the fourth quadrant.

First method. As before, let AOP' be any angle in the fourth quadrant. The functions of any such angle are the same as the corresponding functions of the positive angle $AOP' = 360° - P'OQ$. Let x be the measure of the acute angle $P'OQ$, and construct $AOP = P'OQ = x$. Now, drawing the lines representing all the functions of the angles x and $360° - x$, we get, just as in the previous cases,

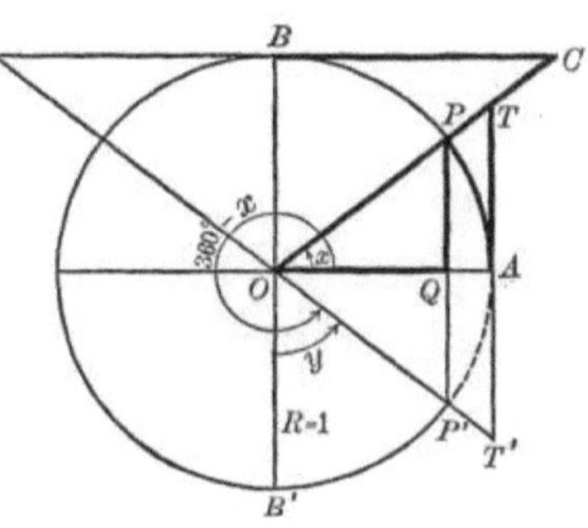

$$\sin (360° - x) = -\sin x; \qquad \csc (360° - x) = -\csc x;$$
$$\cos (360° - x) = \cos x; \qquad \sec (360° - x) = \sec x;$$
$$\tan (360° - x) = -\tan x; \qquad \cot (360° - x) = -\cot x;$$

Hence we have the

Theorem. *The functions of an angle in the fourth quadrant equal numerically the same-named functions of the acute angle between its terminal side and the terminal side of 360°. The algebraic signs, however, are those for an angle in the fourth quadrant.*

Ex. 1. Express $\sin 327°$ as the function of an acute angle, and find its value.

Solution. Since $360° - 327° = 33°$,

$$\sin 327° = \sin (360° - 33°) = -\sin 33° = -.5446. \quad Ans.$$

Ex. 2. Find value of $\cot \dfrac{5\,\pi}{3}$.

Solution. $\cot \dfrac{5\,\pi}{3} = \cot 300° = \cot (360° - 60°) = -\cot 60° = -\dfrac{1}{\sqrt{3}}. \quad Ans.$

Ex. 3. Find value of $\cos 1000°$.

Solution. This is an angle in the fourth quadrant, for $1000° - 720° = 280°$.

Hence $\cos 1000° = \cos 280° = \cos (360° - 80°) = \cos 80° = .1736. \quad Ans.$

Second method. The angle AOP' may also be written $270° + y$, where y measures the acute angle $B'OP'$. Since the angles $B'OP'$ and $P'OQ$ are complementary, we have, from theorem on p. 47, combined with the above results, remembering that $360° - x = 270° + y$,

$$\sin(270° + y) = -\cos y; \qquad \csc(270° + y) = -\sec y;$$
$$\cos(270° + y) = \sin y; \qquad \sec(270° + y) = \csc y;$$
$$\tan(270° + y) = -\cot y; \qquad \cot(270° + y) = -\tan y.$$

Hence we have the

Theorem. *The functions of an angle in the fourth quadrant equal numerically the co-named functions of the acute angle between its terminal side and the terminal side of 270°. The algebraic signs, however, are those of an angle in the fourth quadrant.*

Ex. 4. Find value of $\cos \dfrac{11\pi}{6}$.

Solution. $\cos \dfrac{11\pi}{6} = \cos 330° = \cos(270° + 60°) = \sin 60° = \dfrac{\sqrt{3}}{2}.$ *Ans.*

As before, the first method is generally to be preferred.

EXAMPLES

1. Construct a table of sines, cosines, and tangents of all angles from 180° to 360° at intervals of 30°.

Ans.

	180°	210°	240°	270°	300°	330°	360°
sin	0	$-\dfrac{1}{2}$	$-\dfrac{\sqrt{3}}{2}$	-1	$-\dfrac{\sqrt{3}}{2}$	$-\dfrac{1}{2}$	0
cos	-1	$-\dfrac{\sqrt{3}}{2}$	$-\dfrac{1}{2}$	0	$\dfrac{1}{2}$	$\dfrac{\sqrt{3}}{2}$	1
tan	0	$\dfrac{1}{\sqrt{3}}$	$\sqrt{3}$	∞	$-\sqrt{3}$	$-\dfrac{1}{\sqrt{3}}$	0

2. Construct a table of sines, cosines, and tangents of all angles from 270° to 360° at intervals of 15°, using table on p. 9.

Ans.

	270°	285°	300°	315°	330°	345°	360°
sin	-1.0000	$-.9659$	$-.8660$	$-.7071$	$-.5000$	$-.2588$	0
cos	0	.2588	.5000	.7071	.8660	.9659	1.0000
tan	∞	-3.7321	-1.7321	-1.0000	$-.5774$	$-.2679$	0

3. Construct a table of sines, cosines, and tangents of all angles from 270° to 360° at intervals of 5°.

4. Express the following as functions of an acute angle:

(a) $\sin 289°$.

(b) $\cos 322.4°$.

(c) $\tan 295° \ 43'$.

(d) $\cot 356° \ 11'$.

(e) $\sin 655°$.

(f) $\csc \dfrac{9\,\pi}{5}$.

(g) $\sin 275.5°$.

(h) $\cos \dfrac{18\,\pi}{5}$.

(i) $\sec 246°$.

(j) $\tan \dfrac{3\,\pi}{4}$.

5. Find values of the following:

(a) $\sin 275° = -.9962$.

(b) $\cos 336° = .9135$.

(c) $\tan 687° = -.6494$.

(d) $\cot 1055°$.

(e) $\sec 295° \ 52.6'$.

(f) $\sin \dfrac{17\,\pi}{9}$.

(g) $\csc 5.2$.

(h) $\cos \dfrac{15\,\pi}{4}$.

(i) $\csc \dfrac{11\,\pi}{6}$.

(j) $\tan \dfrac{5\,\pi}{3}$.

(k) $\sin 275° \ 22'$.

(l) $\cot 348°$.

(m) $\tan 660°$.

(n) $\sec \dfrac{13\,\pi}{8}$.

(o) $\sin \dfrac{5\,\pi}{6}$.

28. Reduction of functions of negative angles. Simple relations exist between the functions of the angle x and $-x$ where x is any angle whatever. It is evident that x and $-x$ will lie, one in the first quadrant and the other in the fourth quadrant, as angles AOP and AOP'

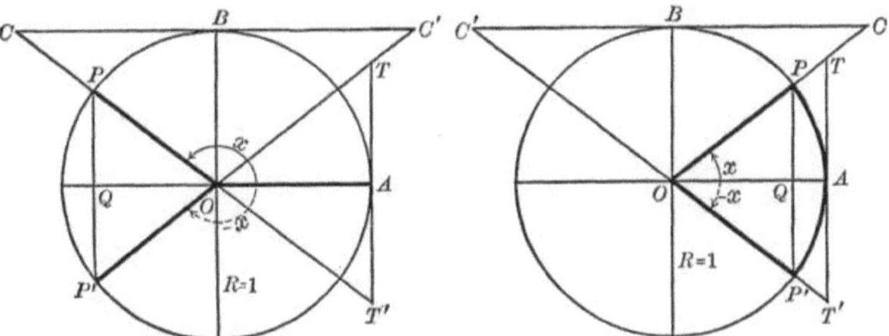

in the right-hand figure; or, one will lie in the second quadrant and the other in the third quadrant, as the angles AOP and AOP' in the left-hand figure. In either figure, remembering the rule for signs (§ 16, p. 29), we get

$$QP = -QP', \qquad \therefore \sin x = -\sin(-x);$$
$$OQ = OQ, \qquad \therefore \cos x = \cos(-x);$$
$$AT = -AT', \qquad \therefore \tan x = -\tan(-x);$$
$$OT = OT', \qquad \therefore \sec x = \sec(-x);$$
$$BC = -BC', \qquad \therefore \cot x = -\cot(-x);$$
$$OC = -OC', \qquad \therefore \csc x = -\csc(-x).$$

We may write these results in the form

$$\sin(-x) = -\sin x; \qquad \csc(-x) = -\csc x;$$
$$\cos(-x) = \cos x; \qquad \sec(-x) = \sec x;$$
$$\tan(-x) = -\tan x; \qquad \cot(-x) = -\cot x.$$

Hence we have the

Theorem. *The functions of $-x$ equal numerically the same-named functions of x. The algebraic sign, however, will change for all functions except the cosine and secant.**

Ex. 1. Express $\tan(-29°)$ as the function of an acute angle, and find its value.
Solution. $\tan(-29°) = -\tan 29° = -.5543.$ *Ans.*

Ex. 2. Find value of $\sec(-135°)$.
Solution. $\sec(-135°) = \sec 135° = \sec(180° - 45°) = -\sec 45° = -\sqrt{2}.$ *Ans.*

Ex. 3. Find value of $\sin(-540°)$.
Solution. $\sin(-540°) = -\sin 540° = -\sin(360° + 180°) = -\sin 180° = 0.$ *Ans.*

EXAMPLES

1. Construct a table of sines, cosines, and tangents of all angles from $0°$ to $-360°$ at intervals of $30°$. *Ans.*

2. Find values of the following:

(a) $\tan(-33°) = -.6494.$

(b) $\sin(-60°) = -\dfrac{\sqrt{3}}{2}.$

(c) $\cos(-145°) = -\dfrac{1}{\sqrt{2}}.$

(d) $\cot(-259°).$

(e) $\sec\left(-\dfrac{\pi}{3}\right).$

(f) $\sin(-1231°).$

(g) $\cos\left(-\dfrac{17\pi}{ }\right).$

(h) $\sin(-1000°).$

(i) $\cos(-2.3).$

(j) $\cot\left(-\dfrac{5\pi}{8}\right).$

(k) $\sin(-176.9°).$

(l) $\cos(-88° 12.7')$

(m) $\tan\left(-\dfrac{\pi}{4}\right).$

(n) $\cot(-842°).$

Angle	sin	cos	tan
$0°$	0	1	0
$-30°$	$-\dfrac{1}{2}$	$\dfrac{\sqrt{3}}{2}$	$-\dfrac{1}{\sqrt{3}}$
$-60°$	$-\dfrac{\sqrt{3}}{2}$	$\dfrac{1}{2}$	$-\sqrt{3}$
$-90°$	-1	0	∞
$-120°$	$-\dfrac{\sqrt{3}}{2}$	$-\dfrac{1}{2}$	$\sqrt{3}$
$-150°$	$-\dfrac{1}{2}$	$-\dfrac{\sqrt{3}}{2}$	$\dfrac{1}{\sqrt{3}}$
$-180°$	0	-1	0
$-210°$	$\dfrac{1}{2}$	$-\dfrac{\sqrt{3}}{2}$	$-\dfrac{1}{\sqrt{3}}$
$-240°$	$\dfrac{\sqrt{3}}{2}$	$-\dfrac{1}{2}$	$-\sqrt{3}$
$-270°$	1	0	∞
$-300°$	$\dfrac{\sqrt{3}}{2}$	$\dfrac{1}{2}$	$\sqrt{3}$
$-330°$	$\dfrac{1}{2}$	$\dfrac{\sqrt{3}}{2}$	$\dfrac{1}{\sqrt{3}}$
$-360°$	0	1	0

* Another method for reducing the functions of a negative angle consists in adding such a multiple of $+360°$ to the negative angle that the sum becomes a positive angle less than $360°$. The functions of this positive angle will be the same as the functions of the given negative angle, since their terminal sides will coincide. To illustrate :

Ex. Find value of $\cos(-240°)$.
Solution. Adding $+360°$ to $-240°$ gives $+120°$.
Hence $\cos(-240°) = \cos 120° = \cos(180° - 60°) = -\cos 60° = -\tfrac{1}{2}.$ *Ans.*

29. General rule for reducing the functions of any angle to the functions of an acute angle. The results of the last seven sections may be stated in compact form as follows, x being an acute angle.*

General Rule.

I. *Whenever the angle is $180° \pm x$ or $360° \pm x$, the functions of the angle are numerically equal to the **same-named** functions of x.*

II. *Whenever the angle is $90° \pm x$ or $270° \pm x$, the functions of the angle are numerically equal to the **co-named** functions of x.*

III. *In any case the sign of the result is the same as the sign of the given function taken in the quadrant where the given angle lies.*

The student is advised to use I wherever possible, since the liability of making a mistake is less when the name of the function remains unchanged throughout the operation. Work out examples from pp. 50–56, applying the above general rule.

EXAMPLES

1. Construct a table for every five degrees from $90°$ to $180°$.

	Angle	sin	cos	tan	cot	sec	csc
Ans.	$90°$	1.0000	0.0000	∞	0.0000	∞	1.0000
	$95°$	.9962	$-.0872$	-11.430	$-.0875$	-11.474	1.0038
	$100°$	.9848	$-.1736$	-5.6713	$-.1763$	-5.7588	1.0154
	. . .	. . .	. . .	. . .	. . .	. . .	. . .

2. Construct a table as in Ex. 1 for every $15°$ from $180°$ to $270°$.

3. Construct a table as in Ex. 1 for every $10°$ from $0°$ to $-90°$.

4. Reduce the following to functions of x :

(a) $\sin(x - 90°) = -\cos x.$†

(b) $\cos(x - \pi) = -\cos x.$

(c) $\tan\left(-x - \dfrac{3\pi}{2}\right) = \cot x.$

(d) $\cot(x - 2\pi) = \cot x.$

(e) $\sec(x - 180°) = -\sec x.$

(f) $\csc\left(-x - \dfrac{\pi}{2}\right) = -\sec x.$

(g) $\sin(x - 270°).$

(h) $\cos(-x - \pi).$

(i) $\tan\left(x - \dfrac{\pi}{2}\right).$

(j) $\cot(-x - 8\pi).$

(k) $\sec(x - 630°).$

(l) $\csc(x - 720°).$

---

* In case the given angle is greater than $360°$ we assume that it has first been reduced to a positive angle less than $360°$ by the subtraction of some multiple of $360°$. Or, if the given angle is negative, we assume that it has been reduced to a positive angle by the theorem on p. 56.

† Since x is acute, $x - 90°$ is a negative angle. Hence $\sin(x - 90°) = -\sin(90° - x) = -\cos x.$

5. Find values of the following :

(a) $\cos 420°$.

(b) $\sin 768°$.

(c) $\sec\left(-\dfrac{4\,\pi}{3}\right)$.

(d) $\cot(-240°)$.

(e) $\csc\dfrac{13\,\pi}{3}$.

(f) $\tan 7.5$.

(g) $\sin(-2.8)$.

(h) $\cos 952.8°$.

(i) $\cot 549° \ 39'$.

(j) $\csc 387° \ 58'$.

(k) $\sec\left(-\dfrac{11\,\pi}{6}\right)$.

(l) $\tan\left(-\dfrac{3\,\pi}{4}\right)$.

(m) $\sin(-830°)$.

(n) $\cos\dfrac{9\,\pi}{4}$.

(o) $\cot 1020°$.

( ) $\sec\dfrac{13\,\pi}{6}$.

(q) $\sin(-5.3)$.

(r) $\cos\left(-\dfrac{}{12}\right)$.

(s) $\tan\left(-\dfrac{7\,\pi}{2}\right)$.

(t) $\sec(-123.8°)$.

(u) $\sin(-256° \ 19.6')$

(v) $\cos(-98° \ 31')$.

6. Prove the following :

(a) $\sin 420° \cdot \cos 390° + \cos(-300°) \cdot \sin(-330°) = 1$.

(b) $\cos 570° \cdot \sin 510° - \sin 330° \cdot \cos 390° = 0$.

(c) $a \cos(90° - x) + b \cos(90° + x) = (a - b)\sin x$.

(d) $m \cos\left(\dfrac{\pi}{2}\ x\right) \cdot \sin\left(\dfrac{\pi}{2} - x\right) = m \sin x \cos x$.

(e) $(a - b)\tan(90° - x) + (a + b)\cot(90° + x) = (a - b)\cot x - (a + b)\tan x$.

(f) $\sin\left(\dfrac{\pi}{2} + x\right)\sin(\pi + x) + \cos\left(\dfrac{\pi}{2} + x\right)\cos(\pi - x) = 0$.

(g) $\cos(\pi + x)\cos\left(\dfrac{3\,\pi}{2} - y\right) - \sin(\pi + x)\sin\left(\dfrac{3\,\pi}{2} - y\right) = \cos x \sin y - \sin x \cos y$.

(h) $\tan x + \tan(-y) - \tan(\pi - y) - \tan x$.

(i) $\cos(90° + a)\cos(270° - a) - \sin(180° - a)\sin(360° - a) = 2\sin^2 a$.

(j) $\dfrac{\sin(180° - y)}{\sin(270° - y)}\cdots(90° + y) + \dfrac{1}{\sin^2(270° - y)} = 1 - \sec^2 y$.

(k) $3 \tan 210° + 2 \tan 120° = -\sqrt{3}$.

(l) $5 \sec^2 135° - 6 \cot^2 300° = -8$.

(m) $\cos\tfrac{1}{3}(x - 270°) = \sin\tfrac{1}{3}x$.

(n) $\tan^1(2\,\pi + x) = \tan^1 x$.

(o) $\csc^1(x - 2\,\pi) = -\sec^1 x$.

(p) $\cos\tfrac{1}{3}(y - 810°) = -\sin\tfrac{1}{3}y$.

CHAPTER III

RELATIONS BETWEEN THE TRIGONOMETRIC FUNCTIONS

30. Fundamental relations between the functions. From the definitions (and footnote) on p. 29 we have at once the *reciprocal relations*

$$(19) \qquad \sin x = \frac{1}{\csc x}, \qquad \csc x = \frac{1}{\sin x};$$

$$(20) \qquad \cos x = \frac{1}{\sec x}, \qquad \sec x = \frac{1}{\cos x};$$

$$(21) \qquad \tan x = \frac{1}{\cot x}, \qquad \cot x = \frac{1}{\tan x}.$$

Making use of the unit circle, we shall now derive five more very important relations between the functions.

In the right triangle QPO

$$\tan x = \frac{QP}{OQ}, \quad \text{and} \quad \cot x = \frac{OQ}{QP}.$$

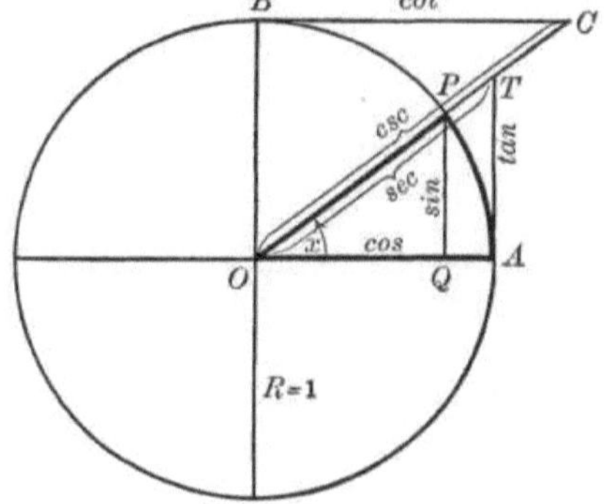

Substituting the functions equivalent to QP and OQ, we get

$$(22) \quad \tan x = \frac{\sin x}{\cos x}, \quad \cot x = \frac{\cos x}{\sin x}.$$

Again, in the same triangle,

$$\overline{QP}^2 + \overline{OQ}^2 = \overline{OP}^2, \quad \text{or,}$$

$$(23) \qquad \sin^2 x + \cos^2 x = 1.$$

In triangle OAT, $\quad \overline{OA}^2 + \overline{AT}^2 = \overline{OT}^2, \quad$ or,

$$(24) \qquad 1 + \tan^2 x = \sec^2 x.$$

In triangle OCB, $\quad \overline{OB}^2 + \overline{BC}^2 = \overline{OC}^2, \quad$ or

$$(25) \qquad 1 + \cot^2 x = \csc^2 x.$$

While in the above figure the angle x has been taken in the first quadrant, the results hold true for any angle whatever, for the above proofs apply to any one of the figures on p. 36 without the change of a single letter.

While it is of the utmost importance to memorize formulas (**19**) to (**25**), p. 59, as they stand, the student should also learn the following formulas where each one of the functions is expressed explicitly in terms of other functions.

$$(26)\quad \sin x = \frac{1}{\csc x}. \hspace{4cm} (19),\ \text{p. }59$$

$$(27)\quad \sin x = \pm \sqrt{1 - \cos^2 x}.^* \hspace{2cm} \text{Solving (23), p. 59, for } \sin x$$

$$(28)\quad \cos x = \frac{1}{\sec x}. \hspace{4cm} (20),\ \text{p. }59$$

$$(29)\quad \cos x = \pm \sqrt{1 - \sin^2 x}. \hspace{2cm} \text{Solving (23), p. 59, for } \cos x$$

$$(30)\quad \tan x = \frac{1}{\cot x}. \hspace{4cm} (21),\ \text{p. }59$$

$$(31)\quad \tan x = \pm \sqrt{\sec^2 x - 1}. \hspace{2cm} \text{Solving (24), p. 59, for } \tan x$$

$$(32)\quad \tan x = \frac{\sin x}{\cos x} = \frac{\sin x}{\pm \sqrt{1 - \sin^2 x}} = \frac{\pm \sqrt{1 - \cos^2 x}}{\cos x}.$$

[From (22), p. 59; also (29) and (27).]

$$(33)\quad \csc x = \frac{1}{\sin x}. \hspace{4cm} (19),\ \text{p. }59$$

$$(34)\quad \csc x = \pm \sqrt{1 + \cot^2 x}. \hspace{2cm} \text{Solving (25), p. 59, for } \csc x$$

$$(35)\quad \sec x = \frac{1}{\cos x}. \hspace{4cm} (20),\ \text{p. }59$$

$$(36)\quad \sec x = \pm \sqrt{1 + \tan^2 x}. \hspace{2cm} \text{Solving (24), p. 59, for } \sec x$$

$$(37)\quad \cot x = \frac{1}{\tan x}. \hspace{4cm} (21),\ \text{p. }59$$

$$(38)\quad \cot x = \pm \sqrt{\csc^2 x - 1}. \hspace{2cm} \text{Solving (25), p. 59, for } \cot x$$

$$(39)\quad \cot x = \frac{\cos x}{\sin x} = \frac{\cos x}{\pm \sqrt{1 - \cos^2 x}} = \frac{\pm \sqrt{1 - \sin^2 x}}{\sin x}.$$

[From (22), p. 59; also (27) and (29).]

31. Any function expressed in terms of each of the other five functions. By means of the above formulas we may easily find any function in terms of each one of the other five functions as follows:

---

* The double sign means that we get two values for some of the functions unless a condition is given which determines whether to choose the plus or minus sign. The reason for this is that there are two angles less than 360° for which a function has a given value.

Ex. 1. Find $\sin x$ in terms of each of the other five functions of x.

(a) $\sin x = \dfrac{1}{\csc x}$, $\hspace{4em}$ from **(26)**

(b) $\sin x = \pm \sqrt{1 - \cos^2 x}$, $\hspace{4em}$ from **(27)**

(c) $\sin x = \dfrac{1}{\pm \sqrt{1 + \cot^2 x}}$, $\hspace{3em}$ substitute **(34)** in (a)

(d) $\sin x = \pm \sqrt{1 - \dfrac{1}{\sec^2 x}} = \dfrac{\pm \sqrt{\sec^2 x - 1}}{\sec x}$, $\hspace{2em}$ substitute **(28)** in (b)

(e) $\sin x = \dfrac{1}{\pm \sqrt{1 + \dfrac{1}{\tan^2 x}}} = \dfrac{\tan x}{\pm \sqrt{\tan^2 x + 1}}$. $\hspace{2em}$ Substitute **(37)** in (c)

Ex. 2. Find $\cos x$ in terms of each of the other five functions.

(a) $\cos x = \dfrac{1}{\sec x}$, $\hspace{4em}$ from **(28)**

(b) $\cos x = \pm \sqrt{1 - \sin^2 x}$, $\hspace{4em}$ from **(29)**

(c) $\cos x = \dfrac{1}{\pm \sqrt{1 + \tan^2 x}}$, $\hspace{3em}$ substitute **(36)** in (a)

(d) $\cos x = \pm \sqrt{1 - \dfrac{1}{\csc^2 x}} = \dfrac{\pm \sqrt{\csc^2 x - 1}}{\csc x}$, $\hspace{2em}$ substitute **(26)** in (b)

(e) $\cos x = \dfrac{1}{\pm \sqrt{1 + \dfrac{1}{\cot^2 x}}} = \dfrac{\cot x}{\pm \sqrt{\cot^2 x + 1}}$. $\hspace{2em}$ Substitute **(30)** in (c)

Ex. 3. Find $\tan x$ in terms of each of the other five functions.

(a) $\tan x = \dfrac{1}{\cot x}$, $\hspace{4em}$ from **(30)**

(b) $\tan x = \pm \sqrt{\sec^2 x - 1}$, $\hspace{4em}$ from **(31)**

(c) $\tan x = \dfrac{\sin x}{\pm \sqrt{1 - \sin^2 x}}$, $\hspace{4em}$ from **(32)**

(d) $\tan x = \dfrac{\pm \sqrt{1 - \cos^2 x}}{\cos x}$, $\hspace{4em}$ from **(32)**

(e) $\tan x = \dfrac{1}{\pm \sqrt{\csc^2 x - 1}}$. $\hspace{3em}$ Substituting **(38)** in (a)

Ex. 4. Prove that $\sec x - \tan x \cdot \sin x = \cos x$.

Solution. Let us take the first member and reduce it by means of the formulas **(26)** to **(39)**, p. 60, until it becomes identical with the second member. Thus

$$\sec x - \tan x \cdot \sin x = \frac{1}{\cos x} - \frac{\sin x}{\cos x} \cdot \sin x$$

$$\left[\text{Since } \sec x = \frac{1}{\cos x} \text{ and } \tan x = \frac{\sin x}{\cos x}^* \right]$$

$$= \frac{1 - \sin^2 x}{\cos x} = \frac{\cos^2 x}{\cos x} \hspace{3em} \textbf{(23)}, \text{ p. 59}$$

$$= \cos x. \ \textit{Ans.}$$

---

* Usually it is best to change the given expression into one containing sines and cosines only, and then change this into the required form. Any operation is admissible that does not change the value of the expression. Use radicals only when unavoidable.

Ex. 5. Prove that $\sin x(\sec x + \csc x) - \cos x(\sec x - \csc x) = \sec x \csc x$.

Solution. $\sin x(\sec x + \csc x) - \cos x(\sec x - \csc x)$

$$= \sin x\left(\frac{1}{\cos x} + \frac{1}{\sin x}\right) - \cos x\left(\frac{1}{\cos x} - \frac{1}{\sin x}\right)$$

$$\left[\text{Since } \sec x = \frac{1}{\cos x} \text{ and } \csc x = \frac{1}{\sin x}\right]$$

$$= \frac{\sin x}{\cos x} + 1 - 1 + \frac{\cos x}{\sin x}$$

$$= \frac{\sin x}{\cos x} + \frac{\cos x}{\sin x}$$

$$= \frac{\sin^2 x + \cos^2 x}{\cos x \sin x} = \frac{1}{\cos x \sin x} \qquad \textbf{(23)}, \text{ p. 59}$$

$$= \frac{1}{\cos x}\,\frac{1}{\sin x} = \sec x \csc x. \quad \textit{Ans.}$$

EXAMPLES

1. Find $\sec x$ in terms of each of the other five functions of x.

$$\textit{Ans.}\quad \frac{1}{\cos x}, \ \pm\sqrt{1 + \tan^2 x}, \ \frac{1}{\pm\sqrt{1 - \sin^2 x}}, \ \frac{\pm\sqrt{\cot^2 x + 1}}{\cot x}, \ \frac{\csc x}{\pm\sqrt{\csc^2 x - 1}}.$$

2. Find $\cot x$ in terms of each of the other five functions of x.

$$\textit{Ans.}\quad \frac{1}{\tan x}, \ \pm\sqrt{\csc^2 x - 1}, \ \frac{\pm\sqrt{1 - \sin^2 x}}{\sin x}, \ \frac{\cos x}{\pm\sqrt{1 - \cos^2 x}}, \ \frac{1}{\pm\sqrt{\sec^2 x - 1}}.$$

3. Find $\csc x$ in terms of each of the other five functions of x.

$$\textit{Ans.}\quad \frac{1}{\sin x}, \ \pm\sqrt{1 + \cot^2 x}, \ \frac{1}{\pm\sqrt{1 - \cos^2 x}}, \ \frac{\pm\sqrt{\tan^2 x + 1}}{\tan x}, \ \frac{\sec x}{\pm\sqrt{\sec^2 x - 1}}.$$

4. Prove the following:

(a) $\cos x \tan x = \sin x$.

(b) $\sin x \sec x = \tan x$.

(c) $\sin y \cot y = \cos y$.

(d) $(1 + \tan^2 y)\cos^2 y = 1$.

(e) $\sin^2 A + \sin^2 A \tan^2 A = \tan^2 A$.

(f) $\cot A - \cos^2 A = \cot^2 A \cos^2 A$.

(g) $\tan A + \cot A = \sec A \csc A$.

(h) $\cos A \csc A = \cot A$.

(i) $\cos^2 A - \sin^2 A = 1 - 2\sin^2 A$.

(j) $\cos^2 A - \sin^2 A = 2\cos^2 A - 1$.

(k) $(1 + \cot^2 B)\sin^2 B = 1$.

(l) $(\csc^2 A - 1)\csc^2 A = \cos^2 A$.

(m) $\sec^2 A + \csc^2 A = \sec^2 A \csc^2 A$.

(n) $\cos^4 C - \sin^4 C + 1 = 2\cos^2 C$.

(o) $(\sin x + \cos x)^2 + (\sin x - \cos x)^2 = 2$.

(p) $\sin^3 x \cos x + \cos^3 x \sin x = \sin x \cos x$.

(q) $\tan A + \cot A = \sec A \csc A$.

(r) $\cot y + \dfrac{\sin y}{1 + \cos y} = \csc y$.

(s) $\cos B \tan B + \sin B \cot B = \sin B + \cos B$.

(t) $\sec x \csc x(\cos^2 x - \sin^2 x) = \cot x - \tan x$.

(u) $\dfrac{\cos C}{1 - \tan C} + \dfrac{\sin C}{1 - \cot C} = \sin C + \cos C$.

(v) $\dfrac{\sin z}{1 + \cos z} + \dfrac{1 + \cos z}{\sin z} = 2\csc z$.

CHAPTER IV

TRIGONOMETRIC ANALYSIS

32. Functions of the sum and of the difference of two angles. We now proceed to express the trigonometric functions of the sum and difference of two angles in terms of the trigonometric functions of the angles themselves.* The fundamental formulas to be derived are:

$$\text{(40)} \qquad \sin(x + y) = \sin x \cos y + \cos x \sin y.$$
$$\text{(41)} \qquad \sin(x - y) = \sin x \cos y - \cos x \sin y.$$
$$\text{(42)} \qquad \cos(x + y) = \cos x \cos y - \sin x \sin y.$$
$$\text{(43)} \qquad \cos(x - y) = \cos x \cos y + \sin x \sin y.$$

33. Sine and cosine of the sum of two angles. Proofs of formulas (40) and (42). Let the angles x and y be each a positive angle less than 90°. In the unit circle whose center is O, lay off the angle $AOP = x$ and the angle $POQ = y$. Then the angle $AOQ = x + y$.

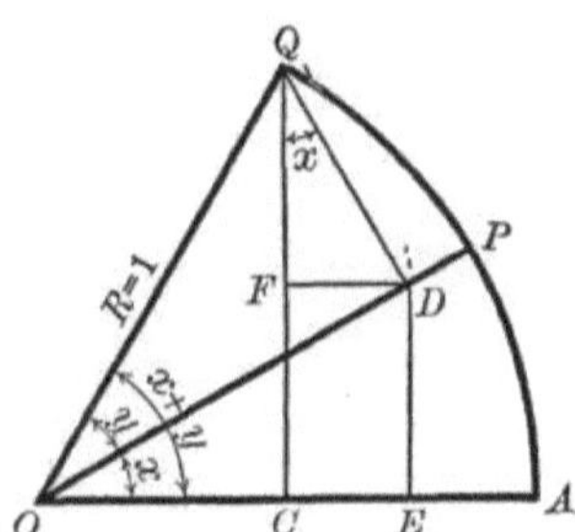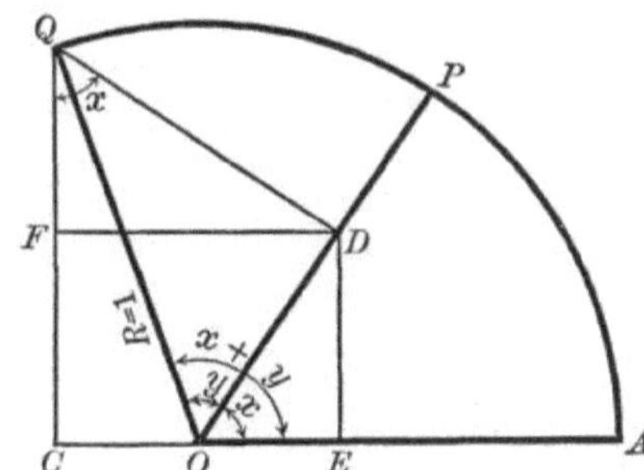

In the first figure the angle $x + y$ is less than 90°, in the second greater than 90°.

Draw QC perpendicular to OA. Then

$$\text{(a)} \qquad \sin(x + y) = CQ, \quad \text{and}$$
$$\text{(b)} \qquad \cos(x + y) = OC.$$

Draw QD perpendicular to OP. Then

$$\text{(c)} \qquad \sin y = DQ, \quad \text{and}$$
$$\text{(d)} \qquad \cos y = OD.†$$

---

* Since x and y are angles, their sum $x+y$ and their difference $x-y$ are also angles. Thus if $x = 61°$ and $y = 23°$, then $x + y = 84°$ and $x - y = 38°$. The student should observe that $\sin(x+y)$ is not the same as $\sin x + \sin y$, or $\cos(x-y)$ the same as $\cos x - \cos y$, etc.

† The student will see this at once if the book is turned until OP appears horizontal.

Draw DE perpendicular and DF parallel to OA. Then angle $DQF =$ angle AOP $(= x)$, having their sides perpendicular each to each. From (a),

(e) $$\sin(x + y) = CQ = CF + FQ = ED + FQ.$$

ED being one side of the right triangle OED, we have

$$ED = OD \cdot \sin x. \qquad \text{from (7), p. 11}$$

But from (d), $OD = \cos y$. Therefore

(f) $$ED = \sin x \cos y.$$

FQ being one side of the right triangle QFD, we have

$$FQ = DQ \cdot \cos x. \qquad \text{from (8), p. 11}$$

But from (c), $DQ = \sin y$. Therefore

(g) $$FQ = \cos x \sin y.$$

Substituting (f) and (g) in (e), we get

(40) $$\sin(x + y) = \sin x \cos y + \cos x \sin y.$$

To derive (42) we use the same figures. From (b),

(h) $$\cos(x + y) = OC = OE - CE = OE - FD.^*$$

OE being one side of the right triangle OED, we have

$$OE = OD \cos x. \qquad \text{from (8), p. 11}$$

But from (d), $OD = \cos y$. Therefore

(i) $$OE = \cos x \cos y.$$

FD being a side of the right triangle QFD, we have

$$FD = DQ \sin x. \qquad \text{from (7), p. 11}$$

But from (c), $DQ = \sin y$. Therefore

(j) $$FD = \sin x \sin y.$$

Substituting (i) and (j) in (h), we get

(42) $$\cos(x + y) = \cos x \cos y - \sin x \sin y.$$

In deriving formulas (40) and (42) we assumed that each of the angles x and y were positive and less than 90°. It is a fact, however, that these formulas hold true for values of x and y of any magnitude whatever, positive or negative. The work which follows will illustrate how this may be shown for any particular case.

---

* When $x + y$ is greater than 90°, OC is negative.

Show that (**42**) *is true when x is a positive angle in the second quadrant and y a positive angle in the fourth quadrant.*

Proof. Let $x = 90° + x'$ and $y = 270° + y'$*; then $x + y = 360° + (x' + y')$ and

(k) $\qquad x' = x - 90°, \qquad y' = y - 270°, \qquad x' + y' = x + y - 360°.$

$$\begin{aligned}
\cos (x + y) &= \cos [360° + (x' + y')] = \cos (x' + y') && \text{by § 29, p. 57}\\
&= \cos x' \cos y' - \sin x' \sin y' && \text{by (42)}\\
&= \cos (x - 90°) \cos (y - 270°) - \sin (x - 90°) \sin (y - 270°) && \text{from (k)}\\
&= \sin x (- \sin y) - (- \cos x \cos y) && \text{by § 29, p. 57}\\
&= \cos x \cos y - \sin x \sin y. && \text{Q. E. D.}
\end{aligned}$$

Show that (**40**) *is true when x is a positive angle in the first quadrant and y a negative angle in the second quadrant.*

Proof. Let $x = 90° - x'$ and $y = - 180° - y'$; then $x + y = - 90° - (x' + y')$ and

(l) $\qquad x' = 90° - x, \qquad y' = - 180° - y, \qquad x' + y' = - 90° - (x + y).$

$$\begin{aligned}
\sin (x + y) &= \sin [- 90° - (x' + y')] = - \cos (x' + y') && \text{by § 28, p. 56}\\
&= - [\cos x' \cos y' - \sin x' \sin y'] && \text{by (40)}\\
&= - [\cos (90° - x) \cos (- 180° - y) - \sin (90° - x) \sin (- 180° - y)] && \text{from (l)}\\
&= - [\sin x (- \cos y) - \cos x \sin y] && \text{by § 29, p. 57}\\
&= \sin x \cos y + \cos x \sin y. && \text{Q. E. D.}
\end{aligned}$$

EXAMPLES

1. Find $\sin 75°$, using (**40**) and the functions of $45°$ and $30°$.

Solution. Since $75° = 45° + 30°$, we get from (**40**)

$$\begin{aligned}
\sin 75° = \sin (45° + 30°) &= \sin 45° \cos 30° + \cos 45° \sin 30°\\
&= \frac{1}{\sqrt{2}} \cdot \frac{\sqrt{3}}{2} + \frac{1}{\sqrt{2}} \cdot \frac{1}{2} && \text{from p. 5}\\
&= \frac{\sqrt{3} + 1}{2 \sqrt{2}}. \quad Ans.
\end{aligned}$$

2. Find $\cos (x + y)$, having given $\sin x = \frac{3}{5}$ and $\sin y = \frac{5}{13}$, x and y being positive acute angles.

Solution. By the method shown on p. 30 we get first

$$\sin x = \tfrac{3}{5}, \qquad \cos x = \tfrac{4}{5}, \qquad \sin y = \tfrac{5}{13}, \qquad \cos y = \tfrac{12}{13}.$$

Substituting these values in (**42**), we get

$$\cos (x + y) = \tfrac{4}{5} \cdot \tfrac{12}{13} - \tfrac{3}{5} \cdot \tfrac{5}{13} = \tfrac{33}{65}. \quad Ans.$$

---

* The student should note that x' and y' are acute angles.

3. Show that $\cos 75° = \dfrac{\sqrt{3}-1}{2\sqrt{2}}$, using the functions of 45° and 30°.

4. Prove that $\sin 90° = 1$ and $\cos 90° = 0$, using the functions of 60° and 30°.

5. If $\tan x = \frac{3}{4}$ and $\tan y = \frac{7}{24}$, find $\sin(x+y)$ and $\cos(x+y)$ when x and y are acute angles.　　　　　　　　*Ans.* $\sin(x+y) = \frac{4}{5}$, $\cos(x+y) = \frac{3}{5}$.

6. By means of **(40)** and **(42)** express the sine and cosine of $90° + x$, $180° + x$, $270° + x$, in terms of $\sin x$ and $\cos x$.

7. Verify the following :

(a) $\sin(45° + x) = \dfrac{\cos x + \sin x}{\sqrt{2}}$.

(b) $\cos(60° + A) = \dfrac{\cos A - \sqrt{3}\sin A}{2}$.

(c) $\sin(y + 30°) = \dfrac{\sqrt{3}\sin y + \cos y}{2}$.

(d) $\cos(B + 45°) = \dfrac{\cos B - \sin B}{\sqrt{2}}$.

8. Find $\sin(A + B)$ and $\cos(A + B)$, having given $\sin A = \frac{1}{2}$ and $\sin B = \frac{2}{3}$.

$$\textit{Ans.}\quad \sin(A+B) = \frac{\pm\sqrt{5} \pm 2\sqrt{3}}{6}, \quad \cos(A+B) = \frac{\pm\sqrt{15}-2}{6}.$$

34. Sine and cosine of the difference of two angles. Proofs of formulas (41) and (43). It was shown in the last section that **(40)** and **(42)** hold true for values of x and y of any magnitude whatever, positive or negative. Hence **(41)** and **(43)** are merely special cases of **(40)** and **(42)** respectively. Thus, from **(40)**,

$$\sin(x+y) = \sin x \cos y + \cos x \sin y.$$

Now replace y by $-y$. This gives

(a)　　　　$\sin(x-y) = \sin x \cos(-y) + \cos x \sin(-y).$

But $\cos(-y) = \cos y$, and $\sin(-y) = -\sin y$.　　　　from p. 55
Substituting back in (a), we get

(41)　　　　$\sin(x-y) = \sin x \cos y - \cos x \sin y.$

Similarly, from **(42)**,

$$\cos(x+y) = \cos x \cos y - \sin x \sin y.$$

Now replace y by $-y$. This gives

(b)　　　　$\cos(x-y) = \cos x \cos(-y) - \sin x \sin(-y).$

But $\cos(-y) = \cos y$, and $\sin(-y) = -\sin y$.　　　　from p. 55
Substituting back in (b), we get

(43)　　　　$\cos(x-y) = \cos x \cos y + \sin x \sin y.$

EXAMPLES

1. Find $\cos 15°$, using **(43)** and the functions of 45° and 30°.

Solution. Since $15° = 45° - 30°$, we get from **(43)**

$$\cos 15° = \cos (45° - 30°) = \cos 45° \cos 30° + \sin 45° \sin 30°$$
$$= \frac{1}{\sqrt{2}} \cdot \frac{\sqrt{3}}{2} \quad \frac{1}{\sqrt{2}} \cdot \frac{1}{2}$$
$$= \frac{\sqrt{3} + 1}{2\sqrt{2}} \cdot \quad Ans.$$

Work out the above example, taking $15° = 60° - 45°$.

2. Prove $\sin (60° + x) - \sin (60° - x) = \sin x$.

Solution. $\sin (60° + x) = \sin 60° \cos x + \cos 60° \sin x.$ by **(40)**

$\sin (60° - x) = \sin 60° \cos x - \cos 60° \sin x.$ by **(41)**

$\therefore \ \sin (60° + x) - \sin (60° - x) = 2 \cos 60° \sin x$ by subtraction

$$= 2 \cdot \tfrac{1}{2} \cdot \sin x = \sin x. \quad Ans.$$

3. Show that $\sin 15° = \dfrac{\sqrt{3} - 1}{2\sqrt{2}}$, using the functions of 45° and 30°.

4. Find $\sin (x - y)$ and $\cos (x - y)$, having given $\sin x = \tfrac{1}{2}$ and $\sin y = \tfrac{1}{3}$, x and y being acute angles.

$$Ans. \ \sin (x - y) = \frac{2\sqrt{2} - \sqrt{15}}{12}, \ \cos (x - y) = \frac{2\sqrt{30} + 1}{12}.$$

5. Find $\sin (x - y)$ and $\cos (x - y)$, having given $\tan x = \tfrac{4}{3}$ and $\tan y = \tfrac{3}{4}$, x and y being acute angles. $Ans. \ \sin (x - y) = \tfrac{7}{25}, \ \cos (x - y) = \tfrac{24}{25}.$

6. By means of **(41)** and **(43)** express the sine and cosine of $90° - x$, $180° - x$, $270° - x$, $360° - x$, in terms of $\sin x$ and $\cos x$.

7. Verify the following:

(a) $\sin (45° - x) = \dfrac{\cos x - \sin x}{\sqrt{2}}.$ (c) $\sin (y - 30°) = \dfrac{\sqrt{3} \sin y - \cos y}{2}.$

(b) $\cos (60° - A) = \dfrac{\cos A + \sqrt{3} \sin A}{2}.$ (d) $\cos (B + 45°) = \dfrac{\cos B - \sin B}{\sqrt{2}}.$

(e) $\sin (60° + x) - \sin x = \sin (60° - x).$

(f) $\cos (30° + y) - \cos (30° - y) = - \sin y.$

(g) $\cos (45° + x) + \cos (45° - x) = \sqrt{2} \cos x.$

(h) $\sin (45° + P) - \sin (45° - P) = \sqrt{2} \sin P.$

(i) $\cos (Q + 45°) + \sin (Q - 45°) = 0.$

(j) $\sin (x + y) \sin (x - y) = \sin^2 x - \sin^2 y.$

(k) $\sin (x + y + z) = \sin x \cos y \cos z + \cos x \sin y \cos z$
$+ \cos x \cos y \sin z - \sin x \sin y \sin z.$

Hint. $\sin (x + y + z) = \sin (x + y) \cos z + \cos (x + y) \sin z.$

35. Tangent and cotangent of the sum and of the difference of two angles. From (**22**), p. 59, and (**40**) and (**42**), p. 63, we get

$$\tan(x+y) = \frac{\sin(x+y)}{\cos(x+y)} = \frac{\sin x \cos y + \cos x \sin y}{\cos x \cos y - \sin x \sin y}.$$

Now divide both numerator and denominator by $\cos x \cos y$. This gives

$$\tan(x+y) = \frac{\dfrac{\sin x \cos y}{\cos x \cos y} + \dfrac{\cos x \sin y}{\cos x \cos y}}{\dfrac{\cos x \cos y}{\cos x \cos y} - \dfrac{\sin x \sin y}{\cos x \cos y}}$$

$$= \frac{\dfrac{\sin x}{\cos x} + \dfrac{\sin y}{\cos y}}{1 - \dfrac{\sin x}{\cos x} \cdot \dfrac{\sin y}{\cos y}}.$$

Since $\dfrac{\sin x}{\cos x} = \tan x$ and $\dfrac{\sin y}{\cos y} = \tan y$, we get

$$(44) \qquad \tan(x+y) = \frac{\tan x + \tan y}{1 - \tan x \tan y}.$$

In the same way, from (**41**) and (**43**) we get

$$(45) \qquad \tan(x-y) = \frac{\tan x - \tan y}{1 + \tan x \tan y}.$$

From (**22**), p. 59, and (**40**) and (**42**), p. 63, we get

$$\cot(x+y) = \frac{\cos(x+y)}{\sin(x+y)} = \frac{\cos x \cos y - \sin x \sin y}{\sin x \cos y + \cos x \sin y}.$$

Now divide both numerator and denominator by $\sin x \sin y$. This gives

$$\cot(x+y) = \frac{\dfrac{\cos x \cos y}{\sin x \sin y} - \dfrac{\sin x \sin y}{\sin x \sin y}}{\dfrac{\sin x \cos y}{\sin x \sin y} + \dfrac{\cos x \sin y}{\sin x \sin y}}$$

$$= \frac{\dfrac{\cos x}{\sin x} \cdot \dfrac{\cos y}{\sin y} - 1}{\dfrac{\cos y}{\sin y} + \dfrac{\cos x}{\sin x}}.$$

Since $\dfrac{\cos x}{\sin x} = \cot x$, and $\dfrac{\cos y}{\sin y} = \cot y$, we get

$$(46) \qquad \cot(x+y) = \frac{\cot x \cot y - 1}{\cot y + \cot x}.$$

In the same way, from **(41)** and **(43)** we get

$$(47) \qquad \cot(x - y) = \frac{\cot x \cot y + 1}{\cot y - \cot x}.$$

Formulas **(40)** to **(47)** may be written in a more compact form as follows:

$$\sin(x \pm y) = \sin x \cos y \pm \cos x \sin y,$$
$$\cos(x \pm y) = \cos x \cos y \mp \sin x \sin y,$$
$$\tan(x \pm y) = \frac{\tan x \pm \tan y}{1 \mp \tan x \tan y},$$
$$\cot(x \pm y) = \frac{\cot x \cot y \mp 1}{\cot y \pm \cot x}.$$

The formulas derived in this chapter demonstrate the **Addition Theorem** for trigonometric functions, namely, that *any function of the algebraic sum of two angles is expressible in terms of the functions of those angles.*

EXAMPLES

1. Find $\tan 15°$, using **(45)** and the functions of 60° and 45°.

Solution. Since $15° = 60° - 45°$, we get from **(45)**

$$\tan 15° = \tan(60° - 45°) = \frac{\tan 60° - \tan 45°}{1 + \tan 60° \tan 45°} = \frac{\sqrt{3} - 1}{1 + \sqrt{3}} = 2 - \sqrt{3}. \ \textit{Ans.}$$

Work out above example, taking $15° = 45° - 30°$.

2. Find $\tan(x + y)$ and $\tan(x - y)$, having given $\tan x = \frac{1}{2}$ and $\tan y = \frac{1}{7}$.

$$\textit{Ans.} \ \tan(x + y) = \tfrac{9}{7}, \ \tan(x - y) = \tfrac{2}{9}.$$

3. Find $\tan 75°$ from the functions of 45° and 30°. $\qquad\qquad$ *Ans.* $2 + \sqrt{3}$.

4. Verify the following:

(a) $\tan(45° + x) = \dfrac{1 + \tan x}{1 - \tan x}.$

(b) $\cot(y - 45°) = \dfrac{1 + \cot y}{1 - \cot y}.$

(c) $\tan(A - 60°) = \dfrac{\tan A - \sqrt{3}}{1 + \sqrt{3}\tan A}.$

(d) $\cot(B + 30°) = \dfrac{\sqrt{3}\cot B - 1}{\cot B + \sqrt{3}}.$

(e) $\tan(x \pm 45°) + \cot(x \mp 45°) = 0.$

(f) $\dfrac{\sin(x + y)}{\sin(x - y)} = \dfrac{\tan x + \tan y}{\tan x - \tan y}.$

(g) $\tan x + \tan y = \dfrac{\sin(x + y)}{\cos x \cos y}.$

(h) $\cot A - \cot B = \dfrac{\sin(B - A)}{\sin A \sin B}.$

(i) $1 - \tan x \tan y = \dfrac{\cos(x + y)}{\cos x \cos y}.$

(j) $\cot P \cot Q - 1 = \dfrac{\cos(P + Q)}{\sin P \sin Q}.$

36. Functions of twice an angle in terms of the functions of the angle. Formulas **(40)** to **(47)** were shown to hold true for all possible values of x and y; hence they must hold true when x equals y.

To find $\sin 2x$ we take **(40)**,

$$\sin(x + y) = \sin x \cos y + \cos x \sin y.$$

Replace y by x. This gives

$$\sin(x + x) = \sin x \cos x + \cos x \sin x, \text{ or}$$

(48) $$\sin 2x = 2 \sin x \cos x.$$

To find $\cos 2x$ we take (42),

$$\cos(x + y) = \cos x \cos y - \sin x \sin y.$$

Replace y by x. This gives

$$\cos(x + x) = \cos x \cos x - \sin x \sin x, \text{ or}$$

(49) $$\cos 2x = \cos^2 x - \sin^2 x.$$

Since $\cos^2 x = 1 - \sin^2 x$, (49) may be written

(49a) $$\cos 2x = 1 - 2\sin^2 x.$$

Or, since $\sin^2 x = 1 - \cos^2 x$, (49) may also be written

(49b) $$\cos 2x = 2\cos^2 x - 1.$$

To find $\tan 2x$ we take (44),

$$\tan(x + y) = \frac{\tan x + \tan y}{1 - \tan x \tan y}.$$

Replace y by x. This gives

$$\tan(x + x) = \frac{\tan x + \tan x}{1 - \tan x \tan x}, \text{ or}$$

(50) $$\tan 2x = \frac{2 \tan x}{1 - \tan^2 x}.$$

37. Functions of multiple angles. The method of the last section may readily be extended to finding the functions of nx in terms of the functions of x.

To find $\sin 3x$ in terms of $\sin x$ we take (40),

$$\sin(x + y) = \sin x \cos y + \cos x \sin y.$$

Replace y by $2x$. This gives

$$\sin(x + 2x) = \sin x \cos 2x + \cos x \sin 2x, \text{ or}$$

$$\begin{aligned}
\sin 3x &= \sin x\,(\cos^2 x - \sin^2 x) + \cos x\,(2 \sin x \cos x) \text{ by (49), (48)}\\
&= 3 \sin x \cos^2 x - \sin^3 x\\
&= 3 \sin x\,(1 - \sin^2 x) - \sin^3 x \qquad\qquad \text{by (23), p. 59}\\
&= 3 \sin x - 4 \sin^3 x. \quad Ans.
\end{aligned}$$

To find $\tan 4x$ in terms of $\tan x$, we take (44), (50),

$$\tan 4x = \tan(2x + 2x) = \frac{2 \tan 2x}{1 - \tan^2 2x} = \frac{4 \tan x\,(1 - \tan^2 x)}{1 - 6 \tan^2 x + \tan^4 x}. \quad Ans.$$

EXAMPLES

1. Given $\sin x = \dfrac{2}{\sqrt{5}}$, x lying in the second quadrant; find $\sin 2x$, $\cos 2x$, $\tan 2x$.

Solution. Since $\sin x = \dfrac{2}{\sqrt{5}}$ and x lies in the second quadrant, we get, using method on p. 30,

$$\sin x = \frac{2}{\sqrt{5}}, \qquad \cos x = -\frac{1}{\sqrt{5}}, \qquad \tan x = -2.$$

Substituting in **(48)**, we get

$$\sin 2x = 2 \sin x \cos x = 2 \cdot \frac{2}{\sqrt{5}}\left(-\frac{1}{\sqrt{5}}\right) = -\frac{4}{5}.$$

Similarly, we get $\cos 2x = -\frac{3}{5}$ by substituting in **(49)**, and $\tan 2x = \frac{4}{3}$ by substituting in **(50)**.

2. Given $\tan x = 2$, x lying in the third quadrant; find $\sin 2x$, $\cos 2x$, $\tan 2x$.

$Ans.$ $\sin 2x = \frac{4}{5}$, $\cos 2x = -\frac{3}{5}$, $\tan 2x = -\frac{4}{3}$.

3. Given $\tan x = \dfrac{a}{b}$; find $\sin 2x$, $\cos 2x$, $\tan 2x$.

$Ans.$ $\pm\dfrac{2ab}{a^2 + b^2}, \dfrac{b^2 - a^2}{b^2 + a^2}, \dfrac{2ab}{b^2 - a^2}.$

4. Show that $\cos 3x = 4 \cos^3 x - 3 \cos x$.

5. Show that $\tan 3A = \dfrac{3 \tan A - \tan^3 A}{1 - 3 \tan^2 A}$.

6. Show that $\sin 4x = 4 \sin x \cos x - 8 \sin^3 x \cos x$.

7. Show that $\cos 4B = 1 - 8 \sin^2 B + 8 \sin^4 B$.

8. Show that $\sin 5x = 5 \sin x - 20 \sin^3 x + 16 \sin^5 x$.

9. Show that $\tan(45° + x) - \tan(45° - x) = 2 \tan 2x$.

10. Show that $\tan(45° + C) + \tan(45° - C) = 2 \sec 2C$.

11. Verify the following:

(a) $\sin 2x = \dfrac{2 \tan x}{1 + \tan^2 x}$.

(b) $\cos 2x = \dfrac{1 - \tan^2 x}{1 + \tan^2 x}$.

(c) $\tan P + \cot P = 2 \csc 2P$.

(d) $\cos 2x = \cos^4 x - \sin^4 x$.

(e) $(\sin x + \cos x)^2 = 1 + \sin 2x$.

(f) $\sec 2x = \dfrac{\csc^2 x}{\csc^2 x - 2}$.

(g) $2 \csc 2s = \sec s \csc s$.

(h) $\cot t - \tan t = 2 \cot 2t$.

(i) $\cos 2x = \dfrac{2 - \sec^2 x}{\sec^2 x}$.

(j) $(\sin x - \cos x)^2 = 1 - \sin 2x$.

12. In a right triangle, C being the right angle, prove the following:

(a) $\cos 2B = \dfrac{\sin^2 A - \sin^2 B}{\sin^2 A + \sin^2 B}$.

(b) $\sin(A - B) + \cos 2A = 0$.

(c) $\sin(A - B) + \sin(2A + C) = 0$.

(d) $(\sin A - \sin B)^2 + (\cos A + \cos B)^2 = 2$.

(e) $\sqrt{\dfrac{a + b}{a - b}} + \sqrt{\dfrac{a - b}{a + b}} = \dfrac{2 \sin A}{\sqrt{\cos 2B}}$.

(f) $\tan B = \cot A + \cos C$.

(g) $\cos 2A + \cos 2B = 0$.

(h) $\tan 2A = \dfrac{2ab}{b^2 - a^2}$.

(i) $\sin 2A = \sin 2B$.

(j) $\sin 2A = \dfrac{2ab}{c^2}$.

(k) $\cos 2A = \dfrac{b^2 - a^2}{c^2}$.

(l) $\sin 3A = \dfrac{3ab^2 - a^3}{c^3}$.

38. Functions of an angle in terms of functions of half the angle.
From **(48)**, p. 70,
$$\sin 2x = 2 \sin x \cos x.$$

Replace $2x$ by x, or, what amounts to the same thing, replace x by $\dfrac{x}{2}$. This gives

(51)
$$\sin x = 2 \sin \frac{x}{2} \cos \frac{x}{2}.$$

From **(49)**, p. 70, $\quad \cos 2x = \cos^2 x - \sin^2 x.$

Replace $2x$ by x, or, what amounts to the same thing, replace x by $\dfrac{x}{2}$. This gives

(52)
$$\cos x = \cos^2 \frac{x}{2} - \sin^2 \frac{x}{2}.$$

From **(50)**, p. 70, $\quad \tan 2x = \dfrac{2 \tan x}{1 - \tan^2 x}.$

Replace $2x$ by x, or, what amounts to the same thing, replace x by $\dfrac{x}{2}$. This gives

(53)
$$\tan x = \frac{2 \tan \frac{x}{2}}{1 - \tan^2 \frac{x}{2}}.$$

39. Functions of half an angle in terms of the cosine of the angle.
From **(49 a)** and **(49 b)**, p. 70, we get

$$2 \sin^2 x = 1 - \cos 2x,$$
and
$$2 \cos^2 x = 1 + \cos 2x.$$

Solving for $\sin x$ and $\cos x$,

$$\sin x = \pm \sqrt{\frac{1 - \cos 2x}{2}},$$
and
$$\cos x = \pm \sqrt{\frac{1 + \cos 2x}{2}}.$$

Replace $2x$ by x, or, what amounts to the same thing, replace x by $\dfrac{x}{2}$. This gives

(54)
$$\sin \frac{x}{2} = \pm \sqrt{\frac{1 - \cos x}{2}},$$

(55)
$$\cos \frac{x}{2} = \pm \sqrt{\frac{1 + \cos x}{2}}.$$

To get $\tan\frac{x}{2}$ we divide (54) by (55). This gives

$$\tan\frac{x}{2} = \frac{\sin\frac{x}{2}}{\cos\frac{x}{2}} = \frac{\pm\sqrt{\dfrac{1-\cos x}{2}}}{\pm\sqrt{\dfrac{1+\cos x}{2}}}, \text{ or,}$$

(56)
$$\tan\frac{x}{2} = \pm\sqrt{\frac{1-\cos x}{1+\cos x}}.$$

Multiplying both numerator and denominator of the right-hand member by $\sqrt{1+\cos x}$, we get *

(57)
$$\tan\frac{x}{2} = \frac{\sin x}{1+\cos x};$$

or, multiplying both numerator and denominator by $\sqrt{1-\cos x}$, we get

(58)
$$\tan\frac{x}{2} = \frac{1-\cos x}{\sin x}.$$

Since tangent and cotangent are reciprocal functions, we have at once from (56), (57), and (58),

(59)
$$\cot\frac{x}{2} = \pm\sqrt{\frac{1+\cos x}{1-\cos x}}.$$

(60)
$$\cot\frac{x}{2} = \frac{1+\cos x}{\sin x}.$$

(61)
$$\cot\frac{x}{2} = \frac{\sin x}{1-\cos x}.$$

40. Sums and differences of functions. From p. 63,

(40)
$$\sin(x+y) = \sin x \cos y + \cos x \sin y.$$

(41)
$$\sin(x-y) = \sin x \cos y - \cos x \sin y.$$

(42)
$$\cos(x+y) = \cos x \cos y - \sin x \sin y.$$

(43)
$$\cos(x-y) = \cos x \cos y + \sin x \sin y.$$

---

* $\sqrt{\dfrac{1-\cos x}{1+\cos x}}\cdot\dfrac{\sqrt{1+\cos x}}{\sqrt{1+\cos x}} = \dfrac{\sqrt{1-\cos^2 x}}{1+\cos x} = \dfrac{\sin x}{1+\cos x}.$

The positive sign only of the radical is taken since $1+\cos x$ can never be negative and $\tan\frac{x}{2}$ and $\sin x$ always have like signs.

First add and then subtract (**40**) and (**41**). Similarly, (**42**) and (**43**). This gives

(a) $\sin(x+y)+\sin(x-y) = 2\sin x \cos y.$ Adding (**40**) and (**41**)

(b) $\sin(x+y)-\sin(x-y) = 2\cos x \sin y.$ Subtracting (**41**) from (**40**)

(c) $\cos(x+y)+\cos(x-y) = 2\cos x \cos y.$ Adding (**42**) and (**43**)

(d) $\cos(x+y)-\cos(x-y) = -2\sin x \sin y.$ Subtracting (**43**) from (**42**)

Let $x+y=A$ $x+y=A$

and $\underline{x-y=B}$ $\underline{x-y=B}$

Adding, $2x=A+B$ Subtracting, $2y=A-B$

         $x=\tfrac{1}{2}(A+B).$ $y=\tfrac{1}{2}(A-B).$

Now replacing the values of $x+y$, $x-y$, x, y in terms of A and B in (a) to (d) inclusive, we get

(**62**) $\sin A + \sin B = 2\sin\tfrac{1}{2}(A+B)\cos\tfrac{1}{2}(A-B).$

(**63**) $\sin A - \sin B = 2\cos\tfrac{1}{2}(A+B)\sin\tfrac{1}{2}(A-B).$

(**64**) $\cos A + \cos B = 2\cos\tfrac{1}{2}(A+B)\cos\tfrac{1}{2}(A-B).$

(**65**) $\cos A - \cos B = -2\sin\tfrac{1}{2}(A+B)\sin\tfrac{1}{2}(A-B).$

Dividing (**62**) by (**63**), member for member, we obtain

$$\frac{\sin A + \sin B}{\sin A - \sin B} = \frac{2\sin\tfrac{1}{2}(A+B)\cos\tfrac{1}{2}(A-B)}{2\cos\tfrac{1}{2}(A+B)\sin\tfrac{1}{2}(A-B)}$$

$$= \frac{\sin\tfrac{1}{2}(A+B)}{\cos\tfrac{1}{2}(A+B)}\cdot\frac{\cos\tfrac{1}{2}(A-B)}{\sin\tfrac{1}{2}(A-B)}$$

$$= \tan\tfrac{1}{2}(A+B)\cot\tfrac{1}{2}(A-B).$$

But $\cot\tfrac{1}{2}(A-B) = \dfrac{1}{\tan\tfrac{1}{2}(A-B)}.$ Hence

(**66**) $\dfrac{\sin A + \sin B}{\sin A - \sin B} = \dfrac{\tan\tfrac{1}{2}(A+B)}{\tan\tfrac{1}{2}(A-B)}.$

EXAMPLES

1. Find $\sin 22\tfrac{1}{2}°$, having given $\cos 45° = \dfrac{1}{\sqrt{2}}$.

Solution. From (**54**), $\sin\dfrac{x}{2} = \pm\sqrt{\dfrac{1-\cos x}{2}}.$

Let $x=45°$, then $\dfrac{x}{2}=22\tfrac{1}{2}°$, and we get

$$\sin 22\tfrac{1}{2}° = \sqrt{\frac{1-\dfrac{1}{\sqrt{2}}}{2}} = \frac{1}{2}\sqrt{2-\sqrt{2}}. \quad Ans.$$

2. Reduce the sum $\sin 7x + \sin 3x$ to the form of a product.

Solution. From **(62)**,

$$\sin A + \sin B = 2 \sin \tfrac{1}{2}(A + B) \cos \tfrac{1}{2}(A - B).$$

Let $A = 7x$, $B = 3x$. Then $A + B = 10x$, and $A - B = 4x$. Substituting back, we get

$$\sin 7x + \sin 3x = 2 \sin 5x \cos 2x. \quad Ans.$$

3. Find cosine, tangent, and cosecant of $22\tfrac{1}{2}^\circ$.

$$Ans. \quad \frac{1}{2}\sqrt{2 + \sqrt{2}}, \quad \sqrt{2} - 1, \quad \frac{2}{\sqrt{2 - \sqrt{2}}}.$$

4. Find sine, cosine, and tangent of 15°, having given $\cos 30^\circ = \dfrac{\sqrt{3}}{2}$.

$$Ans. \quad \frac{\sqrt{3} - 1}{2\sqrt{2}}, \quad \frac{\sqrt{3} + 1}{2\sqrt{2}}, \quad \frac{\sqrt{3} - 1}{\sqrt{3} + 1}.$$

5. Verify the following:

(a) $\sin 32^\circ + \sin 28^\circ = \cos 2^\circ$.

(b) $\sin 50^\circ + \sin 10^\circ = 2 \sin 30^\circ \cos 20^\circ$.

(c) $\cos 80^\circ - \cos 20^\circ = -2 \sin 50^\circ \sin 30^\circ$.

(d) $\cos 5x + \cos 9x = 2 \cos 7x \cos 2x$.

(e) $\dfrac{\sin 7x - \sin 5x}{\cos 7x + \cos 5x} = \tan x$.

(f) $\dfrac{\sin 33^\circ + \sin 3^\circ}{\cos 33^\circ + \cos 3^\circ} = \tan 18^\circ$.

(g) $\left(\sin \dfrac{x}{2} - \cos \dfrac{x}{2}\right)^2 = 1 - \sin x$.

(h) $\tan \dfrac{x}{4} = \dfrac{\sin \tfrac{1}{2}x}{1 + \cos \tfrac{1}{2}x}$.

(i) $\cot \dfrac{x}{4} = \dfrac{\sin \tfrac{1}{2}x}{1 - \cos \tfrac{1}{2}x}$.

(j) $\dfrac{\sin A + \sin B}{\cos A + \cos B} = \tan \dfrac{1}{2}(A + B)$.

6. Find sine, cosine, and tangent of $\dfrac{x}{2}$, if $\cos x = \dfrac{1}{3}$ and x lies in the first quadrant.

$$Ans. \quad \frac{\sqrt{3}}{3}, \quad \frac{\sqrt{6}}{3}, \quad \frac{\sqrt{2}}{2}.$$

7. Find sine, cosine, and tangent of $\dfrac{x}{2}$, if $\cos x = a$.

$$Ans. \quad \pm\sqrt{\frac{1 - a}{2}}, \quad \pm\sqrt{\frac{1 + a}{2}}, \quad \pm\sqrt{\frac{1 - a}{1 + a}}.$$

8. Express sine, cosine, and tangent of $3x$ in terms of $\cos 6x$.

$$Ans. \quad \pm\sqrt{\frac{1 - \cos 6x}{2}}, \quad \pm\frac{\sqrt{1 + \cos 6x}}{2}, \quad \pm\sqrt{\frac{1 - \cos 6x}{1 + \cos 6x}}.$$

9. In a right triangle, C being the right angle and c the hypotenuse, prove the following:

(a) $\sin^2 \dfrac{B}{2} = \dfrac{c - a}{2c}$.

(b) $\left(\cos \dfrac{A}{2} + \sin \dfrac{A}{2}\right)^2 = \dfrac{a + c}{c}$.

(c) $\cos^2 \dfrac{A}{2} = \dfrac{b + c}{2c}$.

(d) $\dfrac{a - b}{a + b} = \tan \dfrac{A - B}{2}$.

(e) $\tan \dfrac{A}{2} = \dfrac{a}{b + c}$.

41. Trigonometric identities. A trigonometric identity is an expression which states in the form of an equation a relation which holds true for all values of the angles involved. Thus, formulas **(26)** to **(39)**, p. 60, are all trigonometric identities, since they hold true for all values of x; also formulas (62) to (66), p. 74, are identities, since they hold true for all values of A and B. In fact, a large part of the work of this chapter has consisted in learning how to prove identities by reducing one member to the form of the other, using any known identities (as in Ex. 4, p. 61).

Another method for proving an identity is to reduce both members simultaneously, step by step, using known identities, *until both members are identical in form*. No general method can be given that will be the best to follow in all cases, but the following general directions will be found useful.

General directions for proving a trigonometric identity. *

First step. *If multiple angles, fractional angles, or the sums or differences of angles are involved, reduce all to functions of single angles only † and simplify.*

Second step. *If the resulting members are not readily reducible to the same form, change all the functions into sines and cosines.*

Third step. *Clear of fractions and radicals.*

Fourth step. *Change all the functions to a single function. In case the second step has been taken, this means that we change to sines only or to cosines only. The two members may now easily be reduced to the same form.*

Ex. 1. Prove the identity

$$1 + \tan 2x \tan x = \sec 2x.$$

Solution. Since $\tan 2x = \dfrac{2\tan x}{1 - \tan^2 x}$ and $\sec 2x = \dfrac{1}{\cos 2x} = \dfrac{1}{\cos^2 x - \sin^2 x}$, the equation becomes :

First step.
$$1 + \frac{2\tan^2 x}{1 - \tan^2 x} = \frac{1}{\cos^2 x - \sin^2 x},$$

or, simplifying,
$$\frac{1 + \tan^2 x}{1 - \tan^2 x} = \frac{1}{\cos^2 x - \sin^2 x}.$$

Second step.
$$\frac{1 + \dfrac{\sin^2 x}{\cos^2 x}}{1 - \dfrac{\sin^2 x}{\cos^2 x}} = \frac{1}{\cos^2 x - \sin^2 x}, \qquad \text{(22), p. 59}$$

or, simplifying,
$$\frac{\cos^2 x + \sin^2 x}{\cos^2 x - \sin^2 x} = \frac{1}{\cos^2 x - \sin^2 x}.$$

Third step.
$$\cos^2 x + \sin^2 x = 1,$$

or
$$1 = 1. \qquad \text{from (23), p. 59}$$

---

* In working out examples under this head it will appear that it is not necessary to take all of the steps in every case, nor will it always be found the best plan to take the steps in the order indicated.

† For instance, replace $\sin 2x$ by $2\sin x \cos x$, $\tan 2x$ by $\dfrac{2\tan x}{1 - \tan^2 x}$, $\cos (x + y)$ by $\cos x \cos y - \sin x \sin y$, etc.

Ex. 2. Prove $\dfrac{\sin(x+y)}{\sin(x-y)} = \dfrac{\tan x + \tan y}{\tan x - \tan y}$.

Solution. Since $\sin(x+y) = \sin x \cos y + \cos x \sin y$, and $\sin(x-y) = \sin x \cos y - \cos x \sin y$, the equation becomes:

First step. $\dfrac{\sin x \cos y + \cos x \sin y}{\sin x \cos y - \cos x \sin y} = \dfrac{\tan x + \tan y}{\tan x - \tan y}$.

Second step. $\dfrac{\sin x \cos y + \cos x \sin y}{\sin x \cos y - \cos x \sin y} = \dfrac{\dfrac{\sin x}{\cos x} + \dfrac{\sin y}{\cos y}}{\dfrac{\sin x}{\cos x} - \dfrac{\sin y}{\cos y}}$.

Simplifying, $\dfrac{\sin x \cos y + \cos x \sin y}{\sin x \cos y - \cos x \sin y} = \dfrac{\sin x \cos y + \cos x \sin y}{\sin x \cos y - \cos x \sin y}$,

or, $$1 = 1.$$

EXAMPLES

Prove the following identities :

1. $\tan x \sin x + \cos x = \sec x$.

2. $\cot x - \sec x \csc x (1 - 2\sin^2 x) = \tan x$.

3. $(\tan x + \cot x)\sin x \cos x = 1$.

4. $\dfrac{\sin y}{1 + \cos y} = \dfrac{1 - \cos y}{\sin y}$.

5. $\sqrt{\dfrac{1 - \sin A}{1 + \sin A}} = \sec A - \tan A$.

6. $\tan x \sin x \cos x + \sin x \cos x \cot x = 1$.

7. $\cot^2 x = \cos^2 x + (\cot x \cos x)^2$.

8. $(\sec y + \csc y)(1 - \cot y) = (\sec y - \csc y)(1 + \cot y)$.

9. $\sin^2 z \tan z + \cos^2 z \cot z + 2\sin z \cos z = \tan z + \cot z$.

10. $\sin^6 x + \cos^6 x = \sin^4 x + \cos^4 x - \sin^2 x \cos^2 x$.

11. $\sin B \tan^2 B + \csc B \sec^2 B = 2\tan B \sec B + \csc B - \sin B$.

12. Work out (a) to (v) under Ex. 4, p. 62, following the above general directions.

13. $\cos(x+y)\cos(x-y) = \cos^2 x - \sin^2 y$.

14. $\sin(A+B)\sin(A-B) = \cos^2 B - \cos^2 A$.

15. $\sin(x+y)\cos y - \cos(x+y)\sin y = \sin x$.

16. $\dfrac{\cos(x-y)}{\cos(x+y)} = \dfrac{1 + \tan x \tan y}{1 - \tan x \tan y}$.

17. $\tan A - \tan B = \dfrac{\sin(A-B)}{\cos A \cos B}$.

18. $\cot x + \cot y = \dfrac{\sin(x+y)}{\sin x \sin y}$.

19. $\sin x \cos(y+z) - \sin y \cos(x+z) = \sin(x-y)\cos z$.

20. $\dfrac{\tan(\theta - \phi) + \tan\phi}{1 - \tan(\theta - \phi)\tan\phi} = \tan\theta.$

21. $\cos(x - y + z) = \cos x \cos y \cos z + \cos x \sin y \sin z - \sin x \cos y \sin z$
$\qquad\qquad\qquad + \sin x \sin y \cos z.$

22. $\sin(x + y - z) + \sin(x + z - y) + \sin(y + z - x)$
$\qquad\qquad = \sin(x + y + z) + 4\sin x \sin y \sin z.$

23. $\cos x \sin(y - z) + \cos y \sin(z - x) + \cos z \sin(x - y) = 0.$

24. Work out (a) to (k) under Ex. 7, p. 67, and (a) to (j), under Ex. 4, p. 69, following the above general directions.

25. $\dfrac{1 + \sin 2x}{1 - \sin 2x} = \left(\dfrac{\tan x + 1}{\tan x - 1}\right)^2.$

26. $\tan x = \dfrac{\sin 2x}{1 + \cos 2x}.$

27. $\cot x = \dfrac{\sin 2x}{1 - \cos 2x}.$

28. $\dfrac{\cot A - 1}{\cot A + 1} = \sqrt{\dfrac{1 - \sin 2A}{1 + \sin 2A}}.$

29. $\cot(A + 45°) = \dfrac{1 - \sin 2A}{\cos 2A}.$

30. $\dfrac{\cos^3 x + \sin^3 x}{\cos x + \sin x} = \dfrac{2 - \sin 2x}{2}.$

31. $\dfrac{\sin 3x - \sin x}{\cos 3x + \cos x} - \tan x.$

32. $\dfrac{\sin 3x - \sin x}{\cos x - \cos 3x} = \cot 2x.$

33. $\sin 3x = 4\sin x \sin(60° + x)\sin(60° - x).$

34. $\dfrac{\sin 4x}{\sin 2x} = 2\cos 2x.$

35. $\sin 2x + \sin 2y + \sin 2z = 4\sin x \sin y \sin z.$

36. Work out (a) to (j) under Ex. 11, p. 71, following the above general directions.

37. $\sin 9x - \sin 7x = 2\cos 8x \sin x.$

38. $\cos 7x + \cos 5x = 2\cos 6x \cos x.$

39. $\dfrac{\sin 5x - \sin 2x}{\cos 2x - \cos 5x} = \cot\dfrac{7x}{2}.$

40. $\left(\sin\dfrac{x}{2} + \cos\dfrac{x}{2}\right)^2 = 1 + \sin x.$

41. $\dfrac{1 + \sec y}{\sec y} = 2\cos^2\dfrac{y}{2}.$

42. $\dfrac{\sin A + \sin B}{\cos A - \cos B} = -\cot\dfrac{1}{2}(A - B).$

43. $\dfrac{\cos\theta}{1 - \sin\theta} = \dfrac{1 + \tan\dfrac{\theta}{2}}{1 - \tan\dfrac{\theta}{2}}.$

44. $\cos 3\alpha - \cos 7\alpha = 2\sin 5\alpha \sin 2\alpha.$

45. $\cot\dfrac{x}{2} + \tan\dfrac{x}{2} = 2\csc x.$

46. $\dfrac{1 - \tan^2\dfrac{x}{2}}{1 + \tan^2\dfrac{x}{2}} = \cos x.$

47. $\dfrac{2\tan\dfrac{x}{2}}{1 + \tan^2\dfrac{x}{2}} = \sin x.$

48. $1 + \tan x \tan\dfrac{x}{2} = \sec x.$

49. $\tan\dfrac{x}{2} + 2\sin^2\dfrac{x}{2}\cot x = \sin x.$

50. $1 + \cot^2\dfrac{x}{2} = \dfrac{2}{\sin x \tan\dfrac{x}{2}}.$

51. $\dfrac{\tan^2\dfrac{x}{2} + \cot^2\dfrac{x}{2}}{\tan^2\dfrac{x}{2} - \cot^2\dfrac{x}{2}} = -\dfrac{1 + \cos^2 x}{2\cos x}.$

52. Work out (a) to (f) under Ex. 5, p. 75, following the above general directions.

CHAPTER V

GENERAL VALUES OF ANGLES. INVERSE TRIGONOMETRIC FUNCTIONS. TRIGONOMETRIC EQUATIONS

42. General value of an angle. Since all angles having the same initial and terminal sides have the same functions, it follows that we can add 2π to the angle or subtract 2π from the angle as many times as we please without changing the value of any function. Hence all functions of the angle x equal the corresponding functions of the angle

$$2n\pi + x,$$

where n is zero or any positive or negative integer.

The *general value* of an angle having a given trigonometric function is the expression or formula that includes *all angles* having this trigonometric function. Such general values will now be derived for all the trigonometric functions.

43. General value for all angles having the same sine or the same cosecant. Let x be the least positive angle whose sine has the given value a, and consider first the case when a is positive.

Construct the angle $x\,(= XOP)$, as on p. 31, and also the angle $\pi - x\,(= XOP')$, having the same value a for its sine. Then every angle whose terminal side is either OP or OP' has its sine equal to a, and it is evident that all such angles are found by adding even multiples of π to, or subtracting them from, x and $\pi - x$.

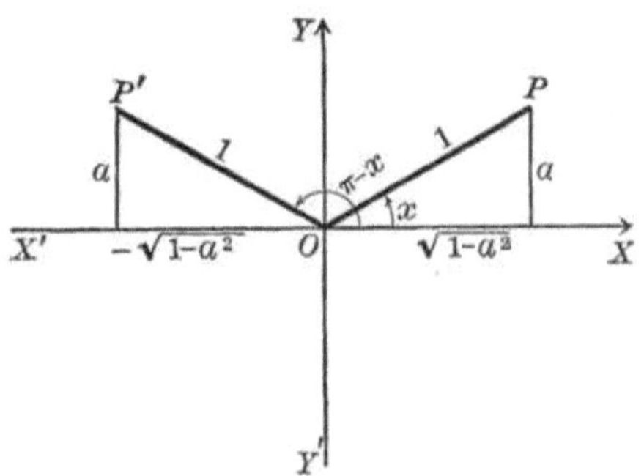

Let n denote zero, or any positive or negative integer. When n *is even*, $n\pi + x$ includes all the angles, and only those, which have the same initial and terminal sides as $x\,(= XOP)$. Therefore, when n is even,

$$(A) \qquad n\pi + x = n\pi + (-1)^n x.^*$$

---

* The factor $(-1)^n$ is positive for all even values of n and negative for all odd values of n.

Again, when n *is odd*, $n-1$ *is even*, and $(n-1)\pi + (\pi - x)$ includes all the angles, and only those, which have the same initial and terminal sides as $\pi - x$ $(= XOP')$. But when n is odd,

$$(B) \qquad (n-1)\pi + (\pi - x) = n\pi - x = n\pi + (-1)^n x.$$

From (A) and (B) it follows that the expression $n\pi + (-1)^n x$ for all values of n includes all the angles, and only those, which have the same initial and terminal sides as x and $\pi - x$.

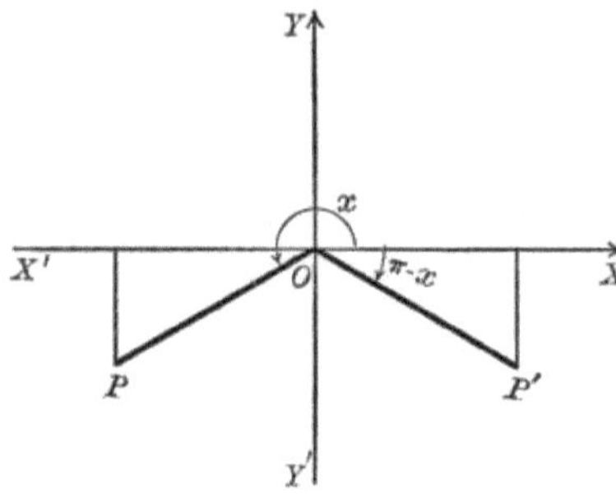

In case a is negative, $\pi - x$ will be negative, as shown in the figure, but the former line of reasoning will still hold true in every particular.

Since sine and cosecant are reciprocal functions, it follows that the expression for all angles having the same cosecant is also $n\pi + (-1)^n x$. Hence

$$(67) \qquad n\pi + (-1)^n x$$

is the general value of all the angles, and only those, which have the same sine or cosecant as x.*

This result may also be expressed as follows:

$$\sin x = \sin[n\pi + (-1)^n x],$$
$$\csc x = \csc[n\pi + (-1)^n x].$$

Ex. 1. Find the general value of all angles having the same sine as $\dfrac{3\pi}{4}$.

Solution. Let $x = \dfrac{3\pi}{4}$ in (67). This gives

$$n\pi + (-1)^n \frac{3\pi}{4}. \quad Ans.$$

Ex. 2. Find the four least positive angles whose cosecant equals 2.

Solution. The least positive angle whose cosecant $= 2$ is $\dfrac{\pi}{6}$. Let $x = \dfrac{\pi}{6}$ in (67). This gives

$$n\pi + (-1)^n \frac{\pi}{6}.$$

---

* In deriving this rule we have assumed x to be the least positive angle having the given sine. It follows immediately from the discussion, however, that the rule holds true if we replace x by an angle of any magnitude whatever, positive or negative, which has the given sine. The same observation applies to the rules derived in the next two sections.

When $n = 0$, we get $\qquad\qquad \dfrac{\pi}{6} = 30°.$

When $n = 1$, we get $\qquad\qquad \pi - \dfrac{\pi}{6} = 150°.$

When $n = 2$, we get $\qquad\qquad 2\pi + \dfrac{\pi}{6} = 390°.$

When $n = 3$, we get $\qquad\qquad 3\pi - \dfrac{\pi}{6} = 510°.$ *Ans.*

44. General value for all angles having the same cosine or the same secant. Let x be the least positive angle whose cosine has the given value a, and consider first the case when a is positive. Construct the angle $x\,(= XOP)$, and also the angle $-x\,(= XOP')$, having the same value a for its cosine. Then every angle whose terminal side is either OP or OP' has its cosine equal to a, and it is evident that all such angles are found by adding even multiples of π to, or subtracting them from, x and $-x$.

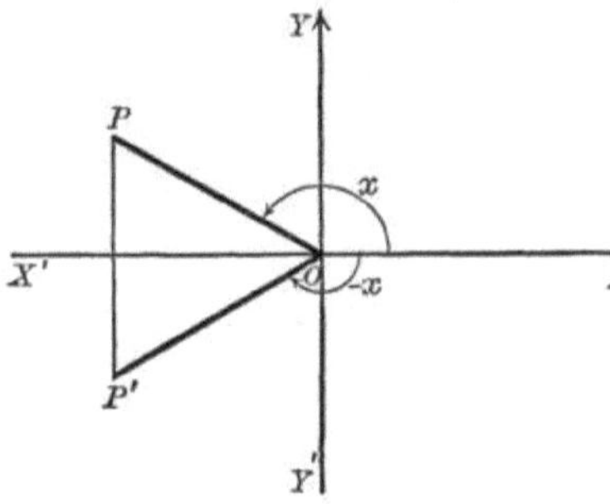

Let n denote zero, or any positive or negative integer. For any value of n,

$$(A) \qquad\qquad 2\,n\pi + x\,{}^*$$

includes all the angles, and only those, which have the same initial and terminal sides as $x\,(= XOP)$. Similarly,

$$(B) \qquad 2\,n\pi - x$$

includes all the angles, and only those, which have the same initial and terminal sides as $-x\,(= XOP')$.

In case a is negative, the same line of reasoning still holds true.

Since cosine and secant are reciprocal functions, it follows that the same discussion holds for the secant. Hence, from (A) and (B),

$$(68) \qquad\qquad 2\,n\pi \pm x$$

is the general value of all the angles, and only those, which have the same cosine or secant as x.

* $2n$ is even and $2n - 1$ is odd for all values of n.

This result may also be expressed as follows:

$$\cos x = \cos (2\,n\pi \pm x),$$
$$\sec x = \sec (2\,n\pi \pm x).$$

Ex. 1. Given $\cos A = -\dfrac{1}{\sqrt{2}}$; find the general value of A. Also find the five least positive values of A.

Solution. The least positive angle whose cosine $= -\dfrac{1}{\sqrt{2}}$ is $\dfrac{3\,\pi}{4}$. If we let $x = \dfrac{3\,\pi}{4}$ in (68), we get

$$A = 2\,n\pi \pm \frac{3\,\pi}{4}.$$

When $n = 0$, $\qquad A = \dfrac{3\,\pi}{4} = 135°.$

When $n = 1$, $\qquad A = 2\,\pi \pm \dfrac{3\,\pi}{4} = 225°$ or $495°.$

When $n = 2$, $\qquad A = 4\,\pi \pm \dfrac{3\,\pi}{4} = 585°$ or $855°.$ *Ans.*

45. General value for all angles having the same tangent or the same cotangent. Let x be the least positive angle whose tangent has the given value a, and consider first the case when a is positive.

Construct the angle $x\ (= XOP)$, and also the angle $\pi + x\ (= XOP')$ having the same value a for its tangent. Then every angle whose

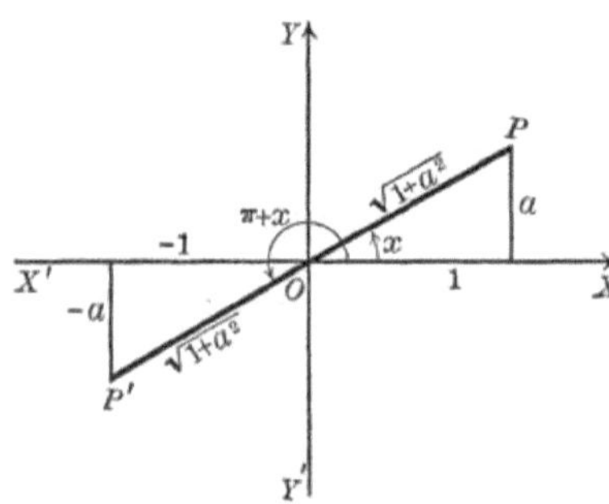

terminal side is either OP or OP' has its tangent equal to a, and it is evident that all such angles are found by adding even multiples of π to, or subtracting them from, x and $\pi + x$.

Let n denote zero, or any positive or negative integer. When n *is even*,

$$(A)\qquad n\pi + x$$

includes all the angles, and only those, which have the same initial and terminal sides as $x\ (= XOP)$.

Again, when n *is odd*, $n - 1$ *is even*, and

$$(B)\qquad (n-1)\pi + (\pi + x) = n\pi + x$$

includes all the angles, and only those, which have the same initial and terminal sides as $\pi + x\ (= XOP')$.

In case a is negative, the same line of reasoning still holds true.

Since tangent and cotangent are reciprocal functions, it follows that the same discussion holds for the cotangent. Hence, from (A) and (B), for all values of n,

(69) $n\pi + x$

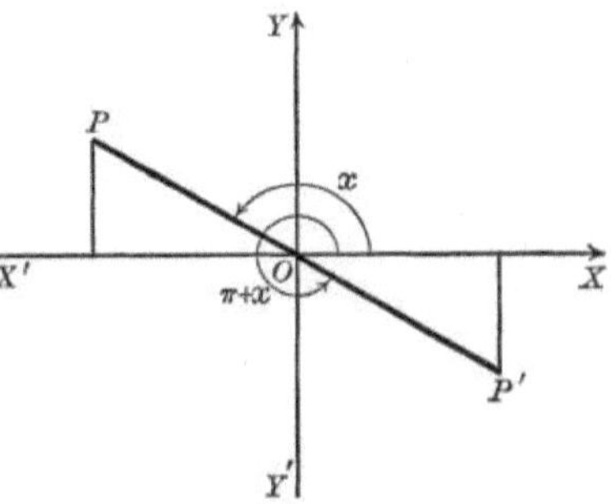

is the general value of all the angles, and only those, which have the same tangent or cotangent as x.

This result may also be stated as follows:

$$\tan x = \tan (n\pi + x),$$
$$\cot x = \cot (n\pi + x).$$

EXAMPLES

1. Given $\sin A = \tfrac{1}{2}$; find the general value of A. Also find the four least positive values of A.

$$Ans. \quad n\pi + (-1)^n \frac{\pi}{6};\ 30°,\ 150°,\ 390°,\ 510°.$$

2. Given $\cos A = \dfrac{\sqrt{3}}{2}$; find the general value of A. Also find all values of A numerically less than 2π.

$$Ans. \quad 2n\pi \pm \frac{\pi}{6};\ \pm\frac{\pi}{6},\ \pm\frac{11\pi}{6}.$$

3. Given $\tan A = 1$; find the general value of A. Also find the values of A numerically less than 4π.

$$Ans. \quad n\pi + \frac{\pi}{4};\ \frac{\pi}{4},\ -\frac{3\pi}{4},\ \frac{5\pi}{4},\ -\frac{7\pi}{4},\ \frac{9\pi}{4},\ -\frac{11\pi}{4},\ \frac{13\pi}{4},\ -\frac{15\pi}{4}.$$

4. Given $\sin 2x = \dfrac{1}{2}$; show that $x = \dfrac{n\pi}{2} + (-1)^n \dfrac{\pi}{12}.$

5. Given $\cos 3x = -\dfrac{1}{2}$; show that $x = \dfrac{2n\pi}{3} \pm \dfrac{\pi}{9}.$

6. In each of the following examples find the general values of the angles, having given

(a) $\sin A = \pm 1.$

$$Ans. \quad A = n\pi + (-1)^n \left(\pm \frac{\pi}{2}\right) - n\pi \pm \frac{\pi}{2}.$$

(b) $\cot x = \pm \sqrt{3}.$

$$x = n\pi \pm \frac{\pi}{6}.$$

(c) $\cos y = \pm \dfrac{1}{2}.$

$$y = n\pi \pm \frac{\pi}{3}.$$

(d) $\tan B = \pm 1.$

$$B = n\pi \pm \frac{\pi}{4}.$$

(e) $\csc C = \pm \sqrt{2}.$

$$C = n\pi \pm \frac{\pi}{4}.$$

(f) $\sec A = \pm \dfrac{2}{\sqrt{3}}.$

$$A = n\pi \pm \frac{\pi}{6}.$$

7. Given $\sin x = \dfrac{1}{2}$ and $\tan x = \dfrac{1}{\sqrt{3}}$; find the general value of x.

Solution. Since $\sin x$ is $-$ and $\tan x$ is $+$, x must lie in the third quadrant. The smallest positive angle in the third quadrant which satisfies the condition $\sin x = -\dfrac{1}{2}$ is $\dfrac{7\pi}{6}$, and this angle also satisfies $\tan x = \dfrac{1}{\sqrt{3}}$.

Hence
$$x = 2n\pi + \frac{7\pi}{6}. \quad Ans.$$

8. In each of the following examples find all the positive values of x less than 2π which satisfy the given equations.

(a) $\cos x = \dfrac{1}{\sqrt{2}}$.   *Ans.* $\dfrac{\pi}{4}, \dfrac{7\pi}{4}$.

(b) $\sin x = \pm \dfrac{1}{\sqrt{2}}$. $\dfrac{\pi}{4}, \dfrac{3\pi}{4}, \dfrac{5\pi}{4}, \dfrac{7\pi}{4}$.

(c) $\tan x = \pm \sqrt{3}$. $\dfrac{\pi}{3}, \dfrac{2\pi}{3}, \dfrac{4\pi}{3}, \dfrac{5\pi}{3}$.

(d) $\cot x = \pm 1$. $\dfrac{\pi}{4}, \dfrac{3\pi}{4}, \dfrac{5\pi}{4}, \dfrac{7\pi}{4}$.

(e) $\cos x = -\dfrac{\sqrt{3}}{2}$. $\dfrac{5\pi}{6}, \dfrac{7\pi}{6}$.

(f) $\sec x = 2$. $\dfrac{\pi}{3}, \dfrac{5\pi}{3}$.

(g) $\sin x = \pm \dfrac{1}{2}$. $\dfrac{\pi}{6}, \dfrac{5\pi}{6}, \dfrac{7\pi}{6}, \dfrac{11\pi}{6}$.

46. Inverse trigonometric functions. The value of a trigonometric function of an angle depends on the value of the angle; and conversely, the value of the angle depends on the value of the function. If the angle is given, the sine of the angle can be found; if the sine is given, the angle can be expressed. It is often convenient to represent an angle by the value of one of its functions. Thus, instead of saying that *an angle is 30°*, we may say (what amounts to the same thing) that *it is the least positive angle whose sine is $\frac{1}{2}$*. We then consider the angle as a function of its sine, and the angle is said to be an *inverse trigonometric function*, or *anti-trigonometric*, or *inverse circular function*, and is denoted by the symbol

$$\sin^{-1}\tfrac{1}{2}, \text{ or, } \operatorname{arc} \sin \tfrac{1}{2} *,$$

read "*inverse sine of $\frac{1}{2}$*," or, "*arc (or angle) whose sine is $\frac{1}{2}$.*" Similarly, $\cos^{-1}x$ is read "*inverse cosine of x*," $\tan^{-1}y$ is read "*inverse tangent of y*," etc. If a is the value of the tangent of the angle x, we are now in a position to express the relation between a

---

* Symbol generally used in Continental books.

and x in two different ways. Thus, *tan x = a*, meaning *the tangent of the angle x is a ;* or, *x = tan⁻¹a*, meaning *x is an angle whose tangent is a.* The student should note carefully that in this connection —*1 is not an algebraic exponent,* but is merely a part of the mathematical symbol denoting an inverse trigonometric function. tan⁻¹a *does not denote*

$$(\tan a)^{-1} = \frac{1}{\tan a},$$

but *does denote each and every angle whose tangent is a.*

The trigonometric functions (ratios) are pure numbers, while the inverse trigonometric functions are measures of angles, expressed in degrees or radians.

Consider the expressions

$$\tan x = a, \qquad x = \tan^{-1}a.$$

In the first we know that for a given value of the angle x, $\tan x$ (or a) has *a single definite value.*

In the second we know from (**69**), p. 83, that for a given value of the tangent a, $\tan^{-1}a$ (or x) has *an infinite number of values.*

Similarly, for each of the other inverse trigonometric functions. Hence:

The trigonometric functions are single valued, and the inverse trigonometric functions are many valued.

The smallest value numerically of an inverse trigonometric function is called its *principal value.** For example, if

$$\tan x = 1,$$

then the *general value* of x is, by (**69**), p. 83,

$$x = \tan^{-1}1 = n\pi + \frac{\pi}{4},$$

where n denotes zero or any positive or negative integer, and

$$\frac{\pi}{4} = 45°$$

is the *principal value* of x.

---

* Hence, if $\sin x$, $\csc x$, $\tan x$, or $\cot x$ is positive, the principal value of x lies between 0 and $\frac{\pi}{2}$; if $\sin x$, $\csc x$, $\tan x$, or $\cot x$ is negative, the principal value of x lies between 0 and $-\frac{\pi}{2}$.

If $\cos x$ or $\sec x$ is positive, the principal value of x lies between $\frac{\pi}{2}$ and $-\frac{\pi}{2}$, *preference being given to the positive angle.*

Similarly, if $\cos x = \frac{1}{2}$, then by (**68**), p. 81,

$$x = \cos^{-1}\frac{1}{2} = 2\,n\pi \pm \frac{\pi}{3},$$

where the principal value of x is $\frac{\pi}{3} = 60°$.

Since the sine and cosine of an angle cannot be less than -1 nor greater than $+1$, it follows that the expressions

$$\sin^{-1}a \quad \text{and} \quad \cos^{-1}a$$

have no meaning unless a lies between -1 and $+1$ inclusive. Similarly, it is evident that the expressions

$$\sec^{-1}a \quad \text{and} \quad \csc^{-1}a$$

have no meaning for values of a lying between -1 and $+1$.

Any relation that has been established between trigonometric functions may be expressed by means of the inverse notation. Thus, we know that

$$\cos x = \sqrt{1 - \sin^2 x}. \qquad (\mathbf{29}),\ \text{p. } 60$$

This may be written

$$(A) \qquad x = \cos^{-1}\sqrt{1 - \sin^2 x}.$$

Placing $\sin x = a$, then $x = \sin^{-1}a$, and (A) becomes

$$\sin^{-1}a - \cos^{-1}\sqrt{1 - a^2}.$$

Similarly, since $\quad \cos 2x = 2\cos^2 x - 1, \qquad (\mathbf{49\,b}),\ \text{p. } 70$

we may write

$$(B) \qquad 2x = \cos^{-1}(2\cos^2 x - 1).$$

Placing $\cos x = c$, then $x = \cos^{-1}c$, and (B) becomes

$$2\cos^{-1}c = \cos^{-1}(2c^2 - 1).$$

Since we know that the co-functions of complementary angles are equal, we get for the principal values of the angles that

$$\sin^{-1}a + \cos^{-1}a = \frac{\pi}{2}, \qquad\qquad 0 < a \leqq 1$$

$$\tan^{-1}b + \cot^{-1}b = \frac{\pi}{2}, \qquad\qquad 0 < b$$

$$\sec^{-1}c + \csc^{-1}c = \frac{\pi}{2}. \qquad\qquad 1 < c$$

We shall now show how to prove identities involving inverse trigonometric functions for the principal values of the angles.

Ex. 1. Prove the identity

(a) $$\tan^{-1}m + \tan^{-1}n = \tan^{-1}\frac{m+n}{1-mn}.$$

Proof. Let

(b) $$A = \tan^{-1}m \quad \text{and} \quad B = \tan^{-1}n.$$

(c) Then $\quad \tan A = m \quad$ and $\tan B = n.$

Substituting from (b) in first member of (a), we get

$$A + B = \tan^{-1}\frac{m+n}{1-mn},$$

or, what amounts to the same thing,

(d) $$\tan(A + B) = \frac{m+n}{1-mn}.$$

But from (**44**), p. 68,

(e) $$\tan(A + B) = \frac{\tan A + \tan B}{1 - \tan A \tan B}.$$

Substituting from (c) in second member of (e), we get

(f) $$\tan(A + B) = \frac{m+n}{1-mn}.$$

Since (d) and (f) are identical, we have proven (a) to be true.

Ex. 2. Prove that

(g) $$\sin^{-1}\tfrac{3}{5} + \cos^{-1}\tfrac{15}{17} = \sin^{-1}\tfrac{77}{85}.$$

Proof. Let

(h) $$A = \sin^{-1}\tfrac{3}{5} \quad \text{and} \quad B = \cos^{-1}\tfrac{15}{17}.$$

(i) Then $\quad \sin A = \tfrac{3}{5} \quad$ and $\cos B = \tfrac{15}{17}.$

(j) Also $\quad \cos A = \tfrac{4}{5} \quad$ and $\sin B = \tfrac{8}{17}.$*

Substituting from (h) in first member of (g), we get

$$A + B = \sin^{-1}\tfrac{77}{85},$$

or, what amounts to the same thing,

(k) $$\sin(A + B) = \tfrac{77}{85}.$$

But from (**40**), p. 63,

(l) $$\sin(A + B) = \sin A \cos B + \cos A \sin B.$$

Substituting from (i) and (j) in second member of (l), we get

(m) $$\sin(A + B) = \tfrac{3}{5} \cdot \tfrac{15}{17} + \tfrac{4}{5} \cdot \tfrac{8}{17} = \tfrac{77}{85}.$$

Since (k) and (m) are identical, we have proven (g) to be true.

The following example illustrates how some equations involving inverse trigonometric functions may be solved.

* Found by method explained on p. 30.

Ex. 3. Solve the following equation for x :

$$\tan^{-1} 2\,x + \tan^{-1} 3\,x = \frac{\pi}{4}.$$

Solution. Take the tangent of both sides of the equation. Thus *

$$\tan\left(\tan^{-1} 2\,x + \tan^{-1} 3\,x\right) = \tan\frac{\pi}{4},$$

or,
$$\frac{\tan\left(\tan^{-1} 2\,x\right) + \tan\left(\tan^{-1} 3\,x\right)}{1 - \tan\left(\tan^{-1} 2\,x\right)\tan\left(\tan^{-1} 3\,x\right)} = 1, \qquad \text{from (44), p. 68}$$

or,
$$\frac{2\,x + 3\,x}{1 - 2\,x \cdot 3\,x} = 1.$$

Clearing of fractions and solving for x, we get

$$x = \tfrac{1}{6} \ \text{ or } \ -1.$$

$x = \tfrac{1}{6}$ satisfies the equation for the principal values of $\tan^{-1} 2\,x$ and $\tan^{-1} 3\,x$.
$x = -1$ satisfies the equation for the values

$$\tan^{-1}(-2) = 116.57^\circ,$$
$$\tan^{-1}(-3) = -71.57^\circ.$$

EXAMPLES

1. Express in radians the general values of the following functions :

(a) $\sin^{-1}\dfrac{1}{\sqrt{2}}$.   *Ans.* $n\pi + (-1)^n\dfrac{\pi}{4}$.   (e) $\tan^{-1}\dfrac{1}{\sqrt{3}}$.   *Ans.* $n\pi + \dfrac{\pi}{6}$.

(b) $\sin^{-1}\left(-\dfrac{\sqrt{3}}{2}\right)$. $n\pi - (-1)^n\dfrac{\pi}{3}$.   (f) $\tan^{-1}(\pm\sqrt{3})$. $n\pi \pm \dfrac{\pi}{3}$.

(c) $\cos^{-1}\dfrac{\sqrt{3}}{2}$. $2\,n\pi \pm \dfrac{\pi}{6}$ (g) $\cot^{-1}(\pm 1)$. $n\pi \pm \dfrac{\pi}{4}$.

(d) $\cos^{-1}\left(-\dfrac{1}{2}\right)$. $2\,n\pi \pm \dfrac{2\,\pi}{3}$.   (h) $\cot^{-1}\left(\dfrac{1}{\sqrt{3}}\right)$. $n\pi + \dfrac{\pi}{3}$.

2. Prove the following :

(a) $\tan^{-1}a - \tan^{-1}b = \tan^{-1}\dfrac{a-b}{1+ab}$.   (h) $2\tan^{-1}a = \tan^{-1}\dfrac{2\,a}{1-a^2}$.

(b) $2\tan^{-1}a = \sin^{-1}\dfrac{2\,a}{1+a^2}$.   (i) $\sin^{-1}a = \cos^{-1}\sqrt{1-a^2}$.

(c) $2\sin^{-1}a = \cos^{-1}(1 - 2\,a^2)$.   (j) $\sin^{-1}a = \tan^{-1}\dfrac{a}{\sqrt{1-a^2}}$.

(d) $\tan^{-1}a = \sin^{-1}\dfrac{a}{\sqrt{1+a^2}}$.   (k) $\tan^{-1}a = \cos^{-1}\dfrac{1}{\sqrt{1+a^2}}$.

(e) $\tan^{-1}\dfrac{m}{n} - \tan^{-1}\dfrac{m-n}{m+n} = \dfrac{\pi}{4}$.   (l) $\sin^{-1}\dfrac{3}{5} + \sin^{-1}\dfrac{8}{17} = \sin^{-1}\dfrac{77}{85}$.

(f) $\cos^{-1}\dfrac{4}{5} + \tan^{-1}\dfrac{3}{5} = \tan^{-1}\dfrac{27}{11}$.   (m) $\cos^{-1}\dfrac{4}{5} + \cos^{-1}\dfrac{12}{13} = \cos^{-1}\dfrac{33}{65}$.

(g) $2\tan^{-1}\dfrac{2}{3} = \tan^{-1}\dfrac{12}{5}$.   (n) $\tan^{-1}\dfrac{1}{7} + \tan^{-1}\dfrac{1}{13} = \tan^{-1}\dfrac{2}{9}$.

---

* The student should remember that $\tan^{-1} 2\,x$ and $\tan^{-1} 3\,x$ are measures of angles.

3. Solve the following equations :

(a) $\tan^{-1}x + \tan^{-1}(1-x) = \tan^{-1}\left(\frac{4}{3}\right)$. $\qquad$ *Ans.* $x = \frac{1}{2}$.

(b) $\tan^{-1}x + 2\cot^{-1}x = \frac{2\pi}{3}$. $\qquad x = \sqrt{3}$.

(c) $\tan^{-1}\dfrac{x-1}{x+2} + \tan^{-1}\dfrac{x+1}{x+2} = \dfrac{\pi}{4}$. $\qquad x = \pm\sqrt{\dfrac{5}{2}}$.

(d) $\cos^{-1}\dfrac{x^2-1}{x^2+1} + \tan^{-1}\dfrac{2x}{x^2-1} = \dfrac{2\pi}{3}$. $\qquad x = \sqrt{3}$.

(e) $\tan^{-1}\dfrac{x+1}{x-1} + \tan^{-1}\dfrac{x-1}{x} = \tan^{-1}(-7)$. $\qquad x = 2$.

(f) $\tan^{-1}(x+1) + \tan^{-1}(x-1) = \tan^{-1}\dfrac{8}{61}$. $\qquad x = -8, \dfrac{1}{4}$.

(g) $\sin^{-1}x + \sin^{-1}2x = \dfrac{\pi}{3}$. $\qquad x = \pm\dfrac{\sqrt{21}}{14}$.

(h) $\sin^{-1}\dfrac{5}{x} + \sin^{-1}\dfrac{12}{x} = \dfrac{\pi}{2}$. $\qquad x = 13$.

4. Find the values of the following:

(a) $\sin\left(\tan^{-1}\dfrac{5}{12}\right)$. $\qquad$ *Ans.* $\pm\dfrac{5}{13}$. $\qquad$ (d) $\cos(2\cos^{-1}a)$. *Ans.* $2a^2-1$.

(b) $\cot\left(2\sin^{-1}\dfrac{3}{5}\right)$. $\qquad \pm\dfrac{7}{24}$. $\qquad$ (e) $\tan(2\tan^{-1}a)$. $\qquad \dfrac{2a}{1-a^2}$.

(c) $\sin\left(\tan^{-1}\dfrac{1}{2} + \tan^{-1}\dfrac{1}{3}\right)$. $\qquad \pm\dfrac{1}{\sqrt{2}}$. $\qquad$ (f) $\cos(2\tan^{-1}a)$. $\qquad \dfrac{1-a^2}{1+a^2}$.

47. Trigonometric equations. By these we mean equations involving one or more trigonometric functions of one or more angles. For instance,

$$2\cos^2x + \sqrt{3}\sin x + 1 = 0$$

is a trigonometric equation involving the unknown angle x. We have already worked out many problems in trigonometric equations. Thus, Examples 1–8, pp. 83, 84, are in fact examples requiring the solution of trigonometric equations.

To solve a trigonometric equation involving one unknown angle is to find an expression (the student should look up the general value of an angle, p. 85) for all values of the angle which satisfy the given equation.

No general method can be given for solving trigonometric equations that would be the best to follow in all cases, but the following general directions (which are similar to those given on p. 76 for proving identities) will be found useful.

48. General directions for solving a trigonometric equation. [*]

First step. *If multiple angles, fractional angles, or the sums or differences of angles are involved, reduce all to functions of a single angle,† and simplify.*

Second step. *If the resulting expressions are not readily reducible to the same function, change all the functions into sines and cosines.*

Third step. *Clear of fractions and radicals.*

Fourth step. *Change all the functions to a single function.*

Fifth step. *Solve algebraically (by factoring or otherwise) for the one function now occurring in the equation, and express the general value of the angle thus found by* (**67**), (**68**), *or* (**69**). *Only such values of the angle which satisfy the given equation are solutions.*

Ex. 1. Solve the equation

$$\cos 2x \sec x + \sec x + 1 = 0.$$

Solution. Since $\cos 2x = \cos^2 x - \sin^2 x$, we get

First step.　　　　$(\cos^2 x - \sin^2 x)\sec x + \sec x + 1 = 0.$

Second step. Since $\sec x = \dfrac{1}{\cos x}$, this becomes

$$\frac{\cos^2 x - \sin^2 x}{\cos x} + \frac{1}{\cos x} + 1 = 0.$$

Third step.　　　　$\cos^2 x - \sin^2 x + 1 + \cos x = 0.$

Fourth step. Since $\sin^2 x = 1 - \cos^2 x$, we have

$$\cos^2 x - 1 + \cos^2 x + 1 + \cos x = 0,$$

or,　　　　　　　$2\cos^2 x + \cos x = 0.$

Fifth step.　　　　$\cos x (2\cos x + 1) = 0.$

Placing each factor equal to zero, we get

$$\cos x = 0,$$

or, from (**68**), p. 81,　　　　$x = \cos^{-1} 0 = 2n\pi \pm \dfrac{\pi}{2}.$

Also,　　　　　　$2\cos x + 1 = 0,$

$$\cos x = -\tfrac{1}{2},$$

or,　　　　　$x = \cos^{-1}\left(-\dfrac{1}{2}\right) = 2n\pi \pm \dfrac{2\pi}{3}.$

Hence the general values of the angles which satisfy the equation are

$$2n\pi \pm \frac{\pi}{2} \quad \text{and} \quad 2n\pi \pm \frac{2\pi}{3}.$$

The positive angles less than 2π which satisfy the equation are then

$$\frac{\pi}{2}, \quad \frac{2\pi}{3}, \quad \frac{4\pi}{3}, \quad \frac{3\pi}{2}.$$

---

[*] In working out examples under this head it will appear that it is not necessary to take all of the steps in every case, nor will it always be found the best plan to take the steps in the order indicated.

† For instance, replace $\cos 2x$ by $\cos^2 x - \sin^2 x$, $\sin\left(x + \dfrac{\pi}{4}\right)$ by $\dfrac{\sin x + \cos x}{\sqrt{2}}$, etc.

Ex. 2. Solve the equation

$$2 \sin^2 x + \sqrt{3} \cos x + 1 = 0.$$

Solution. Since $\sin^2 x = 1 - \cos^2 x$, we get

Fourth step. $\qquad 2 - 2 \cos^2 x + \sqrt{3} \cos x + 1 = 0,$

or, $\qquad\qquad 2 \cos^2 x - \sqrt{3} \cos x - 3 = 0.$

Fifth step. This is a quadratic in $\cos x$. Solving, we get

$$\cos x = \sqrt{3} \text{ or } -\frac{\sqrt{3}}{2}.$$

Since no cosine can be greater than 1, the first result, $\cos x = \sqrt{3}$, cannot be used. From the second result,

$$x = \cos^{-1}\left(-\frac{\sqrt{3}}{2}\right) = 2\,n\pi \pm \frac{5\,\pi}{6}. \quad Ans.$$

EXAMPLES

Solve each of the following equations :

1. $\sin^2 x = 1.$ $\qquad\qquad\qquad Ans.\ \ x = n\pi + (-1)^n\left(\pm\frac{\pi}{2}\right)^* = n\pi \pm \frac{\pi}{2}.$

2. $\csc^2 x = 2.$ $\qquad\qquad\qquad x = n\pi + (-1)^n\left(\pm\frac{\pi}{4}\right) = n\pi \pm \frac{\pi}{4}.$

3. $\tan^2 x = 1.$ $\qquad\qquad\qquad x = n\pi \pm \frac{\pi}{4}.$

4. $\cot^2 x = 3.$ $\qquad\qquad\qquad x = n\pi \pm \frac{\pi}{6}.$

5. $\cos^2 x = \frac{1}{4}.$ $\qquad\qquad\qquad x = n\pi \pm \frac{\pi}{3}.$

6. $\sec^2 x - \frac{4}{3}.$ $\qquad\qquad\qquad x = n\pi \pm \frac{\pi}{6}.$

7. $2 \sin^2 x + 3 \cos x = 0.$ $\qquad\qquad x = 2\,n\pi \pm \frac{2\,\pi}{3}.$

8. $\cos^2 \alpha - \sin^2 \alpha = \frac{1}{2}.$ $\qquad\qquad \alpha = n\pi + (-1)^n \cdot \frac{\pi}{6}.$

9. $2\sqrt{3} \cos^2 \alpha = \sin \alpha.$ $\qquad\qquad \alpha = n\pi + (-1)^n \cdot \frac{\pi}{3}.$

10. $\sin^2 y - 2 \cos y + \frac{1}{4} = 0.$ $\qquad y = 2\,n\pi \pm \frac{\pi}{3}.$

11. $\sin A + \cos A = \sqrt{2}.$ $\qquad\qquad A = 2\,n\pi + \frac{\pi}{4}.$

12. $4 \sec^2 y - 7 \tan^2 y = 3.$ $\qquad\qquad y = n\pi \pm \frac{\pi}{6}.$

13. $\tan B + \cot B = 2.$ $\qquad\qquad B = n\pi + \frac{\pi}{4}.$

14. $\tan^2 x - (1 + \sqrt{3}) \tan x + \sqrt{3} = 0.$ $\qquad x = n\pi + \frac{\pi}{4},\ n\pi + \frac{\pi}{3}.$

---

* Since the principal value of $x = \sin^{-1} 1 = \frac{\pi}{2}$ and of $x = \sin^{-1}(-1) = -\frac{\pi}{2}$.

15. $\cot^2 x + \left(\sqrt{3} + \dfrac{1}{\sqrt{3}}\right)\cot x + 1 = 0.$ *Ans.* $x = n\pi + \dfrac{5\pi}{6},\ n\pi + \dfrac{2\pi}{3}.$

16. $\tan^2 x + \cot^2 x = 2.$ $x = n\pi \pm \dfrac{\pi}{4}.$

17. $\tan\left(x + \dfrac{\pi}{4}\right) = 1 + \sin 2x.$ $x = n\pi,\ n\pi - \dfrac{\pi}{4}.$

18. $\csc x \cot x = 2\sqrt{3}.$ $x = 2\,n\pi \pm \dfrac{\pi}{6}.$

19. $\sin\dfrac{x}{2} = \csc x - \cot x.$ $x = 2\,n\pi.$

20. $\csc y + \cot y = \sqrt{3}.$ $y = 2\,n\pi + \dfrac{\pi}{3}.$

21. $3\,(\sec^2 \alpha + \cot^2 \alpha) = 13.$ $\alpha = n\pi \pm \dfrac{\pi}{6},\ n\pi \pm \dfrac{\pi}{3}.$

Find all the positive angles less than 360° which satisfy the following equations :

22. $\cos 2x + \cos x = -1.$ *Ans.* $x = 90°,\ 120°,\ 240°,\ 270°.$

23. $\sin 2x - \cos 2x - \sin x + \cos x = 0.$ $x = 0°,\ 90°,\ 180°.$

24. $\sin(60° - x) - \sin(60° + x) = \dfrac{\sqrt{3}}{2}.$ $x = 240°,\ 300°.$

25. $\sin(30° + x) - \cos(60° + x) = -\dfrac{\sqrt{3}}{2}.$ $x = 210°,\ 330°.$

26. $\tan(45° - x) + \cot(45° - x) = 4.$ $x = 30°,\ 150°,\ 210°,\ 330°.$

27. $\cos 2x = \cos^2 x.$ $x = 0°,\ 180°.$

28. $2\sin y = \sin 2y.$ $y = 0°,\ 180°.$

29. $\sin x + \sin 2x + \sin 3x = 0.$ $x = 0°,\ 90°,\ 120°,\ 180°,\ 240°,\ 270°.$

30. $\tan x + \tan 2x = \tan 3x.$ $x = 0°,\ 60°,\ 120°,\ 180°,\ 240°,\ 300°.$

·31. $\sec x - \cot x = \csc x - \tan x.$ $x = 45°,\ 135°,\ 225°,\ 315°.$

32. $\sin 4x - \cos 3x = \sin 2x.$ $x = 30°,\ 90°,\ 150°,\ 210°,\ 270°,\ 330°.$

33. $\sqrt{1 + \sin x} - \sqrt{1 - \sin x} = 2\cos x.$

34. $\sin^4 x + \cos^4 x = \dfrac{5}{8}.$

35. $\sec(x + 120°) + \sec(x - 120°) = 2\cos x.$

36. $\sin(x + 120°) + \sin(x + 60°) = \dfrac{3}{2}.$

37. $\sin y + \sin 3y = \cos y - \cos 3y.$

Find the general value of x that satisfies the following equations :

38. $\cos x = -\dfrac{1}{\sqrt{2}}$ and $\tan x = 1.$ *Ans.* $x = (2\,n + 1)\pi + \dfrac{\pi}{4}.$

39. $\cot x = -\sqrt{3}$ and $\csc x = -2.$ $x = 2\,n\pi - \dfrac{\pi}{6}.$

40. Find positive values of A and B which satisfy the equations

$$\cos(A - B) = \dfrac{1}{2} \text{ and } \sin(A + B) = \dfrac{1}{2}. \qquad \textit{Ans. } \dfrac{7\pi}{12} \text{ and } \dfrac{\pi}{4}.$$

41. Find positive values of A and B which satisfy the equations

$$\tan(A - B) = 1 \text{ and } \sec(A + B) = \dfrac{2}{\sqrt{3}}. \qquad \textit{Ans. } \dfrac{25\pi}{24} \text{ and } \dfrac{19\pi}{24}.$$

CHAPTER VI

GRAPHICAL REPRESENTATION OF TRIGONOMETRIC FUNCTIONS

49. Variables. A *variable* is a quantity to which an unlimited number of values can be assigned. Variables are usually denoted by the later letters of the alphabet, as x, y, z.

50. Constants. A quantity whose value remains unchanged is called a *constant*. *Numerical* or *absolute constants* retain the same values in all problems, as 2, 5, $\sqrt{7}$, π, etc. *Arbitrary constants* are constants whose values are fixed in any particular problem. These are usually denoted by the earlier letters of the alphabet, as a, b, c, etc.

51. Functions. A *function of a variable* is a magnitude whose value depends on the value of the variable. Nearly all scientific problems deal with quantities and relations of this sort, and in the experiences of everyday life we are continually meeting conditions illustrating the dependence of one quantity on another. Thus, the *weight* a man is able to lift depends on his *strength*, other things being equal. Hence we may consider the weight lifted as a function of the strength of the man. Similarly, the *distance* a boy can run may be considered as a function of the *time*. The *area* of a square is a function of the *length* of a side, and the *volume* of a sphere is a function of its *diameter*. Similarly, the trinomial

$$x^2 - 7x - 6$$

is a *function of* x because its value will depend on the value we assume for x, and

$$\sin A, \quad \cos 2A, \quad \tan \frac{A}{2}$$

are *functions of* A.

52. Graphs of functions. The relation between the assumed values of a variable, and the corresponding values of a function depending on that variable, are very clearly shown by a geometrical representation where the assumed values of the variable are taken as the abscissas, and the corresponding values of the function as the ordinates of points in a plane (see § 13, p. 26). A smooth curve drawn through these points in order is called the *graph of the function.* Following are

General directions for plotting the graph of a function.
First step. *Place y equal to the function.*
Second step. *Assume different values for the variable ($= x$) and calculate the corresponding values of the function ($= y$), writing down the results in tabulated form.*
Third step. *Plot the points having the values of x as abscissas and the corresponding values of y as ordinates.*
Fourth step. *A smooth curve drawn through these points in order is called the graph of the function.*

Ex. 1. Plot the graph of $2x - 6$.

Solution. First step. Let $y = 2x - 6$.
Second step. Assume different values for x and compute the corresponding values of y. Thus, if

$$x = 0, \quad y = -6; \qquad x = -1, \quad y = -8;$$
$$x = 1, \quad y = -4; \qquad x = -2, \quad y = -10;$$
$$x = 2, \quad y = -2; \qquad \text{etc.}$$

Arranging these results in tabulated form, the first two columns give the corresponding values of x and y when we assume positive values of x, and the

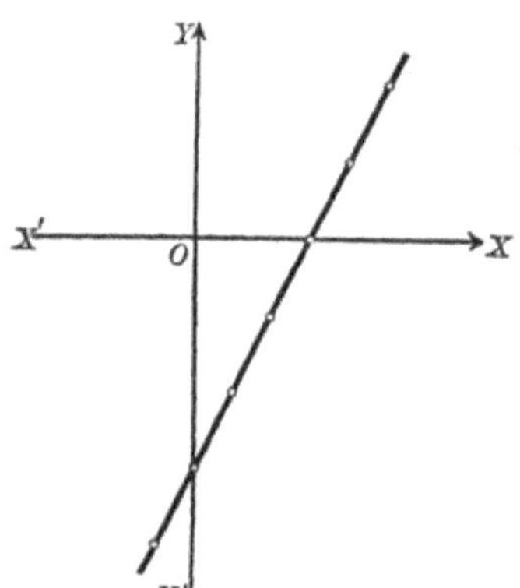

x	y	x	y
0	-6	0	-6
1	-4	-1	-8
2	-2	-2	-10
3	0	-3	-12
4	2	-4	-14
5	4	-5	-16
6	6	-6	-18
etc.	etc.	etc.	etc.

last two columns when we assume negative values of x. For the sake of symmetry $x = 0$ is placed in both pairs of columns.

Third step. Plot the points found.
Fourth step. Drawing a smooth curve through these points gives the graph of the function, which in this case is a straight line.

Ex. 2. Plot the graph of $x^2 - 2x - 3$.

Solution. First step. Let $y = x^2 - 2x - 3$.

Second step. Computing y by assuming values of x, we find the following table of values.

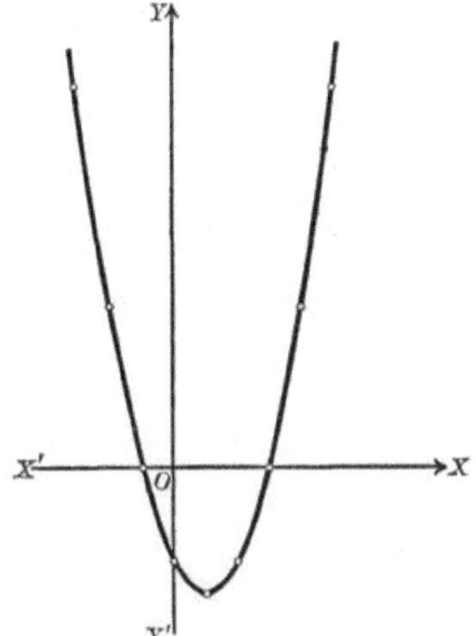

x	y	x	y
0	-3	0	-3
1	-4	-1	0
2	-3	-2	5
3	0	-3	12
4	5	-4	21
5	12	etc.	etc.
6	21		
etc.	etc.		

Third step. Plot the points found.

Fourth step. Drawing a smooth curve through these points gives the graph of the function.

53. Graphs of the trigonometric functions To find the graph of a trigonometric function we assume values for the angle; the circular measures of these angles are taken as the abscissas, and the corresponding values of the function found from the table on p. 9 are taken as the ordinates of points on the graph.

Ex. 1. Plot the graph of $\sin x$.

Solution. First step. Let $y = \sin x$.

Second step. Assuming values of x differing by 30°, we calculate the corresponding values of y from the table on p. 9. In tabulating the results it will be noticed that the angles are expressed both in degree measure and in circular

x		y	x		y
0°	0	0	0°	0	0
30°	$\dfrac{\pi}{6}$	.50	$-30°$	$-\dfrac{\pi}{6}$	$-.50$
60°	$\dfrac{\pi}{3}$	.86	$-60°$	$-\dfrac{\pi}{3}$	$-.86$
90°	$\dfrac{\pi}{2}$	1.00	$-90°$	$-\dfrac{\pi}{2}$	-1.00
120°	$\dfrac{2\pi}{3}$	.86	$-120°$	$-\dfrac{2\pi}{3}$	$-.86$
150°	$\dfrac{5\pi}{6}$	.50	$-150°$	$-\dfrac{5\pi}{6}$	$-.50$
180°	π	0	$-180°$	$-\pi$	0
210°	$\dfrac{7\pi}{6}$	$-.50$	$-210°$	$-\dfrac{7\pi}{6}$	.50
240°	$\dfrac{4\pi}{3}$	$-.86$	$-240°$	$-\dfrac{4\pi}{3}$	.86
270°	$\dfrac{3\pi}{2}$	-1.00	$-270°$	$-\dfrac{3\pi}{2}$	1.00
300°	$\dfrac{5\pi}{3}$	$-.86$	$-300°$	$-\dfrac{5\pi}{3}$	.86
330°	$\dfrac{11\pi}{6}$	$-.50$	$-330°$	$-\dfrac{11\pi}{6}$	.50
360°	2π	0	$-360°$	-2π	0

measure. It is most convenient to use the degree measure of an angle when looking up its function, while in plotting it is necessary to use its circular measure.

Third step. In plotting the points we must use the circular measure of the angles for abscissas. The most convenient way of doing this is to lay off distances $\pi = 3.1416$ to the right and left of the origin and then divide each of these into six equal parts. Then when

$$x = 0, \qquad y = 0;$$
$$x = \frac{\pi}{6}, \qquad y = .50 = AB;$$
$$x = \frac{\pi}{3}, \qquad y = .86 = CD;$$
$$x = \frac{\pi}{2}, \qquad y = 1.00 = EF; \text{ etc.}$$

Also when
$$x = -\frac{\pi}{6}, \qquad y = -.50 = GH; \text{ etc.}$$

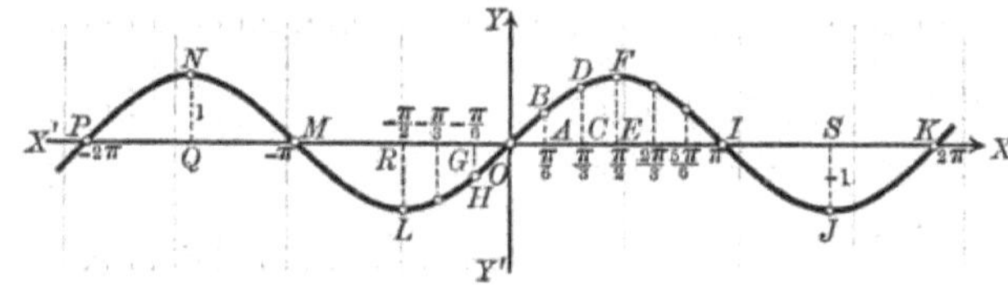

Fourth step. Drawing a smooth curve through these points, we get the graph of sin x for values of x between -2π and 2π. It is called the *sine curve* or *sinusoid*.

Discussion. (a) Since sin $(x \pm 2\pi) = \sin x$, it follows that

$$y = \sin x = \sin (x \pm 2\pi),$$

that is, the graph is *unchanged* if we replace x by $x \pm 2\pi$. This means, however, that every point is moved a distance 2π to the right or left. Hence the arc $PNMLO$ may be moved parallel to XX' until P falls at O, N at F, M at I, etc., that is, into the position $OFIJK$, and it will be a part of the curve in its new position. In the case of the sine curve it is then only necessary to plot points, say, from $x = -\pi$ to $x = \pi$, giving the arc or double undulation $MLOFI$. The sine curve consists of an indefinite number of such arcs extending to the right and left.

(b) From the graph we see that the maximum value of sin $x (= y)$ is $1 (= EF = QN$, etc.) and the minimum value is $-1 (= SJ = RL$, etc.), while x can take on any value whatever.

(c) Since the graph crosses the axis of x an infinite number of times, we see that the equation

$$\sin x = 0$$

has an infinite number of real roots, namely, $x = 0, \pm 2\pi, \pm 4\pi$, etc.

54. Periodicity of the trigonometric functions. From the graph of
$\sin x$ in the above example we saw that as the angle increased from
0 to 2π radians, the sine first increased from 0 to 1, then decreased
from 1 to -1, and finally increased from -1 to 0. As the angle
increased from 2π radians to 4π radians, the sine again went through
the same series of changes, and so on. Thus the sine goes through
all its changes while the angle changes 2π radians in value. This
is expressed by saying that the *period of the sine is 2π.*

Similarly, the cosine, secant, or cosecant passes through all its
changes while the angle changes 2π radians.

The tangent or cotangent, however, passes through all its changes
while the angle changes by π radians. Hence, *the **period** of the sine,
cosine, secant, or cosecant is **2π radians;** while the **period** of the
tangent or cotangent is **π radians.***

As each trigonometric function again and again passes through
the same series of values as the angle increases or decreases uni-
formly, we call them **periodic functions.**

**55. Graphs of trigonometric functions plotted by means of the unit
circle.** The following example will illustrate how we may plot the
graph of a trigonometric function without using any table of
numerical values of the function for different angles such as given
on p. 9.

Ex. 1. Plot the graph of $\sin x$.

Solution. Let $y = \sin x$. Draw a unit circle.

Divide the circumference of the circle into any number of equal parts (12 in

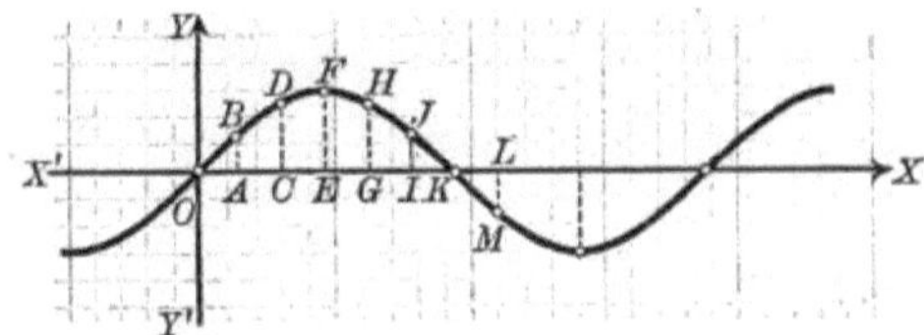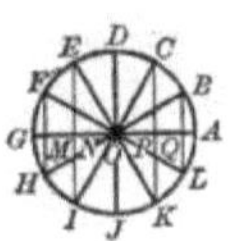

this case). At the several points of division drop perpendiculars to the horizontal diameter. Then the sine of the angle AOB, or, what amounts to the same
thing,

$$\text{sine of arc } AB = QB,$$
$$\text{sine of arc } AE - NE,$$
$$\text{sine of arc } AJ = OJ, \text{ etc.}$$

It is evident that if we take the lengths of the arcs as the abscissas and the
corresponding lengths of the perpendiculars as the ordinates of points in a plane,
these points will lie on the graph of $\sin x$. If we choose the same scale as in

Ex. 1, p. 96, the two graphs could be made to coincide, but in this example the unit of length chosen is larger. The main features of the two graphs of sin x are the same, however, the discussion being the same for both.

	In Circle	In Graph	In Circle	In Graph
When $x = $ arc zero $=$ zero,			$y = $ zero $=$ zero ;		
$x = $ arc AB	$- OA,$		$y = QB$	$= AB$;
$x = $ arc AC	$= OC,$		$y = PC$	$= CD$;
$x = $ arc AD	$= OE,$		$y = OD$	$= EF$;
$x = $ arc AE	$= OG,$		$y = NE$	$= GH$;
$x = $ arc AF	$= OI,$		$y = MF$	$= IJ$;
$x = $ arc AG	$= OK,$		$y = $ zero $=$ zero ;		
$x = $ arc AH	$= OL,$		$y = MH = LM,$ etc.		

EXAMPLES

1. Plot the graphs of the following functions:

(a) $x + 2.$

(b) $3x - 6.$

(c) $2x + 1.$

(d) $x^2.$

(e) $x^3.$

(f) $\dfrac{1}{x}.$

(g) $\dfrac{4}{x-2}.$

(h) $\dfrac{x-2}{x+1}.$

(i) $2^x.$

(j) $\log_{10} x.$

(k) $2x^2 - 4.$

(l) $8 - x^2.$

(m) $6 + 5x + x^2.$

(n) $x^2 - 3x + 2.$

(o) $x^2 - 4x + 3.$

(p) $x^3 - 4x.$

(q) $x^3 - 2x + 1.$

(r) $x^3 - 7x + 6.$

(s) $x^3 - 5x - 12.$

(t) $x^4 - 1.$

(u) $x^5 - 2.$

2. Plot the graph of $\cos x$.

Solution. Let $y = \cos x$. The *cosine curve* is found to be as follows:

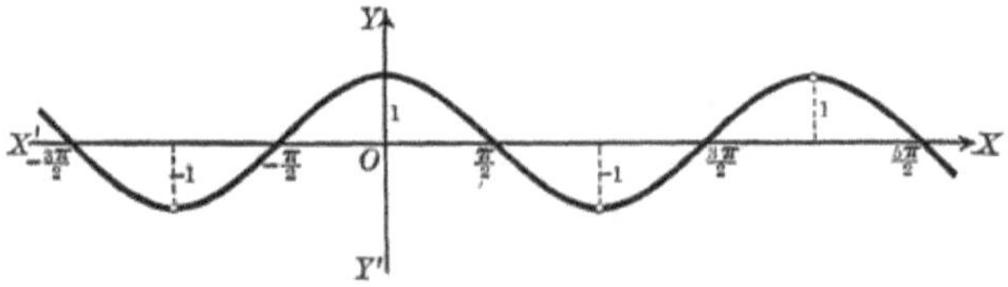

To plot the graph of $\cos x$ by means of the unit circle we may use the circle on p. 97. Taking the abscissas as arcs zero, AB, AC, AD, etc., and the corresponding ordinates as OA, OQ, OP, zero, etc., respectively, we will get points lying on the cosine curve.

3. Plot the graph of $\tan x$.

Solution. The *tangent curve* is shown on next page.

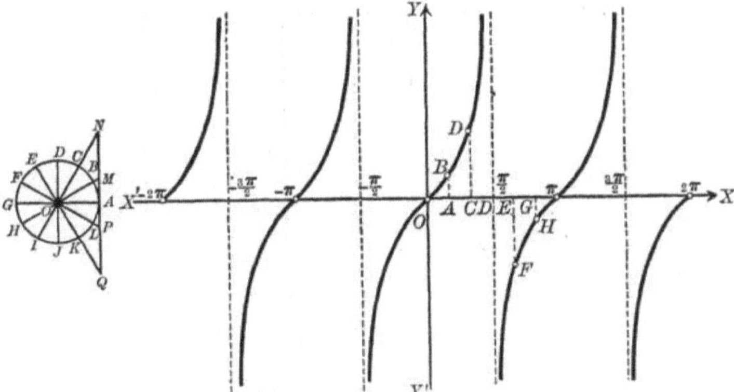

To construct the tangent curve from the unit circle shown, we have

		In Circle	In Graph		In Circle	In Graph
When	$x = \operatorname{arc}$ zero $=$ zero,			$y =$ zero $=$ zero ;		
	$x = \operatorname{arc} AB = OA,$			$y = AM = AB$;		
	$x = \operatorname{arc} AC = OC,$			$y = AN = CD$;		
	$x = \operatorname{arc} AD = OD,$			$y = \infty \ \ = \infty$;		
	$x = \operatorname{arc} AE = OE,$			$y = AQ = EF$, etc.		

4. Plot the graph of sec x.

Solution. The *secant curve* is given below.

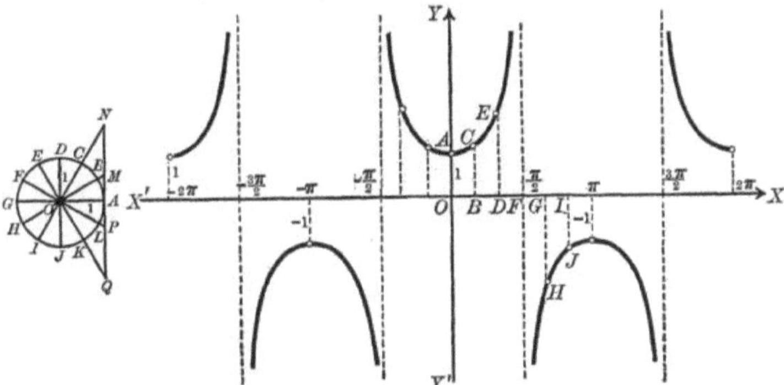

Using the unit circle, we have

		In Circle	In Graph		In Circle	In Graph
When	$x = \operatorname{arc}$ zero $=$ zero,			$y = OA = OA$;		
	$x = \operatorname{arc} AB = OB,$			$y = OM = BC$;		
	$x = \operatorname{arc} AC = OD,$			$y = ON = DE$;		
	$x = \operatorname{arc} AD = OF,$			$y = \infty \ \ = \infty$;		
	$x = \operatorname{arc} AE = OG,$			$y = OQ = GH$, etc.		

5. Plot the *cotangent curve*.

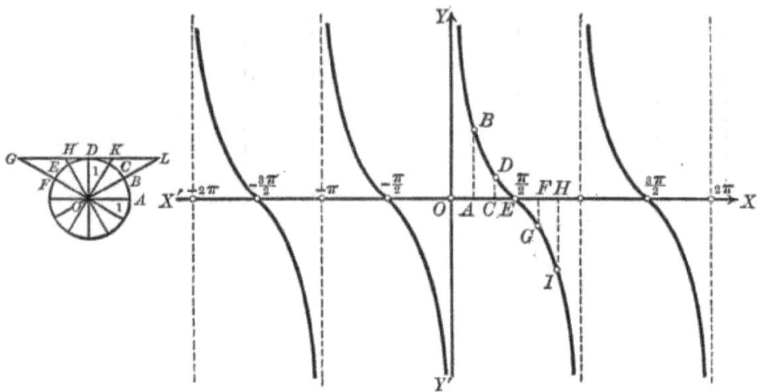

6. Plot the *cosecant curve*.

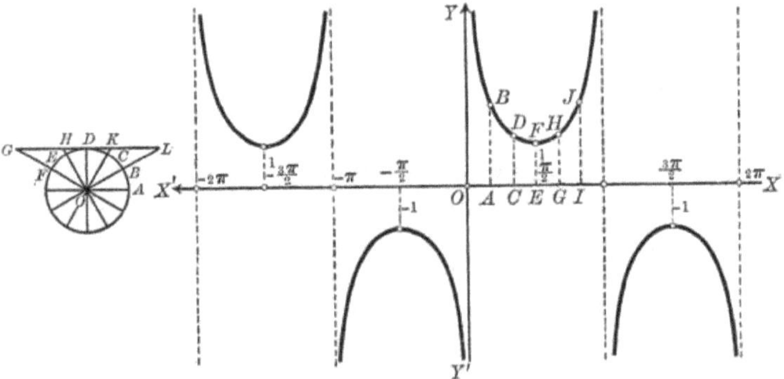

7. Draw graphs of (a) $\sin x + \cos x$, (b) $\cos x - \sin x$, (c) $\sin 2x$, (d) $\tan 2x$, (e) $\sin x \cos x$.

CHAPTER VII

SOLUTION OF OBLIQUE TRIANGLES

56. Relations between the sides and angles of a triangle. One of the principal uses of Trigonometry lies in its application to the solution of triangles. That is, having given three elements of a triangle (sides and angles) at least one of which must be a side, to find the others. In Plane Geometry the student has already been taught how to solve triangles graphically. That is, it has been shown how to construct a triangle, having given

> CASE I. *Two angles and one side.*
> CASE II. *Two sides and an opposite angle.*
> CASE III. *Two sides and the included angle.*
> CASE IV. *Three sides.*

From such a construction of the required triangle the parts not given may be found by actual measurement with a graduated ruler and a protractor. On account of the limitations of the observer and the imperfections of the instruments used, however, the results from such measurements will, in general, be only more or less rough approximations. After having constructed the triangle from the given parts by geometric methods, it will be seen that Trigonometry teaches us how to find the unknown parts of the triangle to any degree of accuracy desired, and the two methods may then serve as checks on each other.

The student should always bear in mind, when solving triangles, the two following geometrical properties which are common to all triangles:

(70) **The sum of the three angles equals 180°.**

(71) **The greater side lies opposite the greater angle, and conversely.**

The trigonometric solution of oblique triangles depends upon the application of three laws, — the law of sines, the law of cosines, and the law of tangents, to the derivation of which we now turn our attention.

57. Law of sines. *The sides of a triangle are proportional to the sines of the opposite angles.*

Proof. Fig. 1 represents a triangle all of whose angles are acute, while Fig. 2 represents a triangle, one angle of which is obtuse (as A).

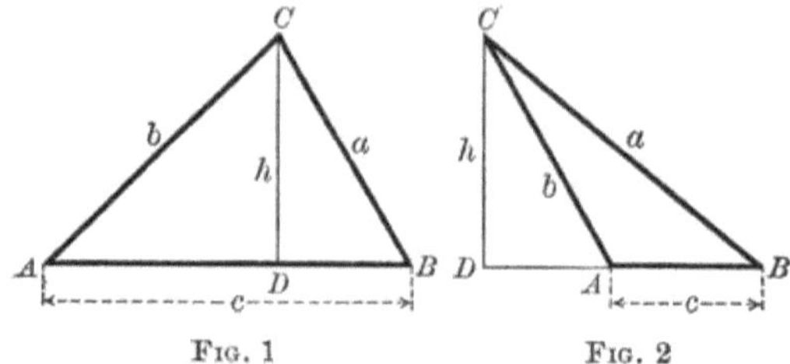

Fig. 1   Fig. 2

Draw the perpendicular $CD(=h)$ on AB or AB produced. From either figure, using the right triangle ACD,

$$(A) \qquad \sin A = \frac{h}{b}.$$

$$\left[\text{In Fig. 2, } \sin A = \sin (180° - A) = \sin CAD = \frac{h}{b}. \right]$$

Also, using the right triangle BCD,

$$(B) \qquad \sin B = \frac{h}{a}.$$

Dividing (A) by (B) gives

$$\frac{\sin A}{\sin B} = \frac{a}{b},$$

or, by alternation in proportion,

$$(C) \qquad \frac{a}{\sin A} = \frac{b}{\sin B}.$$

Similarly, by drawing perpendiculars from A and B we get

$$(D) \qquad \frac{b}{\sin B} = \frac{c}{\sin C}, \text{ and}$$

$$(E) \qquad \frac{c}{\sin C} = \frac{a}{\sin A}, \text{ respectively.}$$

Writing (C), (D), (E) as a single statement, we get the **law of sines.**

$$(72) \qquad \frac{a}{\sin A} = \frac{b}{\sin B} = \frac{c}{\sin C}.$$

Each of these equal ratios has a simple geometrical meaning, as may be shown if the *law of sines* is proved as follows :

Circumscribe a circle about the triangle ABC as shown in the figure, and draw the radii OB, OC. Denote the radius of the circle by R. Draw OM perpendicular to BC.

Since the inscribed angle A is measured by one half of the arc BC and the central angle BOC is measured by the whole arc BC, it follows that the angle $BOC = 2\,A$, or,

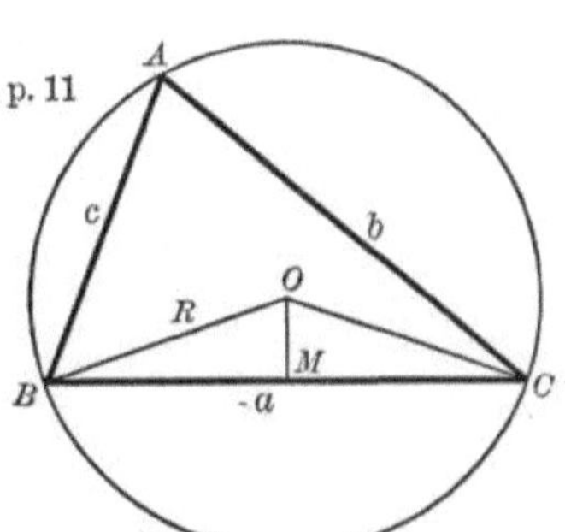

$$\text{angle } BOM = A.$$

Then $BM = R \sin BOM = R \sin A$, by (7), p. 11

and $\qquad a = 2\,BM = 2\,R \sin A,$

or, $\qquad 2R = \dfrac{a}{\sin A}.$

In like manner it may be shown that

$$2R = \frac{b}{\sin B} \text{ and } 2R = \frac{c}{\sin C}.$$

Hence, by equating the results, we get

$$2R = \frac{a}{\sin A} = \frac{b}{\sin B} = \frac{c}{\sin C}.$$

The ratio of any side of a triangle to the sine of the opposite angle is numerically equal to the diameter of the circumscribed circle.

It is evident that a triangle may be solved by the aid of the law of sines *if two of the three known elements are a side and its opposite angle.* The case of two angles and the included side being given, may also be brought under this head, since by (**70**), p. 101, we may find the third angle which lies opposite the given side.

Ex. 1. Given $A = 65°$, $B = 40°$, $a = 50$ ft.; solve the triangle.

Solution. Construct the triangle. Since two angles are given we get the third angle at once from (**70**), p. 101. Thus,

$$C = 180° - (A + B) = 180° - 105° = 75°.$$

Since we know the side a and its opposite angle A we may use the law of sines, but we must be careful to choose such ratios in (**72**) that *only one unknown quantity is involved.* Thus, to find the side b use

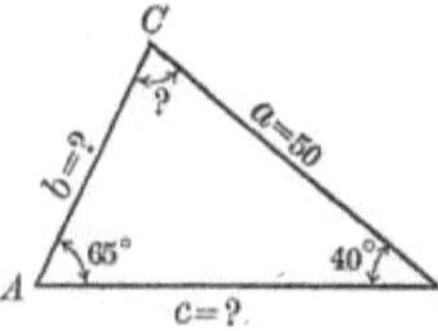

$$\frac{a}{\sin A} = \frac{b}{\sin B}.$$

Clearing of fractions and solving for the only unknown quantity b, we get

$$b = \frac{a \sin B}{\sin A}.$$

Substituting the numerical values of $\sin A$ and $\sin B$ from the table on p. 9, and $a = 50$ ft., we get

$$b = \frac{50 \times 0.6428}{0.9063} = 35.46 \text{ ft.}$$

Similarly, to find the side c, use

$$\frac{a}{\sin A} = \frac{c}{\sin C}.$$

Clearing of fractions and solving for c, we get

$$c = \frac{a \sin C}{\sin A} = \frac{50 \times 0.9659}{0.9063} = 53.29 \text{ ft.}$$

By measurements on the figure we now check the results to see that there are no large errors.

Since we now know all the sides and angles of the triangle, the triangle is said to be solved.

58. The ambiguous case. When two sides and an angle opposite one of them are given, the solution of the triangle depends on the law of sines. We must first find the unknown angle which lies opposite one of the given sides. But when an angle is determined by its sine, it admits of two values which are supplements of each other; hence either value of that angle may be taken unless one is excluded by the conditions of the problem.

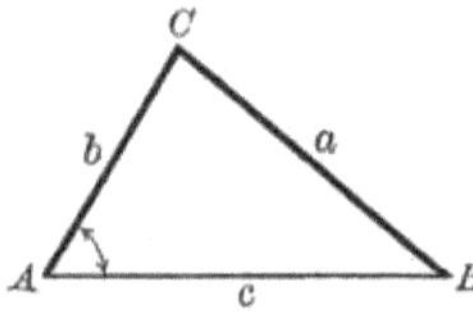

Let a and b be the given sides and A (opposite the side a) the given angle.

If $a > b$, then by Geometry $A > B$, and B must be acute whatever be the value of A, for a triangle can have only one obtuse angle. Hence there is *one, and only one, triangle* that will satisfy the given conditions.

If $a = b$, then by Geometry $A = B$, both A and B must be acute, and *the required triangle is isosceles.*

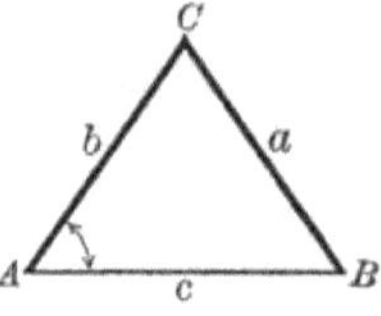

If $a < b$, then by Geometry $A < B$, and A must be acute in order that the triangle shall be possible; and when A is acute it is evident from the figure that *the two triangles ACB and ACB' will satisfy*

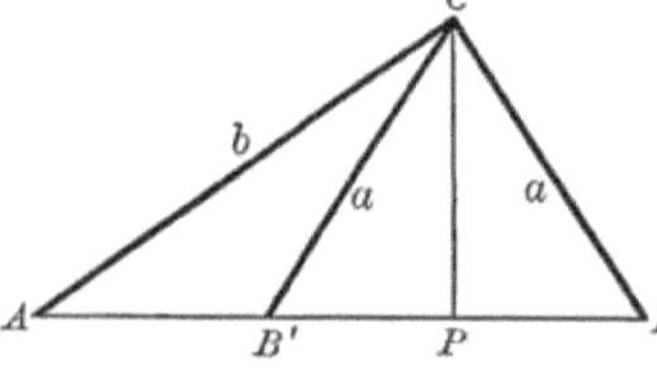

the given conditions provided a is greater than the perpendicular CP; that is, provided

$$a > b \sin A.$$

The angles ABC and $AB'C$ are supplementary (since $\angle B'BC = \angle BB'C$); they are, in fact, the supplementary angles obtained (using the law of sines) from the formula

$$\sin B = \frac{b \sin A}{a}.$$

That is, we get the corresponding acute value B from a table of sines, and the supplementary obtuse value as follows:

$$B' = 180° - B.$$

If, however, $a = b \sin A = CP$, then $\sin B = 1$, $B = 90°$, and *the triangle required is a right triangle.*

If $a < b \sin A$ (that is, greater than CP), then $\sin B > 1$, and *the triangle is impossible.*

These results may be stated in compact form as follows:

Two solutions: *If A is acute and the value of a lies between b and $b \sin A$.*

No solution: *If A is acute and $a < b \sin A$, or if A is obtuse and $a < b$ or $a = b$.*

One solution: *In all other cases.*

The number of solutions can usually be determined by inspection on constructing the triangle. In case of doubt find the value of $b \sin A$ and test as above.

Ex. 1. Given $a = 21$, $b = 32$, $A = 115°$; find the remaining parts.

Solution. In this case $a < b$ and $A > 90°$; hence the triangle is impossible and there is no solution.

Ex. 2. Given $a = 32$, $b = 86$, $A = 30°$; find the remaining parts.

Solution. Here $b \sin A = 86 \times \tfrac{1}{2} = 43$; hence $a < b \sin A$, and there is no solution.

Ex. 3. Given $a = 40$, $b = 30$, $A = 75°$; find the remaining parts.

Solution. Since $a > b$ and A is acute there is one solution only.
By the law of sines,

$$\frac{a}{\sin A} = \frac{b}{\sin B},$$

or,
$$\sin B = \frac{b \sin A}{a} = \frac{30 \times .9659}{40}.$$

$$\therefore \sin B = .7244,$$

or,
$$B = 46.4°,\ \text{the only admissible value of } B.$$

Then
$$C = 180° - (A + B) = 180° - 121.4° = 58.6°.$$

To find C, we get, by the law of sines,

$$\frac{c}{\sin C} = \frac{a}{\sin A},$$

or,
$$c = \frac{a \sin C}{\sin A} = \frac{40 \times .8535}{.9659} = 35.3.$$

Check the results by measurements on the figure.

Ex. 4. Solve the triangle, having given $b = 15$, $a = 12$, $A = 52°$.

Solution. Here $b \sin A = 15 \times .7880 = 11.82$; hence, since A is acute and a lies between b and $b \sin A$, there are two solutions. That is, there are two triangles, ACB_1 and ACB_2, which satisfy the given conditions. By the law of sines,

$$\frac{a}{\sin A} = \frac{b}{\sin B_1},$$

or,
$$\sin B_1 = \frac{b \sin A}{a} = \frac{15 \times .7880}{12} = .9850.$$

This gives $B_1 = 80.07°$, and the supplementary angle $B_2 = 180° - B_1 = 99.93°$. Let us first solve completely the triangle AB_1C.

$$C_1 = 180° - (A + B_1) = 47.93°.$$

By the law of sines,
$$\frac{a}{\sin A} = \frac{c_1}{\sin C_1},$$

or,
$$c_1 = \frac{a \sin C_1}{\sin A} = \frac{12 \times .7423}{.7880} = 11.3.$$

Now, solving the triangle AB_2C,

$$C_2 = 180° - (A + B_2) = 28.07°.$$

By the law of sines, $\dfrac{a}{\sin A} = \dfrac{c_2}{\sin C_2},$

or,
$$c_2 = \frac{a \sin C_2}{\sin A} = \frac{12 \times .4706}{.7880} = 7.2.$$

The solutions then are :

For triangle AB_1C	For triangle AB_2C
$B_1 = 80.07°,$	$B_2 = 99.93°,$
$C_1 = 47.93°,$	$C_2 = 28.07°,$
$c_1 = 11.3.$	$c_2 = 7.2.$

Check the results by measurements on the figure.

In the ambiguous case care should be taken to properly combine the calculated sides and angles.

EXAMPLES

1. Find the number of solutions in the following triangles, having given :

(a) $a = 80,$	$b = 100,$	$A = 30°.$	*Ans.*	Two.
(b) $a = 50,$	$b = 100,$	$A = 30°.$		One.
(c) $a = 40,$	$b = 100,$	$A = 30°.$		None.
(d) $a = 13,$	$b = 11,$	$A = 69°.$		One.
(e) $a = 70,$	$b = 75,$	$A = 60°.$		Two.
(f) $a = 134,$	$b = 84,$	$B = 52°.$		None.
(g) $a = 200,$	$b = 100,$	$A = 30°.$		One.

2. Solve the triangle, having given $a = 50$, $A = 65°$, $B = 40°$.
$$\textit{Ans.} \quad C = 75°, \ b = 35.46, \ c = 53.29.$$

3. Solve the triangle, having given $b = 7.07$, $A = 30°$, $C = 105°$.
$$\textit{Ans.} \quad B = 45°, \ a = 5, \ c = 9.66.$$

4. Solve the triangle, having given $c = 9.56$, $A = 45°$, $B = 60°$.

Ans. $C = 75°$, $a = 7$, $b = 8.57$.

5. Solve the triangle when $c = 60$, $A = 50°$, $B = 75°$.

Ans. $C = 55°$, $b = 70.7$, $a = 56.1$.

6. Solve the triangle when $a = 550$, $A = 10° 12'$, $B = 46° 36'$.

Ans. $C = 123° 12'$, $b = 2257.4$, $c = 2600.2$.

7. Solve the triangle when $a = 18$, $b = 20$, $A = 55.4°$.

Ans. $B_1 = 66.2°$, $C_1 = 58.4°$, $c_1 = 18.6$;

$B_2 = 113.8°$, $C_2 = 10.8°$, $c_2 = 4.1$.

8. Solve the triangle when $a = 3\sqrt{2}$, $b = 2\sqrt{3}$, $A = 60°$.

Ans. $C = 75°$, $B = 45°$, $c = 4.73$.

9. Solve the triangle when $b = 19$, $c = 18$, $C = 15° 49'$.

Ans. $B_1 = 16° 42'$, $A_1 = 147° 28'$, $a_1 = 35.5$;

$B_2 = 163° 18'$, $A_2 = 5.3'$, $a_2 = 1.04$.

10. Solve the triangle when $a = 119$, $b = 97$, $A = 50°$.

Ans. $B = 38.6°$, $C = 91.4°$, $c = 155.3$.

11. Solve the triangle when $a = 120$, $b = 80$, $A = 60°$.

Ans. $B = 35.3°$, $C = 84.7°$, $c = 137.9$.

12. It is required to find the horizontal distance from a point A to an inaccessible point B on the opposite bank of a river. We measure off any convenient horizontal distance as AC, and then measure the angles CAB and ACB.

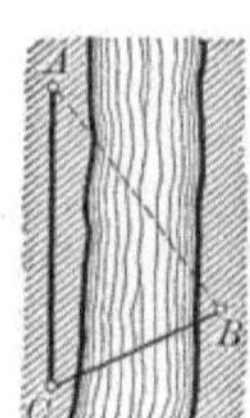

Let $AC = 283$ feet, angle $CAB = 38°$, and angle $ACB = 66.3°$. Solve the triangle ABC for the side AB.   *Ans.* 267.4 ft.

13. A railroad embankment stands on a horizontal plane and it is required to find the distance from a point A in the plane to the top B of the embankment. Select a point C at

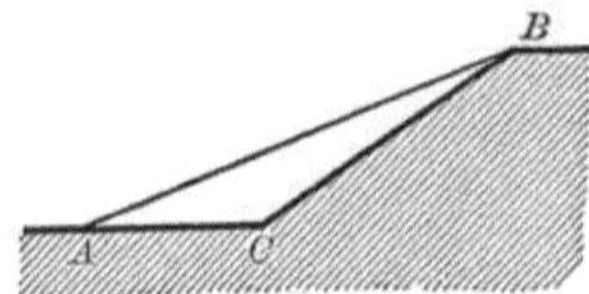

the foot of the embankment lying in the same vertical plane as A and B, and measure the distances AC and CB, and the angle BAC. Let $AC = 48.5$ ft., $BC = 84$ ft., and angle $BAC = 21.5°$. Solve the triangle for the side AB.   *Ans.* 127.2 ft.

14. A tree A is observed from two points B and C, 270 ft. apart, on a straight road. The angle BCA is 55° and the angle $CBA = 65°$. Find the distance from the tree to the nearer point B. *Ans.* 255.4 ft.

15. To determine the distance of a hostile fort A from a place B, a line BC and the angles ABC and BCA were measured and found to be 1006.62 yd., 44°, and 70° respectively. Find the distance AB.   *Ans.* 1035.5 yd.

16. A triangular lot has two sides of lengths 140.5 ft. and 70.6 ft., and the angle opposite the former is 40°. Find the length of a fence around it.

Ans. 353.9 ft., or 529.6 ft.

17. Two buoys are 64.2 yd. apart, and a boat is 74.1 yd. from the nearer buoy. The angle between the lines from the buoys to the boat is 27.3°. How far is the boat from the further buoy?   *Ans.* 120.3 ft.

18. Prove the following for any triangle :

(a) $a = b \cos C + c \cos B,$
$\quad b = a \cos C + c \cos A,$
$\quad c = a \cos B + b \cos A.$

(b) $\sqrt{bc \sin B \sin C} : \dfrac{b^2 \sin C + c^2 \sin B}{b + c}.$

(c) $\dfrac{\sin A + 2 \sin B}{a + 2 b} = \dfrac{\sin C}{c}.$

(d) $\dfrac{\sin^2 A - m \sin^2 B}{a^2 - mb^2} = \dfrac{\sin^2 C}{c^2}.$

(e) $a \sin (B - C) + b \sin (C - A) + c \sin (A - B) = 0.$

19. If R is the radius of the circumscribed circle, prove the following for any triangle $[s = \frac{1}{2}(a + b + c)]$:

(a) $R (\sin A + \sin B + \sin C) = s.$

(b) $bc = 4 R^2 (\cos A + \cos B \cos C).$

(c) $\dfrac{1}{s - a} + \dfrac{1}{s - b} + \dfrac{1}{s - c} - \dfrac{1}{s} = \dfrac{4 R}{\sqrt{s (s - a) (s - b) (s - c)}}.$

20. Show that in any triangle

$$\frac{a + b}{c} = \frac{\cos \frac{1}{2} (A - B)}{\sin \frac{1}{2} C}.$$

59. Law of cosines. *In any triangle the square of any side is equal to the sum of the squares of the other two sides minus twice the product of these two sides into the cosine of their included angle.*

Proof. Suppose we want to find the side a in terms of the other two sides b and c and their included angle A.

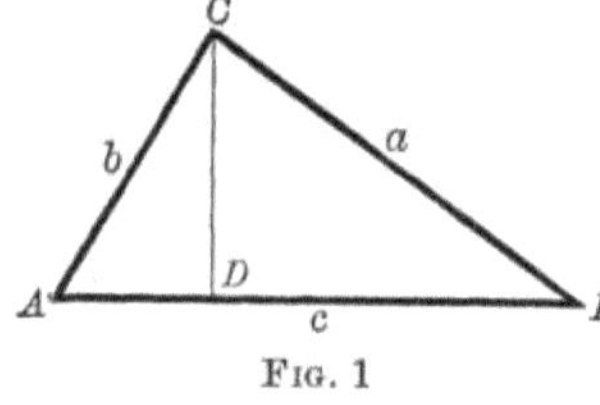

Fig. 1

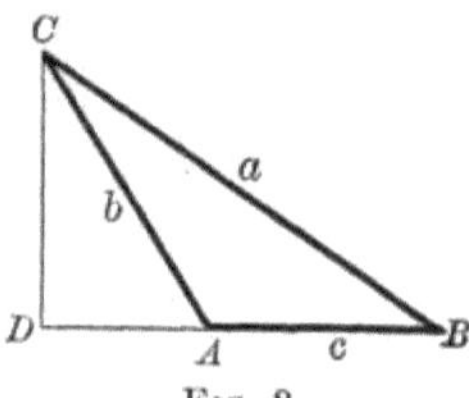

Fig. 2

When the angle A is acute (as in Fig. 1) we have, from Geometry,

$$\overline{CB}^2 = \overline{AC}^2 + \overline{AB}^2 - 2 AB \times AD,$$

$$\left[\begin{array}{l}\text{The square of the side opposite an acute angle equals the sum}\\ \text{of the squares of the other two sides minus twice the product}\\ \text{of one of those sides into the projection of the other upon it.}\end{array}\right]$$

or, $\qquad\qquad\qquad a^2 = b^2 + c^2 - 2 cAD.$

But $\qquad\qquad\qquad AD = b \cos A.$ $\qquad\qquad\qquad$ **(8)**, p. 11

Hence $\qquad\qquad\qquad a^2 = b^2 + c^2 - 2 bc \cos A.$

When the angle A is obtuse (as in Fig. 2) we have, from Geometry,

$$\overline{CB}^2 = \overline{AC}^2 + \overline{AB}^2 + 2\,AB \times AD,$$

> The square of the side opposite an obtuse angle equals the sum
> of the squares of the other two sides plus twice the product
> of one of those sides into the projection of the other upon it.

or,
$$a^2 = b^2 + c^2 + 2\,cAD.$$

But
$$AD = b \cos DAC \qquad\qquad \text{(8), p. 11}$$
$$= b \cos(180^\circ - A)$$
$$= -b \cos A. \quad \text{Hence in any case}$$

(73)
$$a^2 = b^2 + c^2 - 2\,bc \cos A.$$

Similarly, we may find

(74)
$$b^2 = a^2 + c^2 - 2\,ac \cos B.$$

(75)
$$c^2 = a^2 + b^2 - 2\,ab \cos C.*$$

Observe that if $A = 90^\circ$, then $\cos A = 0$, and (73) becomes $a^2 = b^2 + c^2$, which is the known relation between the sides of a right triangle where A is the right angle.

Solving (73), (74), (75) for the cosines of the angles, we get

(76)
$$\cos A = \frac{b^2 + c^2 - a^2}{2\,bc},$$

(77)
$$\cos B = \frac{a^2 + c^2 - b^2}{2\,ac},$$

(78)
$$\cos C = \frac{a^2 + b^2 - c^2}{2\,ab}.$$

These formulas are useful in finding the angles of a triangle, having given its sides.

Formulas (73), (74), (75) may be used for finding the third side of a triangle when two sides and the included angle are given. The other angles may then be found either by the law of sines or by formulas (76), (77), (78).

---

* Since a and A, b and B, c and C stand for *any side* of a triangle and the *opposite angle*, from any formula expressing a general relation between these parts another formula may be deduced by *changing the letters in cyclical order*. Thus, in (73) by changing a to b, b to c, c to a, and A to B we obtain (74); and in (74) by changing b to c, c to a, a to b, and B to C we get (75). This is a great help in memorizing some sets of formulas.

Ex. 1. Having given $A = 47°$, $b = 8$, $c = 10$; solve the triangle.

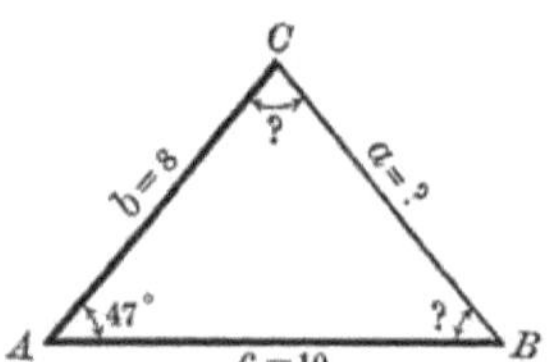

Solution. To find the side a use (**73**).

$$a^2 = b^2 + c^2 - 2\,bc\cos A$$
$$= 64 + 100 - 2 \times 8 \times 10 \times .6820$$
$$= 54.88.$$
$$\therefore a = \sqrt{54.88} = 7.408.$$

To find the angles C and B use the law of sines.

$$\sin B = \frac{b \sin A}{a} = \frac{8 \times .7314}{7.408} = .7898. \quad \therefore B = 52.2°.$$

$$\sin C = \frac{c \sin A}{a} = \frac{10 \times .7314}{7.408} = .9873. \quad \therefore C = 80.8°.$$

To check our work we note the fact that $A + B + C = 47° + 52.2° + 80.8° = 180°$.

Ex. 2. Having given $a = 7$, $b = 3$, $c = 5$; solve the triangle.

Solution. Using formulas (**76**), (**77**), (**78**) in order to find the angles, we get

$$\cos A = \frac{b^2 + c^2 - a^2}{2\,bc} = \frac{3^2 + 5^2 - 7^2}{2 \cdot 3 \cdot 5} = -\frac{1}{2} = -.5000. \quad \therefore A = 120°.$$

$$\cos B = \frac{a^2 + c^2 - b^2}{2\,ac} = \frac{7^2 + 5^2 - 3^2}{2 \cdot 7 \cdot 5} = \frac{13}{14} = .9286. \quad \therefore B = 21.8°.$$

$$\cos C = \frac{a^2 + b^2 - c^2}{2\,ab} = \frac{7^2 + 3^2 - 5^2}{2 \cdot 7 \cdot 3} = \frac{11}{14} = .7857. \quad \therefore C = 38.2°.$$

Check: $A + B + C = 120° + 21.8° + 38.2° = 180°$.

EXAMPLES

1. Having given $a = 30$, $b = 54$, $C = 46°$; solve the triangle.

Ans. $A = 33.1°$, $B = 100.9°$, $c = 39.56$.

2. Having given $A = 60°$, $b = 8$, $c = 5$; find a and the cosines of the angles B and C. *Ans.* $7, \frac{1}{7}, \frac{11}{14}$.

3. Having given $a = 33$, $c = 30$, $B = 35.4°$; find A and C.

Ans. $A = 80.7°$, $C = 63.9°$.

4. Having given $a = 4$, $b = 7$, $c = 10$; solve the triangle.

Ans. $A = 18.2°$, $B = 33.1°$, $C = 128.7°$.

5. Having given $a = 21$, $b = 24$, $c = 27$; solve the triangle.

Ans. $A = 48.2°$, $B = 58.4°$, $C = 73.4°$.

6. Having given $a = 2$, $b = 3$, $c = 4$; find the cosines of the angles A, B, C.

Ans. $\frac{7}{8}, \frac{11}{16}, -\frac{1}{4}$.

7. Having given $a = 77.99$, $b = 83.39$, $C = 72° 15'$; solve the triangle.

Ans. $A = 51° 15'$, $B = 56° 30'$, $c = 95.24$.

8. If two sides of a triangle are 10 and 11 and the included angle is $50°$, find the third side. *Ans.* 8.92.

9. The two diagonals of a parallelogram are 10 and 12 and they form an angle of $49.3°$; find the sides.   *Ans.* 10 and 4.68.

10. In order to find the distance between two objects, A and B, separated by a pond, a station C was chosen, and the distances $CA = 426$ yd., $CB = 322.4$ yd., together with the angle $ACB = 68.7°$, were measured. Find the distance from A to B.

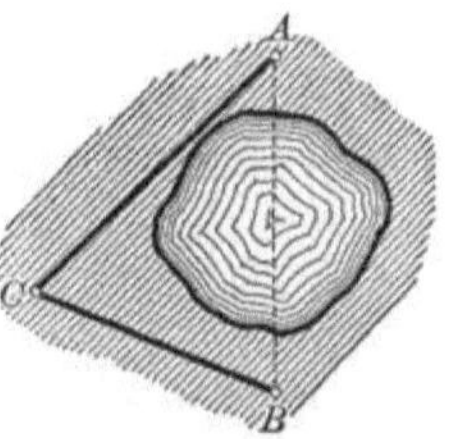

Ans. 430.85 yd.

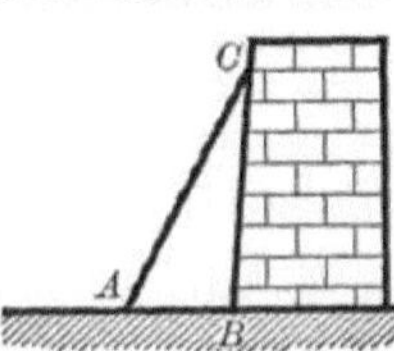

11. A ladder 52 ft. long is set 20 ft. from the foot of an inclined buttress, and reaches 46 ft. up its face. Find the inclination of the face of the buttress.

Ans. $95.9°$.

12. Under what visual angle is an object 7 ft. long seen by an observer whose eye is 5 ft. from one end of the object and 8 ft. from the other end ?

Ans. $60°$.

13. Two stations, A and B, on opposite sides of a mountain, are both visible from a third station C. The distance $AC = 11.5$ mi., $BC = 9.4$ mi., and angle $ACB = 59.5°$. Find the distance between A and B.   *Ans.* 10.5 mi.

14. Prove the following for any triangle :

(a) $a(b^2 + c^2)\cos A + b(c^2 + a^2)\cos B + c(a^2 + b^2)\cos C = 3\,abc.$

(b) $\dfrac{b + c}{a} = \dfrac{\cos B + \cos C}{1 - \cos A}.$

(c) $a + b + c = (b + c)\cos A + (c + a)\cos B + (a + b)\cos C.$

(d) $\dfrac{\cos A}{a} + \dfrac{\cos B}{b} + \dfrac{\cos C}{c} = \dfrac{a^2 + b^2 + c^2}{2\,abc}.$

(e) $a^2 + b^2 + c^2 = 2(ab\cos C + bc\cos A + ca\cos B).$

60. Law of tangents. *The sum of any two sides of a triangle is to their difference as the tangent of half the sum of their opposite angles is to the tangent of half their difference.*

Proof. By the law of sines,

$$\frac{a}{\sin A} = \frac{b}{\sin B},$$

and by division and composition in proportion,

$$(A) \qquad \frac{a + b}{a - b} = \frac{\sin A + \sin B}{\sin A - \sin B}.$$

But from **(66)**, p. 74,

$$(B) \qquad \frac{\sin A + \sin B}{\sin A - \sin B} = \frac{\tan \tfrac{1}{2}(A + B)}{\tan \tfrac{1}{2}(A - B)}.$$

Hence equating (A) and (B), we get

(79)
$$\frac{a+b}{a-b} = \frac{\tan\frac{1}{2}(A+B)}{\tan\frac{1}{2}(A-B)}. \;*$$

Similarly, we get
$$\frac{a+c}{a-c} = \frac{\tan\frac{1}{2}(A+C)}{\tan\frac{1}{2}(A-C)},$$

$$\frac{b+c}{b-c} = \frac{\tan\frac{1}{2}(B+C)}{\tan\frac{1}{2}(B-C)}. \;\dagger$$

When two sides and the included angle are given, as a, b, C, the law of tangents may be employed in finding the two unknown angles A and B.$\ddagger$ Since $a+b$, $a-b$, $A+B$ $(=180° - C)$, and therefore also $\tan\frac{1}{2}(A+B)$, are known, we clear (79) of fractions and solve for the unknown quantity $\tan\frac{1}{2}(A-B)$. This gives

(80)
$$\tan\tfrac{1}{2}(A-B) = \frac{a-b}{a+b}\tan\tfrac{1}{2}(A+B).$$

We shall illustrate the process by means of an example.

Ex. 1. Having given $a = 872.5$, $b = 632.7$, $C = 80°$; solve the triangle.

Solution. $a+b = 1505.2$, $a-b = 239.8$, $A+B = 180° - C = 100°$, and $\frac{1}{2}(A+B) = 50°$.

From (79), since $\tan\frac{1}{2}(A+B) = \tan 50° = 1.1918$,

$$\tan\tfrac{1}{2}(A-B) = \frac{a-b}{a+b}\tan\tfrac{1}{2}(A+B) = \frac{239.8}{1505.2} \times 1.1918 = .1899.$$

$$\therefore \; \tfrac{1}{2}(A-B) = 10.6°.$$

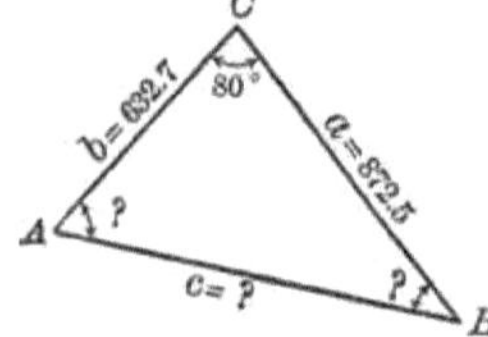

Adding this result to $\frac{1}{2}(A+B) = 50°$ gives

$$A = 60.6°.$$

Subtracting the result from $\frac{1}{2}(A+B) = 50°$ gives

$$B = 39.4°.$$

To find the side c, use the law of sines. Thus,

$$c = \frac{a \sin C}{\sin A} = \frac{872.5 \times .9848}{.8712} = 986.2.$$

We will now derive formulas for solving triangles having three sides given, which are more convenient than (76), (77), (78), p. 109.

---

* If $b > a$, then $B > A$, making $a-b$ and $A-B$ negative. The formula still holds true, but to avoid negative quantities it is better to write the formula in form $\dfrac{b+a}{b-a} = \dfrac{\tan\frac{1}{2}(B+A)}{\tan\frac{1}{2}(B-A)}$.

† These may also be found by changing the letters in cyclical order (see footnote, p. 109).

‡ When logarithms are used in solving triangles, having given two sides and the included angle, the law of tangents, which involves products, is to be preferred to the law of cosines, which involves sums.

61. Trigonometric functions of the half angles of a triangle in terms of its sides. Denote half the sum of the sides of a triangle (i.e. half the perimeter) by s. Then

$$(A) \qquad 2s = a + b + c.$$

Subtracting $2c$ from both sides,

$$2s - 2c = a + b + c - 2c, \text{ or,}$$

$$(B) \qquad 2(s - c) = a + b - c.$$

Similarly,

$$(C) \qquad 2(s - b) = a - b + c,$$

$$(D) \qquad 2(s - a) = - a + b + c.$$

In (**49a**), (**49b**), p. 70, replace $2x$ by A, and, what amounts to the same thing, x by $\tfrac{1}{2}A$. This gives

$$(E) \qquad 2 \sin^2 \tfrac{1}{2} A = 1 - \cos A,$$

$$(F) \qquad 2 \cos^2 \tfrac{1}{2} A = 1 + \cos A.$$

But from (**76**), p. 109, $\cos A = \dfrac{b^2 + c^2 - a^2}{2bc}$; hence (E) becomes

$$(G) \qquad 2 \sin^2 \tfrac{1}{2} A = 1 - \frac{b^2 + c^2 - a^2}{2bc}$$

$$= \frac{2bc - b^2 - c^2 + a^2}{2bc}$$

$$= \frac{a^2 - (b^2 - 2bc + c^2)}{2bc}$$

$$= \frac{a^2 - (b - c)^2}{2bc}$$

$$= \frac{(a + b - c)(a - b + c)}{2bc}$$

[$a^2 - (b - c)^2$ being the product of the sum and difference of a and $b - c$.]

$$= \frac{2(s - c)\, 2(s - b)}{2bc}. \qquad \text{by } (B), (C)$$

$$(81) \qquad \therefore \ \sin \tfrac{1}{2} A = \sqrt{\frac{(s - b)(s - c)}{bc}}.$$

Similarly, (F) becomes

$$2 \cos^2 \tfrac{1}{2} A = 1 + \frac{b^2 + c^2 - a^2}{2\,bc}$$

$$= \frac{2\,bc + b^2 + c^2 - a^2}{2\,bc}$$

$$= \frac{(b + c)^2 - a^2}{2\,bc}$$

$$= \frac{(b + c + a)(b + c - a)}{2\,bc}$$

$$= \frac{2\,s \cdot 2\,(s - a)}{2\,bc}.$$

$$(82) \qquad \therefore \ \cos \tfrac{1}{2} A = \sqrt{\frac{s(s - a)}{bc}}.$$

Since $\tan \tfrac{1}{2} A = \dfrac{\sin \tfrac{1}{2} A}{\cos \tfrac{1}{2} A}$, we get, by substitution from (81) and (82),

$$(83) \qquad \tan \tfrac{1}{2} A = \sqrt{\frac{(s - b)(s - c)}{s(s - a)}}.$$

Since any angle of a triangle must be less than 180°, $\tfrac{1}{2} A$ must be less than 90° and all the functions of $\tfrac{1}{2} A$ must be positive. Hence only the positive signs of the radicals in (81), (82), (83) have been taken.

Similarly, we may get

$$\sin \tfrac{1}{2} B = \sqrt{\frac{(s - a)(s - c)}{ac}}, \qquad \sin \tfrac{1}{2} C = \sqrt{\frac{(s - a)(s - b)}{ab}},$$

$$\cos \tfrac{1}{2} B = \sqrt{\frac{s(s - b)}{ac}}, \qquad \cos \tfrac{1}{2} C = \sqrt{\frac{s(s - c)}{ab}},$$

$$\tan \tfrac{1}{2} B = \sqrt{\frac{(s - a)(s - c)}{s(s - b)}}, \qquad \tan \tfrac{1}{2} C = \sqrt{\frac{(s - a)(s - b)}{s(s - c)}}.^{*}$$

There is then a choice of three different formulas for finding the value of each angle. If half the angle is very near 0°, the formula for the cosine will not give a very accurate result, because the cosines of angles near 0° differ little in value; and the same holds true of the formula for the sine when half the angle is very near 90°. Hence in the first case the formula for the sine, in the second that for the cosine, should be used. In general, however, the formula for the tangent is to be preferred.

---

* Also found by changing the letters in cyclical order.

When two angles, as A and B, have been found, the third angle, C, may be found by the relation $A + B + C = 180°$, but it is best to compute all the angles from the formulas, so that we use the sum of the angles as a test of the accuracy of the results.

It is customary to use a second form of (**83**), found as follows:

$$\tan \tfrac{1}{2} A = \sqrt{\frac{(s-b)(s-c)}{s(s-a)}}$$

$$= \sqrt{\frac{(s-a)(s-b)(s-c)}{s(s-a)^2}}$$

$$\left[\begin{array}{l}\text{Multiplying both numerator and denomina-}\\ \text{tor of the fraction under the radical by } s-a.\end{array}\right]$$

$$= \frac{1}{s-a} \sqrt{\frac{(s-a)(s-b)(s-c)}{s}}.$$

Denoting the radical part of the expression by r,

(**84**) $\qquad r = \sqrt{\dfrac{(s-a)(s-b)(s-c)}{s}}$, and we get

(**85**) $\qquad \tan \tfrac{1}{2} A = \dfrac{r}{s-a}.$ Similarly,

(**86**) $\qquad \tan \tfrac{1}{2} B = \dfrac{r}{s-b}$,

(**87**) $\qquad \tan \tfrac{1}{2} C = \dfrac{r}{s-c}.$ *

By proving one of the last three formulas geometrically it may be shown that r is the radius of the inscribed circle.

Proof. Since angle $NAO = \tfrac{1}{2} A$,

(A) $\qquad \tan \tfrac{1}{2} A = \dfrac{NO}{AN}.$

If s denotes half the perimeter, we have

$2s = AN + NB + BL + LC + CM + MA.$

But $NB = BL$, $CM = LC$, $MA = AN$; therefore

$$2s = 2AN + 2BL + 2LC,$$

or, $\qquad s = AN + (BL + LC) = AN + a.$

This gives $\qquad AN = s - a.$

Substituting in (A), $\tan \tfrac{1}{2} A = \dfrac{NO}{s-a}.$

Comparing this result with (**85**) and (**84**) shows that

$$NO = r = \sqrt{\frac{(s-a)(s-b)(s-c)}{s}}.$$

---

* When logarithms are used in solving triangles, having given the three sides, formulas (**84**), (**85**), (**86**), (**87**), which involve products, are more convenient than the law of cosines, which involves sums.

Ex. 1. Solve the triangle whose sides are 13, 14, 15.

Solution. Let $a = 13$, $b = 14$, $c = 15$.

Then
$$2s = a + b + c = 42,$$

or,
$$s = 21.$$

Also,
$$s - a = 8, \quad s - b = 7, \quad s - c = 6.$$

From (**84**),
$$r = \sqrt{\frac{(s-a)(s-b)(s-c)}{s}} = \sqrt{\frac{8 \cdot 7 \cdot 6}{21}} = \sqrt{16} = 4.$$

From (**85**),
$$\tan \tfrac{1}{2} A = \frac{r}{s-a} = \frac{4}{8} = \frac{1}{2} = .5000.$$

$$\therefore \ \tfrac{1}{2} A = 26.56°, \text{ or } A = 53.12°.$$

From (**86**),
$$\tan \tfrac{1}{2} B = \frac{r}{s-b} = \frac{4}{7} = .5714.$$

$$\therefore \ \tfrac{1}{2} B = 29.74°, \text{ or } B = 59.48°.$$

From (**87**),
$$\tan \tfrac{1}{2} C = \frac{r}{s-c} = \frac{4}{6} = \frac{2}{3} = .6667.$$

$$\therefore \ \tfrac{1}{2} C = 33.69°, \text{ or } C = 67.38°.$$

Check: $A + B + C = 53.12° + 59.48° + 67.38° = 179.98°.$[*]

EXAMPLES

1. Solve Examples 1, 3, 8, p. 110, using the law of tangents.

2. Solve Examples 4, 5, 6, p. 110, using formulas (**84**), (**85**), (**86**), (**87**), p. 115.

3. Prove the following for any triangle :

(a) $(a + b) \sin \tfrac{1}{2} C = c \cos \tfrac{1}{2} (A - B).$

(b) $\tan \tfrac{1}{2} B \tan \tfrac{1}{2} C = \dfrac{b + c - a}{b + c + a}.$

(c) $b \cos^2 \tfrac{1}{2} C + c \cos^2 \tfrac{1}{2} B = s.$

(d) $(b + c - a) \tan \tfrac{1}{2} A = (c + a - b) \tan \tfrac{1}{2} B.$

(e) $c^2 = (a + b)^2 \sin^2 \tfrac{1}{2} C + (a - b)^2 \cos^2 \tfrac{1}{2} C.$

(f) $c (\cos A + \cos B) = 2 (a + b) \sin^2 \tfrac{1}{2} C.$

(g) $\dfrac{\cos^2 \tfrac{1}{2} A}{\cos^2 \tfrac{1}{2} B} = \dfrac{a (s - a)}{b (s - b)}.$

(h) $b \sin^2 \dfrac{C}{2} + c \sin^2 \dfrac{B}{2} = s - a.$

(i) $a \cos \tfrac{1}{2} B \cos \tfrac{1}{2} C \csc \tfrac{1}{2} A = s.$

(j) $\sin A = 2 \sin \dfrac{A}{2} \cos \dfrac{A}{2} = \dfrac{2}{bc} \sqrt{s (s - a)(s - b)(s - c)}.$

[*] The error .02° arises from the fact that we used a four-place table. If we had used a table giving the first five significant figures of the tangent, the error would have been less; if a six-place table, still less, etc. For ordinary purposes, however, the results we get, using a four-place table, are sufficiently accurate.

4. If R and r denote the radii of the circumscribed and inscribed circles respectively, prove the following for any triangle :

(a) $r = \dfrac{a \sin \frac{1}{2} B \sin \frac{1}{2} C}{\cos \frac{1}{2} A}$.

(c) $\dfrac{1}{bc} + \dfrac{1}{ca} + \dfrac{1}{ab} = \dfrac{1}{2\,rR}$.

(b) $R = \dfrac{abc}{4\sqrt{s(s-a)(s-b)(s-c)}}$.

(d) $R = \dfrac{1}{2}\sqrt[3]{\dfrac{abc}{\sin A \sin B \sin C}}$.

(e) $abcr = 4\,R(s-a)(s-b)(s-c)$.

62. Formulas for finding the area of an oblique triangle.

CASE I. *When two sides and the included angle are known.*

Let b, c, and A be known. Take c as the base. Denote the altitude by h and the area by S. Then, by Geometry,

$$S = \tfrac{1}{2}\,ch.$$

But $h = b \sin A$ (from **(7)**, p. 11); hence

(88) $$S = \tfrac{1}{2}\,bc \sin A.$$

Similarly, $$S = \tfrac{1}{2}\,ac \sin B = \tfrac{1}{2}\,ab \sin C.$$

The area of a triangle equals half the product of any two sides multiplied by the sine of the included angle.

Ex. 1. Find the area of a triangle, having given $b = 20$ in., $c = 15$ in., $A = 60°$.
Solution. Substituting in **(88)**,

$$S = \frac{1}{2}\,bc \sin A = \frac{1}{2} \times 20 \times 15 \times \frac{\sqrt{3}}{2} = 75\sqrt{3} \text{ sq. in. } Ans.$$

CASE II. *When the three sides are known.*

$$\sin A = 2 \sin \tfrac{1}{2} A \cos \tfrac{1}{2} A \qquad\qquad \textbf{(51)}, \text{ p. } 72$$

$$= 2\sqrt{\frac{(s-b)(s-c)}{bc}}\,\sqrt{\frac{s(s-a)}{bc}} \qquad \begin{cases} \textbf{(81)}, \textbf{(82)}, \\ \text{pp. } 113, 114 \end{cases}$$

$$= \frac{2}{bc}\sqrt{s(s-a)(s-b)(s-c)}.$$

Substituting this value of $\sin A$ in **(88)**, we get

(89) $$S = \sqrt{s(s-a)(s-b)(s-c)}.$$

Ex. 2. Having given $a = 13$, $b = 14$, $c = 15$; find the area.
Solution. $S = \tfrac{1}{2}(a+b+c) = 21$, $s - a = 8$, $s - b = 7$, $s - c = 6$.
Substituting in **(89)**,

$$S = \sqrt{s(s-a)(s-b)(s-c)} = \sqrt{21 \times 8 \times 7 \times 6} = 84. \quad Ans.$$

Case III. *Problems which do not fall under Cases I or II directly may be solved by Case I, if we first find an additional side or angle by the law of sines.*

Ex. 3. Given $a = 10\sqrt{3}$, $b = 10$, $A = 120°$; find the area of the triangle.

Solution. This does not come directly under either Case I or Case II, but, by the law of sines,

$$\sin B = \frac{b \sin A}{a} = \frac{10 \times \frac{1}{2}\sqrt{3}}{10\sqrt{3}} = \frac{1}{2}.$$

Therefore $B = 30°$ and $C = 180° - (A + B) = 30°$.

Since we now have the two sides a and b and the included angle C, the problem comes under Case I, and we get

$$S = \tfrac{1}{2} ab \sin C = \tfrac{1}{2} \times 10\sqrt{3} \times 10 \times \tfrac{1}{2} = 25\sqrt{3}. \quad Ans.$$

EXAMPLES

1. Find the areas of the following triangles, having given

(a) $a = 40$,	$b = 13$,	$c = 37$.	*Ans.*	240.
(b) $b = 8$,	$c = 5$,	$A = 60°$.		17.32.
(c) $b = 10$,	$c = 40$,	$A = 75°$.		193.18.
(d) $a = 10$,	$b = 12$,	$C = 60°$.		$30\sqrt{3}$.
(e) $a = 40$,	$c = 60$,	$B = 30°$.		600.
(f) $a = 7$,	$c = 5\sqrt{2}$,	$B = 135°$.		$17\tfrac{1}{2}$.
(g) $b = 149$,	$A = 70° 42'$,	$B = 39° 18'$.		15,447.7.
(h) $a = 5$,	$b = 6$,	$c = 7$.		$6\sqrt{6}$.
(i) $a = 409$,	$b = 169$,	$c = 510$.		30,600.
(j) $a = 140.5$,	$b = 170.6$,	$A = 40°$.		11,981 or 2345.8.
(k) $c = 8$,	$B = 100.1°$,	$C = 31.1°$.		46.
(l) $a = 7$,	$c = 3$,	$A = 60°$.		10.4.

2. Show that the area of a parallelogram equals the product of any two adjacent sides multiplied by the sine of the included angle.

3. Find a formula for the area of an isosceles trapezoid in terms of the parallel sides and an acute angle.

4. Show that the area of a quadrilateral equals one half the product of its diagonals into the sine of their included angle.

5. The base of an isosceles triangle is 20, and its area is $100 - \sqrt{3}$; find its angles.

6. Prove the following for any triangle:

(a) $S = \dfrac{abc}{4R}$.

(b) $S = rs$.

(c) $S = Rr(\sin A + \sin B + \sin C)$.

(d) $S = \tfrac{1}{2} a^2 \sin B \sin C \csc A$.

(e) $S = \dfrac{2\,abc}{a+b+c}\left(\cos\dfrac{A}{2}\cos\dfrac{B}{2}\cos\dfrac{C}{2}\right)$.

(f) $S = \dfrac{a^2}{4}\sin 2B + \dfrac{b^2}{4}\sin 2A$.

CHAPTER VIII

THEORY AND USE OF LOGARITHMS

63. Need of logarithms* in Trigonometry. Many of the problems arising in Trigonometry involve computations of considerable length. Since the labor connected with extensive and complicated calculations may be greatly lessened by the use of logarithms, it is advantageous for us to use them in much of our trigonometric work. Especially is this true of the calculations connected with the solution of triangles. We shall now give the fundamental principles of logarithms and explain the use of logarithmic tables.

Definition of a logarithm. *The power to which a given number called the base must be raised to equal a second number is called the logarithm of the second number.*

Thus, if

$$(A) \qquad\qquad b^x = N, \qquad\qquad \textit{(exponential form)}$$

then $x =$ *the logarithm of N to the base b.* This statement is written in abbreviated form as follows:

$$(B) \qquad\qquad x = \log_b N. \qquad\qquad \textit{(logarithmic form)}$$

(A) and (B) are then simply two different ways of expressing the same relation between b, x, and N.

(A) is called the *exponential form.*
(B) is called the *logarithmic form.*

The fact that a logarithm is an exponent may be emphasized by writing (A) in the form

$$(\text{base})^{\log} = \text{number}.$$

For example, the following relations in exponential form, namely,

$$3^2 = 9, \qquad 2^5 = 32, \qquad (\tfrac{1}{2})^3 = \tfrac{1}{8}, \qquad x^y = z,$$

are written respectively in the logarithmic form

$$2 = \log_3 9, \quad 5 = \log_2 32, \qquad 3 = \log_{\frac{1}{2}} \tfrac{1}{8}, \quad y = \log_x z;$$

---

* Logarithms were invented by John Napier (1550–1617), Baron of Merchiston in Scotland, and described by him in 1614.

where $2, 5, 3, y$ are the logarithms (exponents),

$3, 2, \frac{1}{2}, x$ are the bases, and

$9, 32, \frac{1}{8}, z$ are the numbers respectively.

Similarly, the relations

$$25^{\frac{1}{2}} = \sqrt{25} = 5, \qquad\qquad 10^{-3} = \frac{1}{10^3} = \frac{1}{1000} = .001,$$

$$8^{\frac{2}{3}} = \sqrt[3]{8^2} = \sqrt[3]{64} = 4, \qquad\qquad b^0 = \frac{b^n}{b^n} = 1$$

are written in logarithmic form as follows:

$$\tfrac{1}{2} = \log_{25} 5, \quad -3 = \log_{10} .001, \quad \tfrac{2}{3} = \log_8 4, \quad 0 = \log_b 1.$$

EXAMPLES

1. In the following name the logarithm (exponent), the base, and the number, and write each in logarithmic form: $2^3 = 8$, $4^2 = 16$, $5^2 = 25$, $3^3 = 27$, $3^4 = 81$.

Solution. In the first one, $3 = $ logarithm, $2 = $ base, $8 = $ number; hence $\log_2 8 = 3$. *Ans.*

2. Express the following equations in logarithmic form: $(\frac{1}{3})^2 = \frac{1}{9}$, $\sqrt[3]{125} = 5$, $2^{-4} = \frac{1}{16}$, $10^{-2} = .01$, $p^s = q$.

3. Express the following equations in the exponential form: $\log_4 64 = 3$, $\log_7 49 = 2$, $\log_6 216 = 3$, $\log_{10} .0001 = -4$, $\log_4 2 = \frac{1}{2}$, $\log_a a = 1$, $\log_a 1 = 0$, $\log_b a = c$.

4. When the base is 2, what are the logarithms of the numbers $1, 2, \frac{1}{2}, 4, \frac{1}{4}, 8, 64, 128$?

5. When the base is 5, what are the logarithms of the numbers $1, 5, 25, 125, \frac{1}{5}, \frac{1}{25}, \frac{1}{625}$?

6. When the base is 10, what are the logarithms of the numbers $1, 10, 100, 1000, 10,000, .1, .01, .001, .0001$?

7. When the base is 4 and the logarithms are $0, 1, 2, 3, -1, -2, \frac{1}{2}$, what are the numbers?

8. What must be the bases when the following equations are true:

$$\log 64 = 2? \qquad \log 121 = 2? \qquad \log 625 = 4? \qquad \log \tfrac{1}{25} = -2?$$

9. When the base is 10, between what integers do the logarithms of the following numbers lie: 83, 251, 1793?

Solution. Since $\log_{10} 10 = 1$ and $\log_{10} 100 = 2$, and 83 is a number lying between 10 and 100, it follows that $\log_{10} 83 = $ a number lying between 1 and 2.

Similarly, $\log_{10} 251 = $ a number lying between 2 and 3,

$\log_{10} 1793 = $ a number lying between 3 and 4.

10. Verify the following:

(a) $\log_{10} 1000 + \log_{10} 100 + \log_{10} 10 + \log_{10} 1 = 6.$

(b) $\log_{10} \frac{1}{10} + \log_{10} \frac{1}{100} - \log_{10} \frac{1}{1000} = 0.$

(c) $\log_{10} .001 - \log_{10} .01 + \log_{10} .1 = -2.$

(d) $\log_2 8 - 3 \log_8 2 + \log_2 1 = 2.$

(e) $2 \log_a a + 2 \log_a \frac{1}{a} + \log_a 1 = 0.$

(f) $2 \log_4 2 + \tfrac{1}{2} \log_2 4 - \log_2 2 = 1.$

(g) $\log_3 3 + \log_3 \frac{1}{9} - \log_3 81 = -5.$

(h) $3 \log_{27} 3 - 1 \log_3 27 + \log_9 3 = \tfrac{1}{2}.$

(i) $4 \log_{16} 4 + 2 \log_4 \frac{1}{16} + \tfrac{1}{2} \log_2 16 = 0.$

(j) $2 \log_8 64 - \log_7 49 + 1 \log_5 \frac{1}{25} = 1.$

(k) $\log_8 64 + \log_4 64 + \log_2 64 = 11.$

(l) $\log_5 25 - \log_5 125 + 2 \log_5 5 = 1.$

(m) $2 \log_{36} 6 - \log_6 36 + \log_6 \frac{1}{36} = -3.$

64. Properties of logarithms. Since a logarithm is simply a new name for an exponent, it follows that the properties of logarithms must be found from the laws in Algebra governing exponents.

Theorem I. *The logarithm of the* **product** *of two factors equals the* **sum** *of the logarithms of the two factors.*

Proof. Let the two factors be M and N, and let x and y be their logarithms to the common base b. Then

(A) $\qquad\qquad\qquad \log_b M = x, \text{ and } \log_b N = y.$

Writing these in the exponential form,

(B) $\qquad\qquad\qquad b^x = M, \text{ and } b^y = N.$

Multiplying together the corresponding members of equations (B),

$$b^{x+y} = MN.$$

Writing this in the logarithmic form gives

$$\log_b MN = x + y = \log_b M + \log_b N. \qquad \text{from } (A)$$

By successive applications this theorem may evidently be extended to the product of any number of factors as follows:

$$\log_b MNPQ = \log_b M \cdot NPQ = \log_b M + \log_b NPQ \qquad \text{Th. I}$$
$$= \log_b M + \log_b N + \log_b PQ$$
$$= \log_b M + \log_b N + \log_b P + \log_b Q.$$

Theorem II. *The logarithm of the **quotient** of two numbers is equal to the logarithm of the dividend **minus** the logarithm of the divisor.*

Proof. As in Theorem I, let

$$(A) \qquad \log_b M = x, \text{ and } \log_b N = y.$$

Writing these in the exponential form,

$$(B) \qquad b^x = M, \text{ and } b^y = N.$$

Dividing the corresponding members of equations (B), we get

$$b^{x-y} = \frac{M}{N}.$$

Writing this in logarithmic form gives

$$\log_b \frac{M}{N} = x - y = \log_b M - \log_b N. \qquad \text{from } (A)$$

Theorem III. *The logarithm of the **pth power** of a number is equal to **p times** the logarithm of the number.*

Proof. Let $\qquad \log_b N = x.$
Then $\qquad b^x = N.$

Raising both sides to the pth power,

$$b^{px} = N^p.$$

Writing this in logarithmic form gives

$$\log_b N^p = px = p \log_b N.$$

Theorem IV. *The logarithm of the **rth root** of a number is equal to the logarithm of the number **divided by r**.*

Proof. Let $\qquad \log_b N = x.$
Then $\qquad b^x = N.$

Extracting the rth root of both sides,

$$b^{\frac{x}{r}} = N^{\frac{1}{r}}.$$

Writing this in logarithmic form gives

$$\log_b N^{\frac{1}{r}} = \frac{x}{r} = \frac{\log_b N}{r} = \frac{1}{r} \log_b N.$$

From the preceding four theorems it follows that if we use the logarithms of numbers instead of the numbers themselves, then the operations of *multiplication, division, raising to powers*, and *extracting roots* are replaced by those of *addition, subtraction, multiplication*, and *division* respectively.

Ex. 1. Find the value of $\log_{10}\sqrt{.001}$.

Solution. $\log_{10}\sqrt{.001} = \tfrac{1}{2}\log_{10}.001$ Th. IV
$$= \tfrac{1}{2}\log_{10}\tfrac{1}{1000} = \tfrac{1}{2}(-3) = -\tfrac{3}{2}. \quad Ans.$$

Ex. 2. Write $\log_b \sqrt[5]{\dfrac{a^3(c+d)^{\frac{1}{2}}}{c^2}}$ in expanded form.

Solution. $\log_b \sqrt[5]{\dfrac{a^3(c+d)^{\frac{1}{2}}}{c^2}} = \dfrac{1}{5}\log_b \dfrac{a^3(c+d)^{\frac{1}{2}}}{c^2}$ Th. IV
$$= \tfrac{1}{5}\{\log_b a^3 + \log_b(c+d)^{\frac{1}{2}} - \log_b c^2\} \quad \text{Th. I, II}$$
$$= \tfrac{1}{5}\{3\log_b a + \tfrac{1}{2}\log_b(c+d) - 2\log_b c\}. \quad \text{Th. III, IV}$$

When no base is indicated we mean that the same base is to be used throughout. Thus, the relation

$$\log \sqrt[5]{\frac{a^3(c+d)^{\frac{1}{2}}}{c^2}} = \frac{1}{5}\left\{3\log a + \frac{1}{2}\log(c+d) - 2\log c\right\}$$

holds true for any number used as the base. For the sake of convenience we shall call the left-hand member of an equation like the last one the *contracted form* of the logarithmic expression, and the right-hand member the *expanded form*.

Ex. 3. Write $3\log(x+1) + 3\log(x-1) + \tfrac{1}{2}\log x - 2\log(x^2+1)$ in the contracted form.

Solution. $3\log(x+1) + 3\log(x-1) + \tfrac{1}{2}\log x - 2\log(x^2+1)$
$$= \log(x+1)^3 + \log(x-1)^3 + \log x^{\frac{1}{2}} - \log(x^2+1)^2$$
$$= \log\frac{(x+1)^3(x-1)^3 x^{\frac{1}{2}}}{(x^2+1)^2} = \log\frac{\sqrt{x}\,(x^2-1)^3}{(x^2+1)^2}. \quad Ans.$$

Another form of the answer is found as follows:

$$\log\frac{\sqrt{x}\,(x^2-1)^3}{(x^2+1)^2} = \log\left(\frac{x(x^2-1)^6}{(x^2+1)^4}\right)^{\frac{1}{2}} = \frac{1}{2}\log\frac{x(x^2-1)^6}{(x^2+1)^4}.$$

EXAMPLES

1. Verify the following:

(a) $\log_{10}\sqrt{1000} + \log_{10}\sqrt{.01} = \tfrac{1}{2}.$

(b) $\log_{10}(.1)^4 - \log_{10}\sqrt[3]{.001} = -3.$

(c) $\log_{10}\sqrt{\tfrac{1}{10}} + \log_{10}\sqrt{10} = 0.$

(d) $\log_{10}\sqrt[3]{100} - \log_{10}(.01)^2 = 1\tfrac{4}{3}.$

(e) $\log_2\sqrt{8} + \log_3(\tfrac{1}{2})^2 = -\tfrac{1}{2}.$

(f) $\log_2(.5)^3 - \log_4\sqrt[3]{16} = -1\tfrac{1}{1}.$

(g) $\log_5\sqrt{125} + \log_{11}\sqrt[3]{121} = 1\tfrac{3}{6}.$

(h) $\log_8(2)^5 + \log_7(\tfrac{1}{49})^{\frac{1}{2}} = 1.$

2. Write the following logarithmic expressions in expanded form: *

(a) $\log\sqrt{\dfrac{s(s-b)(s-c)}{s-a}}.$

(b) $\log\dfrac{ab\sin C}{z}.$

(c) $\log P(1+r)^n.$

(d) $\log\dfrac{a^3 b^2 c^{\frac{1}{2}}}{4\sqrt[3]{d}}.$

(e) $\log\sqrt[3]{\dfrac{x(x-y)}{z(z+x)}}.$

(f) $\log\dfrac{\sqrt[5]{p^2(1-q)}}{\sqrt{p}(1+q)}.$

(g) $\log\dfrac{(m+n)s^2}{\sqrt{m-n}(1+s)}.$

(h) $\log\sqrt[4]{\left(\dfrac{a^2(b-c)}{c\sqrt{a-b}}\right)^3}.$

* To verify your results, reduce them back to the original form.

3. Write the following logarithmic expressions in contracted form:

(a) $2 \log x + \frac{1}{1} \log y - 3 \log z$.

(b) $3 \log (1 - x) - 2 \log (2 + x) + \log c$.

(c) $\frac{5}{6} \log (x - 1) - \frac{2}{3} \log x - \frac{1}{6} \log (x + 2) + \log c$.

(d) $\log y - \frac{1}{2} \log (y^2 + 4) + \log c$.

(e) $\frac{1}{2} \left\{ 2 \log (x - 1) + 3 \log (x + 1) + \frac{1}{2} \log x - \frac{2}{3} \log (x^2 + 1) \right\}$.

65. Common* system of logarithms. Any positive number except unity may be taken as the base, and to every particular base chosen there corresponds a set or system of logarithms. In the common system the base is 10, being the one most convenient to use with our decimal system of numbers. In what follows the base is usually omitted when writing expressions in the logarithmic form, the base 10 being always understood. Thus $\log_{10} 100 = 2$ is written $\log 100 = 2$, etc.

The logarithm of a given number in the common system is then the answer to the question:

What power of 10 will equal the given number?

The following table indicates what numbers have integers for logarithms in the common system.

	Exponential Form		*Logarithmic Form*
Since	$10^4 = 10,000$	we have	$\log 10,000 = 4$
	$10^3 = 1000$		$\log 1000 = 3$
	$10^2 = 100$		$\log 100 = 2$
	$10^1 = 10$		$\log 10 = 1$
	$10^0 = 1$		$\log 1 = 0$
	$10^{-1} = .1$		$\log .1 = -1$
	$10^{-2} = .01$		$\log .01 = -2$
	$10^{-3} = .001$		$\log .001 = -3$
	$10^{-4} = .0001$		$\log .0001 = -4$
	etc.,		etc.

Assuming that as a number increases its logarithm also increases, we see that a number between 100 and 1000 has a logarithm lying between 2 and 3. Similarly, the logarithm of a number between .1 and .01 has a logarithm lying between -1 and -2. In fact the logarithm of any number not an exact power of 10 consists, in general, of a *whole-number* part and a *decimal* part.

---

* Also called the Briggs System, from Henry Briggs (1556–1631), professor at Gresham College, London, and later at Oxford. He modified the new invention of logarithms so as to make it convenient for practical use.

Thus, since 4587 is a number lying between 10^3 and 10^4, we have

$$\log 4587 = 3 + \text{a decimal.}$$

Similarly, since .0067 is a number lying between 10^{-3} and 10^{-2},

$$\log .0067 = -(2 + \text{a decimal})$$
$$= -2 - \text{a decimal.}$$

For practical reasons the logarithm of a number is always written in such a form that the decimal part is positive. When the logarithm as a whole is negative, the decimal part may be made positive by adding plus unity to it. Then, so as not to change the value of the logarithm, we add minus unity to the whole part. Thus in the last example,

$$\log .0067 = (-2) + (-\text{a decimal})$$
$$= (-1 - 2) + (1 - \text{a decimal})$$
$$= -3 + \text{a new decimal.}$$

To emphasize the fact that only the whole part of a logarithm is negative, the minus sign is usually written over the whole part.

For example, $\quad \log .004712 = -2.3268$
$$= -2 - .3268$$
$$= (-1 - 2) + (1 - .3268)$$
$$= \bar{3}.6732.$$

The whole-number part of a logarithm is called the *characteristic* of the logarithm.

The decimal part of a logarithm is called the *mantissa* of the logarithm.

Thus if $\log 357 = 2.5527$ and $\log .004712 = \bar{3}.6732$, 2 and -3 are the characteristics and .5527 and .6732 the mantissas.

From the previous explanations and by inspection of the table on the opposite page we get the following:

66. Rules for determining the characteristic of a logarithm.

The characteristic of a number greater than unity is positive, and one less than the number of digits in the number to the left of the decimal point.

The characteristic of a number less than unity is negative, and is one greater numerically than the number of zeros between the decimal point and the first significant figure of the number.

Ex. Write down the characteristics of the logarithms of the numbers 27,683, 456.2, 9.67, 436,000, 26, .04, .0000612, .7963, .8, .0012.

$$Ans. \quad 4, 2, 0, 5, 1, -2, -5, -1, -1, -3.$$

Theorem V. *Numbers with the same significant part * (and which therefore differ only in the position of the decimal point) have the same mantissa.*

Proof. Consider, for example, the numbers 54.37 and 5437.

Let $$10^x = 54.37.$$

If we multiply both members of this equation by $100 \, (= 10^2)$, we have
$$10^2 \cdot 10^x = 10^{x+2} = 5437,$$
or, $$x + 2 = \log 5437.$$

Hence the logarithm of one number differs from that of the other merely in its whole part (characteristic).

Thus, if $$\log 47{,}120 = 4.6732,$$
then $$\log 47.12 = 1.6732,$$
and $$\log .004712 = \overline{3}.6732.$$

Special care is necessary in dealing with logarithms because of the fact that the mantissa is always positive, while the characteristic may be either positive or negative. When the characteristic is negative it is best for practical reasons to add 10 to it and write -10 after the logarithm, thus giving the logarithm a new form without change of value. Thus, if

(A) $$\log .0249 = \overline{2}.3962,$$

we add 10 to -2, giving 8 in the place of the characteristic, and counteract this by writing -10 after the logarithm; that is

(B) $$\log .0249 = 8.3962 - 10.$$

In case we wish to divide a logarithm having a negative characteristic by an integer (as is sometimes required in applying Theorem IV, p. 122), it is convenient to add and subtract 10 times that integer. Thus in case we wish to divide such a logarithm by 2, we add and subtract 20; if by 3, we add and subtract 30; and so on. Suppose we want to divide the logarithm of .0249, which is $\overline{2}.3962$, by 3. We would then add and subtract 30, so that

(C) $$\log .0249 = 28.3962 - 30,$$

a form more convenient than (A) or (B) when we wish to divide the logarithm by 3. Thus,

$$\tfrac{1}{3} \log .0249 = \tfrac{1}{3}(28.3962 - 30) = 9.4654 - 10.$$

---

* The *significant part* of a number consists of those figures which remain when we ignore all initial and final zeros. Thus, the significant part of 24,000 is 24; of 6.050 is 605; of .00907 is 907; of .00081070 is 8107.

This result may be written in form (A) by adding the 9 in front to the -10 at the end, giving $-1 = \bar{1}$ as the characteristic. Hence

$$\tfrac{1}{3}\log .0249 = \bar{1}.4654.$$

Another method for dividing a logarithm which has a negative characteristic will now be illustrated. Suppose we wish to divide $\bar{2}.3962\,(=-2 + 0.3962)$ by 2. We get at once

$$2\lfloor -2 + 0.3962$$
$$-1 + 0.1981 = \bar{1}.1981.$$

In case we wish to divide by 3 (as in the above example), we first add and subtract 1 in order to make the negative characteristic exactly divisible by 3. Thus,

$$3\lfloor -3 + 1.3962$$
$$-1 + 0.4654 = \bar{1}.4654.$$

The following examples will illustrate the best methods for performing the four fundamental operations of Arithmetic on logarithms.

CASE I. *Addition of logarithms.*

(a) To add two logarithms having positive characteristics, as 3.2659 and 1.9866.

$$\begin{array}{r} 3.2659 \\ 1.9866 \\ \hline 5.2525 \end{array}$$

This is in no way different from ordinary addition.

(b) To add two logarithms, one having a negative characteristic, as $\bar{4}.2560$ and 2.8711.

$$\begin{array}{lll} \bar{4}.2560 & \text{or,} & 6.2560 - 10 \\ \underline{2.8711} & & \underline{2.8711} \\ \bar{1}.1271 & & 9.1271 - 10 \\ & \text{i.e.} & \bar{1}.1271 \end{array}$$

Since the mantissas (decimal parts) are always positive, the carrying figure 1 from the tenth's place is positive. Hence in adding the first way, $1-4+2=-1=\bar{1}$ will be the characteristic of the sum.

(c) To add two logarithms having negative characteristics, as $\bar{2}.4069$ and $\bar{1}.9842$.

$$\begin{array}{lll} \bar{2}.4069 & \text{or,} & 8.4069 - 10 \\ \underline{\bar{1}.9842} & & \underline{9.9842 - 10} \\ \bar{2}.3911 & & 18.3911 - 20 \\ & \text{i.e.} & \bar{2}.3911 \end{array}$$

Case II. *Subtraction of logarithms.*

(a) To subtract logarithms having positive characteristics.

From	5.6233	From	2.4673	or,	12.4673 — 10
take	3.8890	take	3.7851		3.7851
	1.7343		2.6822		8.6822 — 10
				i.e.	$\bar{2}$.6822

In the first example we have ordinary subtraction. In the second we subtract a greater logarithm from a smaller one and the result as a whole is negative.

(b) To subtract logarithms having negative characteristics.

From	2.1163	or,	12.1163 — 10
take	$\bar{3}$.4492		7.4492 — 10
	4.6671		4.6671
From	$\bar{1}$.6899	or,	9.6899 — 10
take	1.9083		1.9083
	$\bar{3}$.7816		7.7816 — 10
		i.e.	$\bar{3}$.7816
From	$\bar{2}$.1853	or,	18.1853 — 20
take	$\bar{1}$.7442		9.7442 — 10
	$\bar{2}$.4411		8.4411 — 10
		i.e.	$\bar{2}$.4411

Case III. *Multiplication of logarithms by numbers.*

Multiply	0.6842	Multiply	$\bar{2}$.7012	or,	8.7012 — 10
by	5	by	3		3
	3.4210		$\bar{4}$.1036		26.1036 — 30
				i.e.	$\bar{4}$.1036

In the second example the carrying figure from tenth's place is + 2. Adding this + 2 to − 2 × 3 gives 2 − 6 = − 4 = 4 for the characteristic.

Case IV. *Division of logarithms by numbers.*

(a) Divide 3.8530 by 2.

$$2\,\lfloor 3.8530$$
$$1.9265$$

(b) Divide $\bar{2}$.2411 by 3.

Here we first add and then we subtract 30, writing the logarithm in the form 28.2411−30.

$$3\,\lfloor 28.2411 - 30$$
$$9.4137 - 10$$
$$\text{i.e.} \quad \bar{1}.4137$$

67. Tables of logarithms. The common system (having the base 10) of logarithms is the one used in practical computations. For the convenience of the calculator the common logarithms of numbers up to a certain number of significant figures have been computed and arranged in tabulated forms called logarithmic tables. The common system has two great advantages.

(*A*) *The characteristic of the logarithm of a number may be written down on mere inspection by following the rules on p. 125.*

Hence, as a rule, only the mantissas of the logarithms of numbers are printed in the tables.

(*B*) *The logarithms of numbers having the same significant part have the same mantissa (Th. V, p. 126).*

Hence a change in the position of the decimal point in a number affects the characteristic alone, and it is sufficient to tabulate the mantissas * of integers only. Thus,

$$\log \ 3104 = 3.4920, \qquad \log \ 31.04 = 1.4920,$$
$$\log .03104 = 2.4920, \qquad \log 310,400 = 5.4920;$$

in fact, the mantissa of any number whatever having 3104 as its significant part will have .4920 as the mantissa of its logarithm.

Table I, pp. 2, 3,† gives immediately the mantissas of the logarithms of all numbers whose first significant figure is 1 and whose significant part consists of four or fewer digits; and on pp. 4, 5 are found the mantissas of the logarithms of all numbers whose first significant figure is greater than 1 and whose significant part consists of three or fewer digits.

68. To find the logarithms of numbers from Table I, pp. 2–5.

When the first significant figure of the number is 1, and there are four or fewer digits in its significant part, follow

Rule I. **First step.** *Determine the characteristic by inspection, using the rule on p. 125.*

Second step. *Find in the vertical column N, Table I, pp. 2, 3, the first three significant figures of the number. The mantissa required is in the same horizontal row with these figures and in the vertical column having the fourth significant figure at the top (and bottom).*

Ex. 1. Find log 1387.

Solution. First step. From the rule on p. 125 we see that the characteristic will be + 3, that is, one less than the number of digits (four) to the left of the decimal point.

Second step. On p. 2, Table I, we find 138 in column *N*. The required mantissa will be found in the same horizontal row with 138 and in the vertical column which has 7 at the top. This gives the mantissa .1421.

Therefore $\qquad\qquad \log 1387 = 3.1421.$ *Ans.*

---

* In order to save space the decimal point in front of each mantissa is usually omitted in the tables.

† The tables referred to in this book are Granville's *Four-Place Tables of Logarithms* (Ginn & Company).

If the significant part of the number consists of less than four digits, annex zeros until you do have four digits.

Ex. 2. Find log 17.

Solution. *First step.* By the rule on p. 125 the characteristic is found to be 1.

Second step. To find the mantissa of 17 we look up the mantissa of 1700. On p. 3, Table I, we locate 170 in column N. The required mantissa is found in the same horizontal row with 170, and in the vertical column having 0 at the top. This gives the mantissa .2304.

Therefore $\log 17 = 1.2304$. *Ans.*

Ex. 3. Find log .00152.

Solution. *First step.* By the rule on p. 125 we find that the characteristic is -3, that is, negative, and one greater numerically than the number of zeros (two) immediately after the decimal point.

Second step. Locate 152 in column N, Table I, p. 3. In the same horizontal row with 152 and in the vertical column with 0 at the top we find the required mantissa .1818.

Therefore $\log .00152 = \bar{3}.1818 = 7.1818 - 10$. *Ans.*

To find the logarithm of a number *when the first significant figure of the number is greater than 1 and there are three or fewer digits in its significant part, follow*

Rule II. **First step.** *Determine the characteristic by rule on p. 125.*

Second step. *Find in the vertical column N, Table I, pp. 4, 5, the first two significant figures of the number. The mantissa required is in the horizontal row with these figures and in the vertical column having the third significant figure at the top (and bottom).*

Ex. 4. Find log 5.63.

Solution. *First step.* The characteristic here is zero.

Second step. On p. 4, Table I, we locate 56 in column N. In the horizontal row with 56 and in the vertical column with 3 at the top we find the required mantissa .7505.

Therefore $\log 5.63 = 0.7505$. *Ans.*

If the significant part of the number consists of less than three digits, annex zeros until you do have three digits.

Ex. 5. Find log 460,000.

Solution. *First step.* The characteristic is 5.

Second step. On p. 4, Table I, we locate 46 in column N. In the horizontal row with 46 and in the vertical column with 0 at the top we find the required mantissa .6628.

Therefore $\log 460,000 = 5.6628$. *Ans.*

Ex. 6. Find log .08.

Solution. First step. The characteristic is -2.

Second step. Using 800, we find that the mantissa is .9031.

Therefore $\qquad\qquad$ $\log .08 = \bar{2}.9031 = 8.9031 - 10.$ *Ans.*

Ex. 7. Find (a) log 1872, (b) log 5, (c) log .7, (d) log 20,000, (e) log 1.808, (f) log .000032, (g) log .01011, (h) log 9.95, (i) log 17.35, (j) log .1289, (k) log 2500, (l) log 1.002.

Ans. (a) 3.2723, (b) 0.6990, (c) $\bar{1}$.8451, (d) 4.3010, (e) 0.2572, (f) $\bar{5}$.5051, (g) $\bar{2}$.0048, (h) 0.9978, (i) 1.2393, (j) $\bar{1}$.1103, (k) 3.3979, (l) 0.0009.

When the first significant figure of a number is 1 and the number of digits in its significant part is greater than 4, its mantissa cannot be found in Table I; nor can the mantissa of a number be found when its first significant figure is greater than 1 and the number of digits in its significant part be greater than 3.

By *interpolation,** however, we may, in the first case, find the mantissa of a number having a fifth significant figure; and in the second case, of a number having a fourth significant figure. In this book no attempt is made to find the logarithms of numbers with more significant figures, since our four-place tables are in general accurate only to that extent.

We shall now illustrate the process of interpolation by means of examples.

Ex. 8. Find log 2445.

Solution. By rule on p. 125 the characteristic is found to be 3. The required mantissa is not found in our table. But by Rule II, p. 130,

$$\log 2450 = 3.3892$$
and $\qquad\qquad\qquad$ $\log 2440 = 3.3874$
$$\text{Difference in logarithms} = \overline{.0018}$$

Since 2445 lies between 2440 and 2450, it is clear that its logarithm must lie between 3.3874 and 3.3892. Because 2445 is just halfway between 2440 and 2450 we assume that its logarithm is halfway between the two logarithms.† We then take half (or .5) of their difference, .0018 (called the tabular difference), and add this to log 2440 = 3.3874. This gives

$$\log 2445 = 3.3874 + .5 \times .0018 = 3.3883.$$

If we had to find log 2442, we should take not half the difference, but .2 of the difference between the logarithms of 2440 and 2445, since 2442 is not halfway between them but two tenths of the way.

---

* Illustrated by examples on pp. 16-19 in the case of trigonometric functions.

† In this process of interpolation we have assumed and used the principle that the increase of the logarithm is proportional to the increase of the number. This principle is not strictly true, though for numbers whose first significant figure is greater than 1 the error is so small as not to appear in the fourth decimal place of the mantissa. For numbers whose first significant figure is 1 this error would often appear, and for this reason Table I, pp. 2, 3, gives the mantissas of all such numbers exact to four decimal places.

In order to save work in interpolating, when looking up the logarithms of numbers whose mantissas are not found in the table, each tabular difference occurring in the table has been multiplied by .1, .2, .3, ⋯, .9, and the results are printed in the large right-hand column with "Prop. Parts" (proportional parts) at the top. Thus, on p. 4, Table I, the first section in the Prop. Parts column shows the products obtained when multiplying the tabular differences 22 and 21* by .1, .2, .3, ⋯, .9. Thus,

Extra Digit	Difference	
	22	21
1	2.2	2.1
2	4.4	4.2
3	6.6	6.3
4	8.8	8.4
5	11.0	10.5
6	13.2	12.6
7	15.4	14.7
8	17.6	16.8
9	19.8	18.9

$$.1 \times 22 = 2.2 \qquad .1 \times 21 = 2.1$$
$$.2 \times 22 = 4.4 \qquad .2 \times 21 = 4.2$$
$$.3 \times 22 = 6.6 \qquad .3 \times 21 = 6.3$$
$$.4 \times 22 = 8.8 \qquad .4 \times 21 = 8.4$$
$$.5 \times 22 = 11.0 \qquad .5 \times 21 = 10.5$$
$$\text{etc.} \qquad\qquad \text{etc.}$$

Hence

To find the logarithm of a number whose mantissa is not found in the table,† use

RULE III. **First step.** *Find the logarithm of the number, using only the first three (or four) digits of its significant part when looking up the mantissa.‡*

Second step. *Subtract the mantissa just found from the next greater mantissa in the table to find the corresponding tabular difference.*

Third step. *In the Prop. Parts column locate the block corresponding to the tabular difference found. Under this difference and opposite the extra digit § of the number will be found the proportional part of the tabular difference which should be added to the extreme right of the logarithm found in the first step. The sum will be the logarithm of the given number.*

Ex. 9. Find log 28.64.

Solution. Since the mantissa of 2864 is not found in our table, this example comes under Rule III, the extra digit being 4.

First step.	log 28.60 = 1.4564	Rule II
Second step.	log 28.70 = 1.4579	Rule II
	Tabular difference = 15 ‖	

---

* These are really .0022 and .0021, it being customary to drop the decimal point.

† That is, a number whose logarithm cannot be found by Rule I or Rule II, because its significant part contains too many digits.

‡ When the first significant figure is 1, use the first four digits, following Rule I; when the first significant figure is greater than 1, use the first three digits, following Rule II.

§ In finding log 4836, for instance, 6 is called the extra digit, or, in finding log 14,835 the extra digit is 5.

‖ The tabular difference = .0015, but the decimal point is usually omitted in practice.

Third step. About halfway down the Prop. Parts column on p. 4 we find the block giving the proportional parts corresponding to the tabular difference 15. Under 15 and opposite the extra digit 4 of our number we find 6.0. Then

$$\log 28.60 = 1.4564$$
$$\underline{\qquad\qquad 6 \quad \text{Prop. Part}}$$
$$\log 28.64 = \overline{1.4570}. \quad \textit{Ans.}$$

Ex. 10. Find log .12548.

Solution. Since the mantissa of 12,548 is not found in our table, this example comes under Rule III, the extra digit being 8.

First step.	$\log .12540 = \bar{1}.0983$
Second step.	$\log .12550 = \bar{1}.0986$
	Tabular difference = $\quad$ 3

Third step. In the Prop. Parts column on p. 2 we find the block giving the proportional parts corresponding to the tabular difference 3. Under 3 and opposite the extra digit 8 we find 2.4 ($= 2$). Then

$$\log .12540 = \bar{1}.0983$$
$$\underline{\qquad\qquad 2 \quad \text{Prop. Part}}$$
$$\log .12548 = \bar{1}.0985. \quad \textit{Ans.}$$

Ex. 11. Verify the following :

(a) $\log 4583 = 3.6612$.	(e) $\log 1000.7 = 3.0003$.
(b) $\log 16.426 = 1.2155$.	(f) $\log 724{,}200 = 5.8598$.
(c) $\log .09688 = 2.9862$.	(g) $\log 9.496 = 0.9775$.
(d) $\log .10108 = 2.9862$.	(h) $\log .0004586 = 4.6614$.

69. **To find the number corresponding to a given logarithm, use**

RULE IV. *On pp. 2–5, Table I, look for the mantissa of the given logarithm. If the mantissa is found exactly in the table, the first significant figures of the corresponding number are found in the same row under the N column, while the last figure is at the top of the column in which the mantissa was found. Noting what the characteristic in the given logarithm is, place the decimal point so as to agree with the rule on p. 125.*

In case the mantissa of the given logarithm is not found exactly in the table *we must take instead the following steps :*

First step. *Locate the given mantissa between two mantissas in the tables.*

Second step. *Write down the number corresponding to the lesser of the two mantissas. This will give the first three (or four) significant figures of the required number.*

Third step. *Find the tabular difference between the two mantissas from the table, and also the difference between the lesser of the two and the given mantissa.*

Fourth step. *Under the Prop. Parts column find the block corresponding to the tabular difference found. Under this tabular difference pick out the proportional part nearest the difference found between the lesser mantissa and the given mantissa, and to the left of it will be found the last (extra) figure of the number, which figure we now annex.*

Fifth step. *Noting what the characteristic of the given logarithm is, place the decimal point so as to agree with the rule on p. 125.*

Ex. 12. Find the number whose logarithm is 2.1892.

Solution. The problem may also be stated thus : find x, having given

$$\log x = 2.1892.$$

On p. 3, Table I, we find this mantissa, .1892 exactly, in the same horizontal row with 154 in the N column and in the vertical column with 6 at the top. Hence the first four significant figures of the required number are 1546. Since the characteristic is 2, we place the decimal point so that there will be three digits to the left of the decimal point, that is, we place it between 4 and 6. Hence

$$x = 154.6. \quad Ans.$$

Ex. 13. Find the number whose logarithm is 4.8409.

Solution. That is, given $\log x = 4.8409$, to find x. Since the mantissa .8409 is not found exactly in our table, we follow the last part of Rule IV.

First step. The given mantissa, .8409, is found to lie between .8407 and .8414 on p. 4, Table I.

Second step. The number corresponding to the lesser one, that is, to .8407, is 693.

Third step. The tabular difference between .8407 and .8414 is 7, and the difference between .8407 and the given mantissa .8409 is 2.

Fourth step. In the Prop. Parts column under the block corresponding to the tabular difference 7, we find that the proportional part 2.1 is nearest to 2 in value. Immediately to the left of 2.1 we find 3, the (extra) figure to be annexed to the number 693 found in the second step. Hence the first four significant figures of the required number are 6933.

Fifth step. Since the characteristic of the given logarithm is 4, we annex one zero and place the decimal point after it in order to have five digits of the number to the left of the decimal point. Hence

$$x = 69,330. \quad Ans.$$

Ex. 14. Find the numbers whose logarithms are (a) 1.8055, (b) $\bar{1}$.4487, (c) 0.2164, (d) 2.9487, (e) 2.0529, (f) 5.2668, (g) 3.9774, (h) 4.0010, (i) 8.4430 − 10,* (j) 9.4975 − 10.

      Ans. (a) 63.9, (b) .281, (c) 1.646, (d) 888.6, (e) .011295, (f) 184,850, (g) 9493, (h) .00010023, (i) .02773, (j) .3144.

---

* By (A), (B), p. 126, $8.4430 - 10 = \bar{2}.4430$.

70. The use of logarithms in computations. The following examples will illustrate how logarithms are used in actual calculations.

Ex. 1. Calculate 243×13.49, using logarithms.

Solution. Denoting the product by x, we may write

$$x = 243 \times 13.49.$$

Taking the logarithms of both sides, we get

$$\log x = \log 243 + \log 13.49. \qquad \text{Th. I, p. 121}$$

Looking up the logarithms of the numbers,

$$\log 243 = 2.3856 \qquad \text{Rule II, p. 130}$$
$$\log 13.49 = 1.1300 \qquad \text{Rule I, p. 129}$$

Adding,
$$\log x = 3.5156$$

By Rule IV, p. 133,
$$x = 3278. \ \textit{Ans.}$$

Ex. 2. Calculate $\dfrac{1375 \times .06423}{76,420}$.

Solution. Let $x = \dfrac{1375 \times .06423}{76,420}$.

Then
$$\log x = \log 1375 + \log .06423 - \log 76,420$$
$$\text{Th. I, p. 121, and Th. II, p. 122}$$

$$\log 1375 = 3.1383 \qquad \text{Rule I, p. 129}$$
$$\log .06423 = 8.8077 - 10 \qquad \text{Rule III, p. 132}$$

Adding,
$$11.9460 - 10$$

$$\log 76,420 = 4.8832 \qquad \text{Rule III, p. 132}$$

Subtracting,
$$\log x = 7.0628 - 10$$

or,
$$\log x = \bar{3}.0628.$$

By Rule IV, p. 133,
$$x = .0011555. \ \textit{Ans.}$$

Ex. 3. Calculate $(5.664)^3$.

Solution. Let
$$x = (5.664)^3.$$

Then
$$\log x = 3 \log 5.664. \qquad \text{Th. III, p. 122}$$

$$\log 5.664 = 0.7531 \qquad \text{Rule III, p. 132}$$

Multiplying by 3,
$$3$$
$$\log x = 2.2593$$

By Rule IV, p. 133,
$$x = 181.67. \ \textit{Ans.}$$

Ex. 4. Calculate $\sqrt[3]{.7182}$.

Solution. Let
$$x = \sqrt[3]{.7182} = (.7182)^{\frac{1}{3}}.$$

Then
$$\log x = \tfrac{1}{3} \log .7182. \qquad \text{Th. IV, p. 122}$$

$$\log .7182 = \bar{1}.8562 \qquad \text{Rule III, p. 132}$$
$$= 29.8562 - 30. \qquad \text{(b), Case IV, p. 128}$$

Dividing by 3,
$$3 \,|\, \underline{29.8562 - 30}$$
$$\log x = 9.9521 - 10$$
$$= \bar{1}.9521.$$

By Rule IV, p. 133,
$$x = .8956. \ \textit{Ans.}$$

Ex. 5. Calculate $\sqrt[3]{\dfrac{\sqrt{7194} \times 87}{98{,}080{,}000}}$.

Solution. Let
$$x = \sqrt[3]{\frac{\sqrt{7194} \times 87}{98{,}080{,}000}} = \left[\frac{(7194)^{\frac{1}{2}} \times 87}{98{,}080{,}000}\right]^{\frac{1}{3}}.$$

Then
$$\log x = \tfrac{1}{3}\left[\tfrac{1}{2}\log 7194 + \log 87 - \log 98{,}080{,}000\right].$$

$$\log 7194 = 3.8569$$

Dividing by 2,
$$2\,|\,3.8569$$
$$\tfrac{1}{2}\log 7194 = 1.9285$$
$$\log 87 = 1.9395$$

Adding,
$$3.8680$$

or, $13.8680 - 10$ (a), Case II, p. 128
$$\log 98{,}080{,}000 = 7.9916$$

Subtracting, $5.8764 - 10$

or, $25.8764 - 30$ (b), Case IV, p. 128

Dividing by 3, $3\,|\,25.8764 - 30$
$$\log x = 8.6255 - 10$$
$$= 2.6255.$$
$$\therefore\ x = .04222.\ \ \textit{Ans.}$$

Ex. 6. Calculate $\dfrac{8 \times 62.73 \times .052}{56 \times 8.793}$.

Solution. Let
$$x = \frac{8 \times 62.73 \times .052}{56 \times 8.793}.$$

Then
$$\log x = [\log 8 + \log 62.73 + \log .052] - [\log 56 + \log 8.793].$$

$$
\begin{array}{ll}
\log 8 = 0.9031 & \log 56 = 1.7482 \\
\log 62.73 = 1.7975 & \log 8.793 = 0.9442 \\
\log .052 = 8.7160 - 10 & \log \text{denominator} = 2.6924 \\
\log \text{numerator} = 11.4166 - 10 & \\
\log \text{denominator} = 2.6924 & \\
\log x = 8.7242 - 10 & \\
= 2.7242. & \\
\therefore\ x = .05299.\ \ \textit{Ans.}^{*} &
\end{array}
$$

---

* Instead of looking up the logarithms at once when we write down log 8, log 62.73, etc., it is better to write down an outline or skeleton of the computation before using the tables at all. Thus, for above example,

$$
\begin{array}{ll}
\log 8 = 0. & \log 56 = 1. \\
\log 62.73 = 1. & \log 8.793 = 0.\underline{\hphantom{0000000}} \\
\log .052 = 8.\quad\quad - 10 & \log \text{denominator} = \\
\log \text{numerator} = & \\
\log \text{denominator} = \underline{\hphantom{000000}} & \\
\log x = & \\
\therefore\ x = &
\end{array}
$$

It saves time to look up all the logarithms at once, and, besides, the student is not so apt to forget to put down the characteristics.

71. Cologarithms. The logarithm of the reciprocal of a number is called its *cologarithm* (abbreviated *colog*). Hence if N is any positive number,

$$\operatorname{colog} N = \log \frac{1}{N} = \log 1 - \log N \qquad \text{Th. II, p. 122}$$
$$= 0 - \log N = - \log N.$$

That is, the cologarithm of a number equals *minus* the logarithm of the number, the minus sign affecting the entire logarithm, both characteristic and mantissa. In order to avoid a negative mantissa in the cologarithm, it is customary to subtract the logarithm of the number from $10 - 10$. Thus, taking 25 as the number,

$$\operatorname{colog} 25 = \log \tfrac{1}{25} = \log 1 - \log 25.$$

But
$$\log 1 = 0,$$

or, what amounts to the same thing,

$$\log 1 = 10.0000 - 10.$$
Also,
$$\log 25 = 1.3979$$
$$\operatorname{colog} 25 = 8.6021 - 10$$

Since dividing by a number is the same as multiplying by the reciprocal of the number, it is evident that when we are calculating by means of logarithms we may either subtract the logarithm of a divisor or add its cologarithm. When a computation is to be made in which several factors occur in the denominator of a fraction, it is more convenient to add the cologarithms of the factors than to subtract their logarithms. Hence

RULE V. *Instead of subtracting the logarithm of a divisor, we may add its cologarithm. The cologarithm of any number is found by subtracting its logarithm from* $10.0000 - 10$.

Ex. 1. Find colog 52.63.

Solution.
$$10.0000 - 10$$
$$\log 52.63 = 1.7212$$
$$\operatorname{colog} 52.63 = 8.2788 - 10. \quad \textit{Ans.} \qquad \text{Rule V}$$

Ex. 2. Find colog .016548.

Solution.
$$10.0000 - 10$$
$$\log .016548 = 8.2187 - 10$$
$$\operatorname{colog} .016548 = 1.7813. \quad \textit{Ans.} \qquad \text{Rule V}$$

Thus we see that the cologarithm may be obtained from the logarithm by subtracting the last significant figure of the mantissa from 10 and each of the others from 9.

In order to show how the use of cologarithms exhibits the written work in more compact form, let us calculate the expression in Ex. 6, namely,

$$x = \frac{8 \times 62.73 \times .052}{56 \times 8.793}.$$

Solution. Using cologarithms,

$$\log x = \log 8 + \log 62.73 + \log .052 + \operatorname{colog} 56 + \operatorname{colog} 8.793.$$

$$
\begin{aligned}
\log 8 &= 0.9031 \\
\log 62.73 &= 1.7975 \\
\log .052 &= 8.7160 - 10 \\
\operatorname{colog} 56 &= 8.2518 - 10 \qquad \text{since} \qquad \log 56 = 1.7482 \\
\operatorname{colog} 8.793 &= 9.0558 - 10 \qquad \text{since} \qquad \log 8.793 = 0.9442 \\
\log x &= 28.7242 - 30 \\
&= \bar{2}.7242. \\
\therefore\ x &= .05299. \quad \textit{Ans.}
\end{aligned}
$$

Calculate the following expressions, using logarithms:

3. $9.238 \times .9152$. *Ans.* 8.454.

4. $336.8 \div 7984$. .04218.

5. $(.07396)^5$. .000002213.

6. $\dfrac{15.008 \times .0843}{.06376 \times 4.248}$. 4.671.

7. $\sqrt{2}$. 1.414.

8. $\sqrt[4]{5}$. 1.495.

9. $\sqrt[3]{.02305}$. .2846.

10. $\sqrt[3]{\dfrac{.03296}{7.962}}$. .1606.

11. $\left(\dfrac{.08726}{.1321}\right)^{\frac{2}{3}}$. *Ans.* .5010.

12. $(538.2 \times .0005969)^{\frac{1}{4}}$. .8678.

13. $\sqrt[17]{\left(\dfrac{31.63}{429}\right)^{3}}$. .631.

14. $\left(\dfrac{35}{}\right)^{\frac{2}{3}}$. .6443.

15. $\sqrt[8]{2} \times \sqrt[5]{3} \times \sqrt[7]{.01}$. .7035.

16. $\dfrac{-401.8}{52.37}$ *. -7.672.

17. $\dfrac{(-2563) \times .03442}{714.8 \times (-.511)}$ *Ans.* .2415.

18. $\dfrac{121.6 \times (-9.025)}{(-48.3) \times 3662 \times (-.0856)}$. $-.0725$.

72. Change of base in logarithms. We have seen how the logarithm of a number to the base 10 may be found in our tables. It is sometimes necessary to find the logarithm of a number to a base different from 10. For the sake of generality let us assume that the logarithms of numbers to the base a have been computed. We wish to find the logarithm of a number, as N, to a new base b; that is, we seek to express $\log_b N$ in terms of logarithms to the base a.

Suppose $\log_b N = x$,

that is, $b^x = N.$

---

* From the definition of a logarithm, p. 119, it is evident that a *negative number* can have no logarithm. If negative numbers do occur in a computation, they should be treated as if they were positive, and the sign of the result determined by the rules for signs in Algebra, irrespective of the logarithmic work. Thus, in Example 16 above, we calculate the value of $401.8 \div 52.37$ and write a minus sign before the result.

Taking the logarithms of both sides of this equation to the base a, we get

$$\log_a b^x = \log_a N,$$

or, $\qquad x \log_a b = \log_a N.$ Th. III, p. 122

Solving, $\qquad x = \dfrac{\log_a N}{\log_a b}.$

But $\qquad \log_b N = x.$ By hypothesis.

(90) $\qquad \therefore \log_b N = \dfrac{\log_a N}{\log_a b}.$

Theorem VI. *The logarithm of a number to the new base b equals the logarithm of the same number to the original base a, divided by the logarithm of b to the base a.*

This formula is also written in the form

$$\log_b N = M \cdot \log_a N,$$

where $M = \dfrac{1}{\log_a b}$ is called the **modulus** *of the new system with respect to the original one.**

This number M does not depend on the particular number N, but only on the two bases a and b.

In actual computations $a = 10$, since the tables we use are computed to the base 10.

Ex. Find $\log_3 21$.

Solution. Here $N = 21$, $b = 3$, $a = 10$. Substituting in (90),

$$\log_3 21 = \frac{\log_{10} 21}{\log_{10} 3} = \frac{1.3222}{.4771} = 2.771. \quad Ans.$$

EXAMPLES

1. Verify the following:

(a) $\log_2 7 = 2.807.$ (e) $\log_9 8 = 0.9464.$ (i) $\log_3 10 = 2.096.$
(b) $\log_3 4 = 1.262.$ (f) $\log_8 5 = 0.7743.$ (j) $\log_5 100 = 2.86.$
(c) $\log_4 9 = 1.585.$ (g) $\log_7 14 = 1.356.$ (k) $\log_3 .1 = -2.096.$
(d) $\log_5 7 = 1.209.$ (h) $\log_5 102 = 2.873.$ (l) $\log_5 .01 = -2.86.$

2. Find the logarithm of $\frac{7}{11}$ in the system of which 0.5 is the base.

3. Find the base of the system in which the logarithm of 8 is $\frac{2}{3}$.

4. Prove $\log_b a \cdot \log_a b = 1.$

5. Prove $\log_N 10 = \dfrac{1}{\log_{10} N}.$

---

* If, then, we have given the logarithms of numbers to a certain base a, and we wish to find the logarithms of the same numbers to a new base b, we multiply the given logarithms by the constant multiplier (modulus) $M = \dfrac{1}{L}$. Thus, having given the common logarithms (base 10) of numbers, we can reduce them to the logarithms of the same numbers to the base $e\,(= 2.718)$ by multiplying them by $M = \dfrac{1}{\log_{10} e} = 2.3026.$

73. Exponential equations. These are equations in which the unknown quantities occur in the exponents. Such equations may often be solved by the use of logarithms, as illustrated in the following examples:

Ex. 1. Given $81^x = 10$; find the value of x.

Solution. Taking the logarithms of both members,

$$\log 81^x = \log 10,$$

or, $$x \log 81 = \log 10. \qquad \text{Th. III, p. 122}$$

Solving, $$x = \frac{\log 10}{\log 81} = \frac{1.0000}{1.9085} = 0.524. \quad Ans.$$

Ex. 2. Express the solution of

$$a^{2x+3} b^x = c$$

in terms of logarithms.

Solution. Taking the logarithms of both members,

$$\log a^{2x+3} + \log b^x = \log c. \qquad \text{Th. I, p. 121}$$
$$(2x+3) \log a + x \log b = \log c. \qquad \text{Th. III, p. 122}$$
$$2x \log a + 3 \log a + x \log b = \log c.$$
$$x(2 \log a + \log b) = \log c - 3 \log a.$$
$$x = \frac{\log c - 3 \log a}{2 \log a + \log b}. \quad Ans.$$

Ex. 3. Solve the simultaneous equations

(A) $$\qquad\qquad 2^x \cdot 3^y = 100.$$
(B) $$\qquad\qquad x + y = 4.$$

Solution. Taking the logarithms of both members of (A), and multiplying (B) through by $\log 2$, we get

$$x \log 2 + y \log 3 = 2 \qquad \text{Th. I, III, p. 122}$$
$$x \log 2 + y \log 2 = 4 \log 2$$

Subtracting, $$y (\log 3 - \log 2) = 2 - 4 \log 2$$

Solving, $$y = \frac{2 - 4 \log 2}{\log 3 - \log 2} = \frac{2 - 1.2040}{.4771 - .3010}$$

$$y = \frac{.7960}{.1761} = 4.52.$$

Substituting back in (B), we get $x = -.52$.

EXAMPLES

1. Solve the following equations:

(a) $5^x = 12$.	*Ans.* 1.54.	(g) $(1.3)^x = 7.2$.		*Ans.* 7.53.
(b) $7^x = 25$.	1.65.			
(c) $(0.4)^{-x} = 7$.	2.12.	(h) $(0.9)^{\frac{1}{x^2}} = (4.7)^{-\frac{1}{3}}$.		0.45.
(d) $10^{x-1} = 4$.	1.602.	(i) $7^{x+3} = 5$.		-2.1728.
(e) $4^{x-1} = 5^{x+1}$.	13.86.	(j) $2^{2x+3} - 6^{x-1} = 0$.		9.542.
(f) $4^x = 40$.	2.66.			

2. Solve the following simultaneous equations:

(a) $4^x \cdot 3^y = 8,$ *Ans.* $x = .9005,$ (c) $2^x \cdot 2^y = 2^{22},$ *Ans.* $x = 13,$
$\quad 2^x \cdot 8^y = 9.$ $\quad y = .7565.$ $\quad x - y = 4.$ $\quad y = 9.$

(b) $3^x \cdot 4^y = 15{,}552,$ $\quad x = 5,$ (d) $2^x \cdot 3^y = 18,$ $\quad x = 1,$
$\quad 4^x \cdot 5^y = 128{,}000.$ $\quad y = 3.$ $\quad 5^x \cdot 7^y = 245.$ $\quad y = 2.$

3. Indicate the solution of the following in terms of logarithms:

(a) $A = P(r + 1)^x.$ $\quad$ *Ans.* $x = \dfrac{\log A - \log P}{\log (r + 1)}.$

(b) $a^{x^2 + 2x} = b.$ $\quad x = -1 \pm \sqrt{\dfrac{\log ab}{\log a}}.$

(c) $a^x \cdot b^y = m,$ $\quad x = \dfrac{\log d \log m - \log b \log n}{\log a \log d - \log b \log c},$

$\quad c^x \cdot d^y = n.$ $\quad y = \dfrac{\log a \log n - \log c \log m}{\log a \log d - \log b \log c}$

(d) $a^{2x-3} \cdot a^{3y-2} = a^8,$ $\quad x = 5,$
$\quad 3x + 2y = 17.$ $\quad y = 1.$

74. Use of the tables of logarithms of the trigonometric functions. On p. 9 the values of the trigonometric functions of angles from 0° to 90° were given in tabulated form. When we are using logarithms in calculating expressions involving these trigonometric functions it saves much labor to have the logarithms of these functions already looked up for us and arranged in tabulated form.* Two complete sets of such logarithms of the trigonometric functions are given. Table II, pp. 8–16, should be used when the given or required angle is expressed in degrees, minutes, and the decimal part of a minute; and Table III, pp. 20–37, when the given or required angle is expressed in degrees, and the decimal part of a degree.† In both tables the following directions hold true:

Angles between 0° and 45° are in the extreme *left-hand* column on each page,‡ and the logarithm of the function of any angle will be found in the same horizontal row with it and in the vertical column with the name of the function at the top; that is, sines in the first column, tangents in the second, cotangents in the third, and cosines in the fourth, counting from left to right.

---

* To distinguish between the two kinds of tables, that on p. 9 is called a Table of Natural Functions, while the logarithms of these functions arranged in tabulated form is called a Table of Logarithmic Functions.

† The division of the degree into decimal parts, instead of using minutes and seconds, has much to recommend it theoretically, and is also regarded with favor by many expert computers. In fact, a movement towards the adoption of such a system of subdivision is not only gaining headway in France and Germany, but is making itself felt in America.

‡ The angles increase as we read downwards.

Angles between 45° and 90° are in the extreme *right-hand* column on each page,* and the logarithm of the function of any angle will be found in the same horizontal row with it and in the vertical column with the name of the function at the bottom; that is, cosines in the first column, cotangents in the second, tangents in the third, and sines in the fourth, counting from left to right.

In order to avoid the printing of negative characteristics, the number 10 has been added to every logarithm in the first, second, and fourth columns (those having log sin, log tan, and log cos at the top). Hence in writing down any logarithm taken from these three columns — 10 should be written after it. Logarithms taken from the third column, having "log cot" at the top, should be used as printed. Thus,

$$\log \sin 38°\ 30' = 9.7941 - 10 = 1.7941. \qquad \text{p. 16}$$
$$\log \cot\ \ 0°\ 10' = 2.5363 \qquad\quad = 2.5363. \qquad\quad \text{p. 8}$$
$$\log \tan 75.6°\ \ = 0.5905 \qquad\quad = 0.5905. \qquad\quad \text{p. 31}$$
$$\log \cos\ \ 2.94° = 9.9994 - 10 = \bar{1}.9994. \qquad \text{p. 25}$$

75. Use of Table II, pp. 8–16, the given or required angle being expressed in degrees and minutes.† This table gives the logarithms of the sines, cosines, tangents, and cotangents of all angles from 0° to 5° and from 85° to 90° for each minute on pp. 8–12; and on pp. 13–16, from 5° to 85° at intervals of 10 minutes.

The small columns headed "diff. 1'" immediately to the right of the columns headed "log sin" and "log cos" contain the differences, called tabular differences, in the logarithms of the sines and cosines corresponding to a difference of 1' in the angle. Similarly, the small column headed "com. diff. 1'" contains the tabular differences for both tangent and cotangent corresponding to a difference of 1' in the angle. It will be observed that any tabular difference is not in the same horizontal row with a logarithm, but midway between the two particular logarithms whose difference it is. Of course that tabular difference should always be taken which corresponds to the interval in which the angle in question lies. Thus, in finding log sin 78° 16', the tabular difference corresponding to the interval between 78° 10' and 78° 20' is 6.1.

---

* The angles increase as we read upwards.

† In case the given angle involves seconds, first reduce the seconds to the decimal part of a minute by dividing by 60. Thus,

$$88°\ 18'\ 42'' = 88°\ 18.7', \quad \text{since } 42'' = \tfrac{42}{60} = .7';$$
$$2°\ 0'\ 16'' = 2°\ 0.27', \quad \text{since } 16'' = \tfrac{16}{60} = .266' \cdots.$$

If the angle is given in degrees and the decimal parts of a degree, and it is desired to use Table II, the angle may be quickly found in degrees and minutes by making use of the Conversion Table on p. 17.

76. To find the logarithm of a function of an angle when the angle is expressed in degrees and minutes, use

RULE VI. *When the given angle is found exactly in Table II, the logarithm of the given function of the angle is immediately found in the same horizontal row and in the vertical column having the given function at the top when the angle is less than 45°, or at the bottom when the angle is greater than 45°.*

In case the given angle is not found exactly in the table *we should take the following steps :*

(a) Write down the logarithm of the same function of the next less angle found in the table, and also the corresponding tabular difference for 1'.

(b) To find the correction necessary, multiply this tabular difference by the excess in minutes of the given angle over the angle whose logarithm was written down.

(c) If sine or tangent, add
*   If cosine or cotangent, subtract* } *this correction.**

This rule, as well as the next three, assumes that the differences of the logarithms of functions are proportional to the differences of their corresponding angles. Unless the angle is very near 0° or 90°, this is in general sufficiently exact for most practical purposes.

Ex. 1. Find log tan 32° 30'.

Solution. On p. 15, Table II, we find the angle 32° 30' exactly; hence, by Rule VI, we get immediately from the table

$$\log \tan 32° 30' = 9.8042 - 10. \quad Ans.$$

Ex. 2. Find log cot 88° 17'.

Solution. On p. 9, Table II, we find the angle 88° 17' exactly ; hence, by Rule VI, we get at once

$$\log \cot 88° 17' = 8.4767 - 10. \quad Ans.$$

Ex. 3. Find log sin 23° 26'.

Solution. The exact angle 23° 26' is not found in Table II; but then, by Rule VI, from p. 14,

$$
\begin{array}{ll}
\log \sin 23° 20' = 9.5978 - 10 & \text{Tab. diff.} = 2.9 \\
\text{corr. for } \quad 6' = \quad\underline{\quad 17 \quad} & \text{Excess } = \underline{\quad 6} \\
\log \sin 23° 26' = \overline{9.5995 - 10}. \; Ans. & \text{Corr. } \quad = 17.4
\end{array}
$$

---

* The sine and tangent increase as the angle increases, hence we add the correction ; the cosine and cotangent, however, decrease as the angle increases, hence we subtract the correction. Of course this is true only for acute angles.

Ex. 4. Find log cos 54° 42′ 18″.

Solution. Since 18″ is less than half a minute, we drop it, and from p. 16, Table II, by Rule VI,

$$\begin{aligned}
\log \cos 54° 40' &= 9.7622 - 10 \\
\text{corr. for } 2' &= \quad 4 \\
\log \cos 54° 42' &= 9.7618 - 10. \quad \textit{Ans.}
\end{aligned}$$

Tab. diff. = 1.8
Excess   = 2
Corr.   = 3.6
i.e. = 4

Ex. 5. Find log cot 1° 34.42′.

Solution. From p. 9, Table II, by Rule VI,

$$\begin{aligned}
\log \cot 1° 34' &= 1.5630 \\
\text{corr. for } .4' &= \quad 18 \\
\log \cot 1° 34.4' &= 1.5612. \quad \textit{Ans.}
\end{aligned}$$

Tab. diff. = 46
Excess   = .4
Corr.   = 18.4

When the angles are given in the table at intervals of 10′, it is only necessary to take our angle to the nearest minute, while if the angles are given for every minute, we take our angle to the nearest tenth of a minute. Thus, in Ex. 4, we find cos 54° 42′, dropping the seconds; and in Ex. 5 we find log cot 1° 34.4′, dropping the final 2.

Ex. 6. Verify the following:

(a) log tan 35° 50′ = 9.8586 − 10.
(b) log sin 61° 58′ = 9.9458 − 10.
(c) log tan 82° 3′ 20″ = .8550.
(d) log cos 44° 32′ 50″ = 9.8528 − 10.
(e) log tan 1° 53.2′ = 8.5177 − 10.
(f) log tan 87° 15.6′ = 1.3201.
(g) log cos 27° 28′ = 9.9480 − 10.
(h) log cot 51° 49′ = 9.8957 − 10.
(i) log sin 85° 57′ = 9.9989 − 10.
(j) log cot 45° 0′ 13″ = 0.0000.
(k) log sin 120° 24.3′ = 9.9358 − 10.
(l) log tan 243° 42′ 15″ = 0.3060.

77. To find the acute angle in degrees and minutes which corresponds to a given logarithmic function, use

RULE VII. *When the given logarithmic function is found exactly in Table II, then the corresponding angle is immediately found in the same horizontal row, to the left if the given function is written at the top of the column, and to the right if at the bottom.*

In case the given logarithmic function is not found exactly in the table *we should take the following steps:*

(a) Write down the angle corresponding to the next less logarithm of the same function found in the table, and also the corresponding tabular difference for 1′.

(b) To find the necessary correction in minutes divide this tabular difference into the excess of the given logarithmic function over the one written down.

(c) If sine or tangent, add
*   If cosine or cotangent, subtract* } *this correction.* *

* See footnote, p. 143.

In searching the table for the given logarithm, attention must be paid to the fact that the functions are found in different columns according as the angle is less or greater than 45°. If, for example, the logarithmic sine is found in the column with "log sin" at the top, the degrees and minutes must be taken from the *left-hand* column, but if it is found in the column with "log sin" at the bottom, the degrees and minutes must be taken from the *right-hand* column. Similarly, for the other functions. Thus, if the logarithmic cosine is given, we look for it in two columns on each page, the one having "log cos" at the top and also the one having "log cos" at the bottom.

Ex. 7. Find the angle whose log tan = 9.6946 − 10.

Solution. This problem may also be stated as follows: having given log tan x = 9.6946 − 10; to find the angle x. Looking up and down the columns having "log tan" at top or bottom, we find 9.6946 exactly on p. 15, Table II, in the column with "log tan" at top. The corresponding angle is then found in the same horizontal row to the left and is $x = 26°\,20'$.

Ex. 8. Find the angle whose log sin = 9.6652 − 10.

Solution. That is, having given log sin x = 9.6652 − 10; to find the angle x. Looking up and down the columns having "log sin" at top or bottom, we do not find 9.6652 exactly; but (Rule VII) the next less logarithm in such a column is found on p. 15, Table II, to be 9.6644, which corresponds to the angle 27° 30′, and the corresponding tabular difference for 1′ is 2.4. Hence

$$
\begin{array}{rl}
\log \sin x = & 9.6652 - 10 \\
\log \sin 27°\,30' = & \underline{9.6644 - 10} \\
\text{excess} = & 8
\end{array}
\qquad
\begin{array}{c|c|c}
\text{Tab. diff. 1}' & \text{Excess} & \text{Corr.} \\
 & 8.0 & \underline{\quad 3\quad} \\
\hline
 & \dfrac{72}{8} &
\end{array}
$$

Since the function involved is the sine, we add this correction, giving

$$x = 27°\,30' + 3' = 27°\,33'. \quad \textit{Ans.}$$

Ex. 9. Find the angle whose log cos = 9.3705 − 10.

Solution. That is, having given log cos x = 9.3705 − 10; to find the angle x. Looking up and down the columns having "log cos" at top or bottom, we do not find 9.3705 exactly; but (Rule VII) the next less logarithm in such a column is found on p. 13, Table II, to be 9.3682, which corresponds to the angle 76° 30′, and the corresponding tabular difference for 1′ is 5.2. Hence

$$
\begin{array}{rl}
\log \cos x = & 9.3705 - 10 \\
\log \cos 76°\,30' = & \underline{9.3682 - 10} \\
\text{excess} = & 23
\end{array}
\qquad
\begin{array}{c|c|c}
\text{Tab. diff. 1}' & \text{Excess} & \text{Corr.} \\
5.2 & 23.0 & \underline{\quad 4\quad} \\
\hline
 & \dfrac{208}{22} &
\end{array}
$$

Since the function involved is the cosine, we subtract this correction, giving

$$x = 76°\,30' - 4' = 76°\,26'. \quad \textit{Ans.}$$

Ex. 10. Given $\log \tan x = 8.7570 - 10$; find x.

Solution. By Rule VII the next less logarithmic tangent is found on p. 11, Table II.

$$\log \tan x = 8.7570 - 10$$
$$\log \tan 3^\circ 16' = 8.7565 - 10$$
$$\text{excess} = 5$$

Tab. diff. 1'	Excess	Corr.
22	50	.2
	44	
	6	

Hence $\qquad x = 3^\circ 16' + .2' = 3^\circ 16.2'$. *Ans.*

Ex. 11. Given $\cot x = (1.01)^5$; find x.

Solution. Taking the logarithms of both sides,

$$\log \cot x = 5 \log 1.01. \qquad \text{Th. III, p. 122}$$

But $\qquad \log 1.01 = 0.0043$

and, multiplying by 5, $\qquad 5$

$$\log \cot x = 0.0215 \text{; to find } x.$$

By Rule VII the next less logarithmic cotangent is found on p. 16, Table II.

$$\log \cot x = 0.0215$$
$$\log \cot 43^\circ 40' = 0.0202$$
$$\text{excess} = 13$$

Tab. diff. 1'	Excess	Corr.
2.6	13.0	5
	13.0	

Hence $\qquad x = 43^\circ 40' - 5' = 43^\circ 35'$. *Ans.*

Ex. 12. Verify the following :

(a) If $\log \sin x = 9.5443 - 10$, then $x = 20^\circ 30'$.

(b) If $\log \cos x = 9.7531 - 10$, then $x = 55^\circ 30'$.

(c) If $\log \tan x = 9.9570 - 10$, then $x = 42^\circ 10'$.

(d) If $\log \cot x = 1.0034$, then $x = 5^\circ 40'$.

(e) If $\log \sin x = 8.0435 - 10$, then $x = 0^\circ 38'$.

(f) If $\log \cos x = 8.7918 - 10$, then $x = 86^\circ 27'$.

(g) If $\log \tan x = 9.5261 - 10$, then $x = 18^\circ 34'$.

(h) If $\log \cot x = 0.6380$, then $x = 12^\circ 58'$.

(i) If $\log \sin x = 9.9995 - 10$, then $x = 87^\circ 16'$.*

(j) If $\log \cos x = 8.2881 - 10$, then $x = 88^\circ 53.3'$.

(k) If $\log \tan x = 2.1642$, then $x = 89^\circ 36.4'$.

(l) If $\log \tan x = 7.9732 - 10$, then $x = 0^\circ 32.3'$.

(m) If $\log \sin x = 9.8500 - 10$, then $x = 45^\circ 4'$.

(n) If $\log \cos x = 9.9000 - 10$, then $x = 37^\circ 25'$.

(o) If $\log \tan x = 0.0035$, then $x = 45^\circ 14'$.

(p) If $\log \cot x = 1.0000$, then $x = 5^\circ 43'$.

(q) If $\log \cot x = 3.9732$, then $x = 89^\circ 27.7'$.

---

* When there are several angles corresponding to the given logarithmic function, we choose the middle one.

EXAMPLES

Use logarithms when making the calculations in the following examples:

1. Given $184 \sin^3 x = (12.03)^2 \cos 57° 20'$; find x.

Solution. First we solve for $\sin x$, giving

$$\sin x = \sqrt[3]{\frac{(12.03)^2 \cos 57° 20'}{184}}.$$

Taking the logarithms of both sides,

$$\log \sin x = \tfrac{1}{3} [2 \log 12.03 + \log \cos 57° 20' + \operatorname{colog} 184].$$

$2 \log 12.03 = \ 2.1606$	since $\log 12.03 = 1.0803$
$\log \cos 57° 20' = \ 9.7322 - 10$	
$\operatorname{colog} 184 = \ \underline{7.7352 - 10}$	since $\log 184 = 2.2648$
$19.6280 - 20$	
$3\,|\,29.6280 - 30$	
$\log \sin x = \ 9.8760 - 10$	
$\therefore \ x = 48° 44'. \ Ans.$	

2. Given $\cos x = (.9854)^{\frac{1}{3}}$; find x. *Ans.* $77° 37'$.

3. Calculate $\dfrac{4.236 \cos 52° 19'}{13.087 \sin 48° 5'}$. 2.659.

4. Given $1.5 \cot 82° = x^2 \sin 12° 15$, find x. $.9968$.

Hint. First solve for x, giving

$$x = \sqrt{\frac{1.5 \cot 82°}{\sin 12° 15'}}.$$

5. Given $50 \tan x = \sqrt[4]{.2584}$; find x. $0° 49'$.

6. Calculate $\dfrac{\sin 24° 13' \cot 58° 2'}{\cos 33° 17' \tan 19° 58'}$. $.8426$.

7. Calculate $\sqrt{\cos 10° 5' \tan 73° 11'}$. 1.805.

8. Calculate $\dfrac{(\sin 33° 18')^3 \sqrt{\cot 71° 20'}}{10.658 \tan 63° 54'}$. $.04422$.

9. Given $3 \cot x = \sqrt[5]{.7}$; find x. $72° 45'$.

10. Given $\sin x = (.9361)^{10}$; find x. $31° 6'$.

11. Given $2.3 \tan x = (1.002)^{125}$; find x. $29° 24'$.

78. Use of Table III, pp. 20–37, the given or required angle being expressed in degrees and the decimal part of a degree.* This table gives, on pp. 20–29, the logarithms of the sines, cosines, tangents, and cotangents of all angles from 0° to 5°, and from 85° to 90° for every hundredth part of a degree; and on pp. 30–37 from 5° to 85° for every tenth of a degree.

The tabular differences between the logarithms given in the table are given in the same manner as were the tabular differences in Table II, and the general arrangement is the same.

---

* In case the angle is given in degrees, minutes, and seconds, and it is desired to use Table III, we may quickly reduce the angle to degrees and the decimal part of a degree by using the Conversion Table on p. 17.

79. To find the logarithm of the function of an angle when the angle is expressed in degrees and the decimal part of a degree, use

RULE VIII. *When the given angle is found exactly in Table III, the logarithm of the given function of the angle is immediately found in the same horizontal row and in the vertical column having the given function at the top when the angle is less than 45°, or at the bottom when the angle is greater than 45°.*

In case the given angle is not found exactly in the table *we should take the following steps :*

*(a) Write down the logarithm of the same function of the next less angle * found in the table and note the tabular difference which follows.*

(b) In the Prop. Parts column locate the block corresponding to this tabular difference. Under this difference and opposite the extra digit of the given angle will be found the proportional part of the tabular difference (that is, the correction).

(c) If sine or tangent, add
If cosine or cotangent, subtract } *this correction.†*

Ex. 1. Find log sin 27.4°.

Solution. On p. 34, Table III, we find the angle 27.4° exactly; hence, by Rule VIII, we get at once

$$\log \sin 27.4° = 9.6629 - 10. \quad Ans.$$

Ex. 2. Find log cot 3.17°.

Solution. On p. 26, Table III, we find the angle 3.17° exactly; hence, by Rule VIII, we get immediately from the table

$$\log \cot 3.17° = 1.2566. \quad Ans.$$

Ex. 3. Find log tan 61.87°.

Solution. The exact angle 61.87° is not found in our tables. But then, by Rule VIII, the next less angle is 61.8°, the extra digit of the given angle being 7, and we have, from p. 34, Table III,

$$\log \tan 61.8° = 10.2707 - 10.$$

The tabular difference between log tan 61.8° and log tan 61.9° is 18. In the Prop. Parts column under 18 and opposite the extra digit 7 we find the proportional part 12.6 (= 13). Then

$$\log \tan 61.80° = 0.2707$$
$$13 \quad \text{Prop. Part.}$$
$$\log \tan 61.87° = 0.27\overline{2}0. \quad Ans.$$

---

* This "next less angle" will not contain the last (extra) digit of the given angle.
† See footnote, p. 143.

Ex. 4. Find log cot 2.158°.

Solution. The exact angle 2.158° is not found in our tables. But then, by Rule VIII, the next less angle is 2.15°, the extra digit of the given angle being 8, and we have, from p. 24, Table III,

$$\log \cot 2.15° = 1.4255.$$

The tabular difference between log cot 2.15° and log cot 2.16° is 20. In the Prop. Parts column under 20 and opposite the extra digit 8 we find the proportional part 16. Then

$$\log \cot 2.150° = 1.4255$$
$$\underline{16} \quad \text{Prop. Part.}$$
$$\log \cot 2.158° = 1.4239. \quad \textit{Ans.}$$

Ex. 5. Verify the following:

(a) log tan 37.6° = 9.8865 − 10.

(b) log sin 63.87° = 9.9532 − 10.

(c) log cot 1.111° = 1.7123.

(d) log sin 0.335° = 7.7669 − 10.

(e) log cos 45.68° = 9.8443 − 10.

(f) log tan 3.867° = 8.8299 − 10.

(g) log tan 88.564° = 1.6009.

(h) log cos 20.03° = 9.9729 − 10.

(i) log sin 89.97° = 0.0000.

(j) log cot 34.84° = 0.1574.

(k) log sin 155.42° = 9.6191 − 10.

(l) log tan 196.85° = 9.4813 − 10.

80. To find the acute angle in degrees and decimal parts of a degree which corresponds to a given logarithmic function, use

RULE IX. *When the given logarithmic function is found exactly in Table III, then the corresponding angle is immediately found in the same horizontal row ; to the left, if the given function is written at top of the column, and to the right if written at the bottom.*

In case the given logarithmic function is not found exactly in the table we should take the following steps :

(a) Locate the given logarithm between two of the logarithms of the same function given in the tables.

(b) The lesser angle of the two angles corresponding to these logarithms will be the required angle complete except for the last digit. Write this angle down with the corresponding logarithmic function.

(c) Find the difference between the logarithm just written down and the given logarithm, also noting the corresponding tabular difference in the table.

(d) In the Prop. Parts column, under this tabular difference, pick out the proportional part nearest the difference found in (c), and to the left of it will be found the last (extra) digit of the required angle, which we now annex.

Ex. 6. Having given log tan x = 9.5364 − 10; to find the angle x.

Solution. Looking up and down the columns having "log tan" at top or bottom, we do not find 9.5364 exactly. But then, by Rule IX, we locate it between 9.5345 and 9.5370, on p. 32, Table III. Except for the last digit the required angle will be the lesser of the two corresponding angles, that is, 18.9°. Then

$$\begin{aligned} \log \tan 18.9° &= 9.5345 - 10 \\ \log \tan x &= \underline{9.5364 - 10} \\ & \quad\quad 19 = \text{difference.} \end{aligned}$$

The corresponding tabular difference being 25, we find in the Prop. Parts column that 20 is the proportional part under 25 which is nearest 19. To the left of 20 is the last (extra) digit 8 of the required angle. Hence x = 18.98°. *Ans.*

Ex. 7. Having given log cos x = 8.6820 − 10; find x.

Solution. On p. 25, Table III, we locate 8.6820 between 8.6810 and 8.6826. Except for the last digit, the required angle must be the lesser of the two corresponding angles, that is, 87.24°. Then

$$\begin{aligned} \log \cos 87.24° &= 8.6826 - 10 \\ \log \cos x &= \underline{8.6820 - 10} \\ & \quad\quad 6 = \text{difference.} \end{aligned}$$

The corresponding tabular difference being 16, we find in the Prop. Parts column that 6.4 is the proportional part under 16 which is nearest 6. To the left of 6.4 is the last (extra) digit 4 of the required angle. Hence x − 87.244°. *Ans.*

Ex. 8. Verify the following:

(a) If log sin x = 9.6371 − 10, then x = 25.7°.

(b) If log cos x = 9.9873 − 10, then x = 76.2°.

(c) If log tan x = 8.9186 − 10, then x = 4.74°.

(d) If log cot x = 1.1597, then x = 3.96°.

(e) If log sin x = 9.5052 − 10, then x = 18.67°.

(f) If log cos x = 9.9629 − 10, then x = 23.35°.

(g) If log tan x = 9.8380 − 10, then x = 34.55°.

(h) If log cot x = 9.3361 − 10, then x = 77.77°.

(i) If log sin x = 8.6852 − 10, then x = 2.776°.

(j) If log cos x = 9.9995 − 10, then x = 2.74°.

(k) If log tan x = 7.2642 − 10, then x = 0.105°.

(l) If log cot x = 1.7900, then x = 0.929°.

(m) If log sin x = 9.5350 − 10, then x = 20.05°.

(n) If log cos x = 9.8000 − 10, then x = 50.88°.

(o) If log tan x = 0.0035, then x = 45.23°.

(p) If log cot x = 2.0000, then x = 0.573°.

(q) If log sin x = 0.0000, then x = 90°.

(r) If log tan x = 0.0000, then x = 45°.

EXAMPLES

Use logarithms when making the calculations in the following examples:

1. Given $\tan x = (1.018)^{12}$; find x.

Solution. Taking the logarithms of both sides,

$$\log \tan x = 12 \log 1.018. \qquad \text{Th. III, p. 122}$$

But $\qquad\qquad\qquad \log 1.018 = 0.0077$
and, multiplying by 12, $\qquad\qquad\quad \underline{12}$
$$\log \tan x = \overline{0.0924}$$

On p. 36 we locate 0.0924 between 0.0916 and 0.0932. Then

$$\log \tan 51.0° = 0.0916$$
$$\log \tan x = \underline{0.0924}$$
$$8 = \text{difference.}$$

The tabular difference is 16. In the Prop. Parts column under 16 we find 8.0 exactly. To the left of 8.0 we find the last digit 5 of the required angle. Hence $x = 51.05°$. *Ans.*

2. Given $56.4 \tan^3 x = (18.65)^5 \cos 69.8°$; find x.

Solution. First we solve for $\tan x$, giving

$$\tan x = \sqrt[3]{\frac{(18.65)^5 \cos 69.8°}{56.4}}.$$

Taking the logarithms of both sides,

$$\log \tan x = \tfrac{1}{3} [5 \log 18.65 + \log \cos 69.8° + \text{colog } 56.4].$$

$$
\begin{array}{ll}
5 \log 18.65 = & 6.3535 \qquad\qquad \text{since } \log 18.65 = 1.2707 \\
\log \cos 69.8° = & 9.5382 - 10 \\
\text{colog } 56.4 = & \underline{8.2487 - 10} \qquad\quad \text{since } \log 56.4 = 1.7513 \\
& 24.1404 - 20 \\
& 5 \mid 54.1404 - 50 \\
\log \tan x = & 10.8281 - 10. \\
\therefore\ x = & 81.55°. \ Ans.
\end{array}
$$

3. Given $\cos x = \sqrt{.9681}$; find x. $\qquad\qquad\qquad\qquad\qquad$ *Ans.* 10.25°.

4. Calculate $\dfrac{26.52 \tan 33.86°}{100.85 \cot 88.963°}$. $\qquad\qquad\qquad\qquad\qquad$ 9.745.

5. Given $\sqrt{3} \sin 48.06° = x^3 \cos 2.143°$; find x. $\qquad\qquad$ 1.0885.

Hint. First solve for x, giving

$$x = \sqrt[3]{\frac{\sqrt{3} \sin 48.06°}{\cos 2.143°}}.$$

6. Given $5 \cot x = \sqrt[3]{.4083}$; find x. $\qquad\qquad\qquad\qquad$ 81.56°.

7. Given $\sin x = \dfrac{\sqrt{83} \cos 52.82°}{(13.382)^2}$, find x. $\qquad\qquad$ 0.557°.

8. Calculate $\sqrt{361} \tan 87.5° \sin 9.53°$. $\qquad\qquad\qquad\qquad$ 37.

81. Use of logarithms in the solution of right triangles. Since the solutions of right triangles involve the calculation of products and quotients, time and labor may be saved by using logarithms in the computations. From p. 7 we have the following:

General directions for solving right triangles.

First step. *Draw a figure as accurately as possible representing the triangle in question.*

Second step. *When one acute angle is known, subtract it from 90° to get the other acute angle.*

Third step. *To find an unknown part, select from (1) to (6), p. 2, a formula involving the unknown part and two known parts, and then solve for the unknown part.**

Fourth step. *Check the values found by seeing whether they satisfy relations different from those already employed in the third step. A convenient numerical check is the relation*

$$a^2 = c^2 - b^2 = (c + b)(c - b).\dagger$$

Large errors may be detected by measurement.

For reference purposes we give the following formulas from p. 8 and p. 11.

$$\text{Area of a right triangle} = \frac{ab}{2}\ddagger.$$

(7) Side opposite an acute angle = hypotenuse $\times$ sine of the angle.

(8) Side adjacent an acute angle = hypotenuse $\times$ cosine of the angle.

(9) Side opposite an acute angle = adjacent side $\times$ tangent of the angle.

It is best to compute the required parts of any triangle as far as possible from the given parts, so that an error made in determining one part will not affect the computation of the other parts.

---

* This also includes formulas (7), (8), (9), on p. 11.

† When we want the hypotenuse, the other two sides being given, this formula is not well adapted to logarithmic computation, since

$$c = \sqrt{a^2 + b^2},$$

and we have a summation under the radical that cannot be performed by the use of our logarithmic tables. If, however, we have the hypotenuse c and one side (as b) given to find the other side a, then

$$a = \sqrt{c^2 - b^2} = \sqrt{(c - b)(c + b)},$$

and we have a product under the radical. The factors $c - b$ and $c + b$ of this product are easily calculated by inspection, and then we can use logarithms advantageously. Thus

$$\log a = \tfrac{1}{2}[\log(c - b) + \log(c + b)].$$

‡ In case a or b is not given, or both a and b are not given, we first find what we need from the known parts, as when solving the triangle, so that we can use the above formula for finding the area.

In trigonometric computations it sometimes happens that the unknown quantity may be determined in more than one way. When choosing the method to be employed it is important to keep in mind the following suggestions :

(*a*) *An angle is best determined from a trigonometric function which changes rapidly, that is, one having large tabular differences, as the tangent or cotangent.*

(*b*) *When a number is to be found (as the side of a triangle) from a relation involving a given angle, it is best to employ a trigonometric function of the angle which changes slowly, as the sine or cosine.*

As was pointed out on pp. 13, 14, the solution of isosceles triangles and regular polygons depends on the solution of right triangles.

The following examples will illustrate the best plan to follow in solving right triangles by the aid of logarithms.

Ex. 1. Solve the right triangle if $A = 48° 17'$, $c = 324$. Also find the area.

Solution. *First step.* Draw a figure of the triangle indicating the known and unknown parts.

Second step. $\qquad\qquad\qquad B = 90° - A = 41° 43'.$

Third step. To find a use $\qquad a = c \sin A.$ $\qquad\qquad$ by (7), p. 11

Taking the logarithms of both sides,

$$\log a = \log c + \log \sin A.$$

Hence, from Tables I and II,*

$$\begin{aligned}
\log c &= 2.5105 \\
\log \sin A &= 9.8730 - 10 \\
\log a &= \overline{12.3835 - 10} \\
&= 2.3835. \\
\therefore\ a &= 241.8.
\end{aligned}$$

To find b use $\qquad b = c \cos A.$ $\qquad\qquad$ by (8), p. 11

Taking the logarithms of both sides,

$$\log b = \log c + \log \cos A.$$

Hence, from Tables I and II,

$$\begin{aligned}
\log c &= 2.5105 \\
\log \cos A &= 9.8231 - 10 \\
\log b &= \overline{12.3336 - 10} \\
&= 2.3336 \\
\therefore\ b &= 215.6.
\end{aligned}$$

* If we wish to use Table III instead of Table II, we reduce 17′ to the decimal of a degree. Thus, $\qquad\qquad A = 48° 17' = 48.28°.$

Fourth step. To check these results numerically, let us see if a, b, c satisfy the equation

$$a^2 = c^2 - b^2 = (c + b)(c - b),$$

or, using logarithms,

$$2 \log a = \log (c + b) + \log (c - b),$$

that is,

$$\log a = \tfrac{1}{2} [\log (c + b) + \log (c - b)].$$

Here $c + b = 539.6$ and $c - b = 108.4$.

$$\log (c + b) = 2.7321$$
$$\log (c - b) = 2.0350$$
$$2 \log a = 4.7671$$

$$\log a = 2.3835.$$

Since this value of $\log a$ is the same as that obtained above, the answers are probably correct.

To find the area use formula

$$\text{Area} = \frac{ab}{2}.$$

$$\log \text{area} = \log a + \log b - \log 2.$$

$$\log a = 2.3835$$
$$\log b = 2.3336$$
$$4.7171$$
$$\log 2 = 0.3010$$
$$\log \text{area} = 4.4161$$

$$\therefore \text{area} = 26{,}070.$$

Ex. 2. Solve the right triangle, having given $b = 15.12$, $c = 30.81$.

Solution. Here we first find an acute angle ; to find A use

$$\cos A = \frac{b}{c}. \qquad \text{(2), p. 2}$$

$$\log \cos A = \log b - \log c.$$

$$\log b = 11.1796 - 10$$
$$\log c = \underline{1.4887}$$
$$\log \cos A = 9.6909 - 10$$

$$\therefore A = 60° 36'. \qquad \text{from Table II, p. 15}$$

Hence

$$B = 90° - A = 29° 24'.$$

To find a we may use

$$a = b \tan A. \qquad \text{by (9), p. 11}$$
$$\log a = \log b + \log \tan A.$$

$$\log b = 1.1796.$$
$$\log \tan A = 0.2491$$
$$\log a = 1.4287$$

$$\therefore a = 26.84.$$

To check the work numerically, take

$$a^2 = (c + b)(c - b),$$

or,
$$\log a = \tfrac{1}{2}\,[\log (c + b) + \log (c - b)].$$

Here $c + b = 45.93$ and $c - b = 15.69$.

$$\log (c + b) = 1.6621$$
$$\log (c - b) = \underline{1.1956}$$
$$2 \log a = 2.8578$$

$$\log a = 1.4288.$$

This we see agrees substantially with the above result.

Ex. 3. Solve the right triangle, having given $B = 2.325°$, $a = 1875.3$.

Solution.
$$A = 90° - B = 87.675°.$$

$$\sin A = \frac{a}{c}.$$ by (1), p. 2

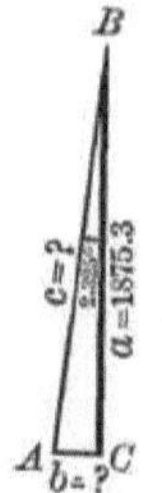

Solving for the unknown side c,

$$c = \frac{a}{\sin A}.$$
$$\log c = \log a - \log \sin A.$$

Hence, from Tables I and III,*

$$\log a = 13.2731 - 10$$
$$\log \sin A = \underline{9.9996 - 10}$$
$$\log c = 3.2735$$

$$\therefore\ c = 1877.$$

$$\tan A = \frac{a}{b}.$$ by (3), p. 2

Solving for the unknown side b,

$$b = \frac{a}{\tan A}.$$
$$\log b = \log a - \log \tan A.$$

$$\log a = 13.2731 - 10$$
$$\log \tan A = \underline{11.3915 - 10}$$
$$\log b = 1.8816$$

$$\therefore\ b = 76.13.$$

To check the work we may use formulas

$$a^2 = (c + b)(c - b),$$

or,
$$b = c \sin B,$$ by (7), p. 11

since neither one was used in the above calculations.

---

* If we wish to use Table II instead of Table III, we reduce $2.325°$ to degrees and minutes. Thus, $B = 2.325° = 2° 19.5'$.

EXAMPLES

Solve the following right triangles ($C = 90°$), using logarithmic Tables I and II.*

No.	Given Parts		Required Parts		
1	$A = 43°\ 30'$	$c = 11.2$	$B = 46°\ 30'$	$a = 7.709$	$b = 8.124$
2	$B = 68°\ 50'$	$a = 729.3$	$A = 21°\ 10'$	$b = 1883.5$	$c = 2019.5$
3	$B = 62°\ 56'$	$b = 47.7$	$A = 27°\ 4'$	$a = 24.37$	$c = 53.56$
4	$a = .624$	$c = .91$	$A = 43°\ 18'$	$B = 46°\ 42'$	$b = .6623$
5	$A = 72°\ 7'$	$a = 83.4$	$B = 17°\ 53'$	$b = 26.91$	$c = 87.64$
6	$b = 2.887$	$c = 5.11$	$B = 34°\ 24'$	$A = 55°\ 36'$	$a = 4.216$
7	$A = 52°\ 41'$	$b = 4247$	$B = 37°\ 19'$	$a = 5571$	$c = 7007$
8	$a = 101$	$b = 116$	$A = 41°\ 2'$	$B = 48°\ 58'$	$c = 153.8$
9	$A = 43°\ 22'$	$a = 158.3$	$B = 46°\ 38'$	$b = 167.6$	$c = 230.5$
10	$a = 204.2$	$c = 275.3$	$A = 47°\ 53'$	$B = 42°\ 7'$	$b = 184.7$
11	$B = 10°\ 51'$	$c = .7264$	$A = 79°\ 9'$	$a = .7133$	$b = .1367$
12	$a = 638.5$	$b = 501.2$	$A = 51°\ 53'$	$B = 38°\ 7'$	$c = 811.7$
13	$b = .02497$	$c = .04792$	$A = 58°\ 36'$	$B = 31°\ 24'$	$a = .0409$
14	$B = 2°\ 19'\ 30''$	$a = 1875.3$	$A = 87°\ 40'\ 30''$	$b = 76.13$	$c = 1877$
15	$B = 21°\ 33'\ 51''$	$a = .8211$	$A = 68°\ 26'\ 9''$	$b = .3246$	$c = .8826$
16	$A = 74°\ 0'\ 18''$	$c = 275.62$	$B = 15°\ 59'\ 42''$	$a = 264.9$	$b = 75.9$
17	$B = 34°\ 14'\ 37''$	$b = 120.22$	$A = 55°\ 45'\ 23''$	$a = 176.57$	$c = 213.6$
18	$a = 10.107$	$b = 17.303$	$A = 30°\ 18'$	$B = 59°\ 42'$	$c = 20.04$
19	$a = 24.67$	$b = 33.02$	$A = 36°\ 46'$	$B = 53°\ 14'$	$c = 41.22$
20	$A = 78°\ 17'$	$a = 203.8$	$B = 11°\ 43'$	$b = 42.27$	$c = 208.15$

21. Find areas of the first five of the above triangles.

$Ans.$ (1) 31.32; (2) 686,900; (3) 581.3; (4) .2067; (5) 1122.5.

Solve the following isosceles triangles where A, B, C are the angles and a, b, c the sides opposite respectively, a and b being the equal sides.

22. Given $A = 68°\ 57'$, $b = 35.09$. $Ans.$ $C = 42°\ 6'$, $c = 25.21$.

23. Given $B = 27°\ 8'$, $c = 3.088$. $Ans.$ $C = 125°\ 44'$, $a = 1.735$.

24. Given $C = 80°\ 47'$, $b = 2103$. $Ans.$ $A = 49°\ 36.5'$, $c = 2726$.

25. Given $a = 79.24$, $c = 106.62$. $Ans.$ $A = 47°\ 43'$, $C = 84°\ 34'$.

26. Given $C = 151°\ 28'$, $c = 95.47$. $Ans.$ $A = 14°\ 16'$, $a = 49.25$.

27. One side of a regular octagon is 24 ft.; find its area and the radii of the inscribed and circumscribed circles. $Ans.$ Area $= 2782$, $r = 28.97$, $R = 31.36$.

---

* For the sake of clearness and simplicity, one set of triangle examples is given which are adapted to practice in using Table II, the given and required angles being expressed in degrees and minutes; and another set is given on p. 157 for practice in the use of Table III, the given and required angles being expressed in degrees and the decimal part of a degree. There is no reason why the student should not work out the examples in the first set, using Table III, and those in the second set, using Table II, if he so desires, except that it may involve a trifle more labor. This extra work of reducing minutes to the decimal part of a degree, or the reverse, may be reduced to a minimum by making use of the Conversion Tables on p. 17. It is possible, however, that an answer thus obtained may differ from the one given here by one unit in the last decimal place. This practice of giving one set of triangle examples for each of the Tables II and III will be followed throughout this book when solving triangles.

Solve the following right triangles ($C = 90°$), using logarithmic Tables I and III.

No.	GIVEN PARTS		REQUIRED PARTS		
28	$a = 5$	$b = 2$	$A = 68.2°$	$B = 21.8°$	$c = 5.385$
29	$B = 32.17°$	$c = .02728$	$A = 57.83°$	$a = .02309$	$b = .01452$
30	$A = 58.65°$	$c = 35.73$	$B = 31.35°$	$a = 30.51$	$b = .18.59$
31	$A = 22.23°$	$b = 13.242$	$B = 67.77°$	$a = 5.413$	$c = 14.31$
32	$b = .02497$	$c = .04792$	$A = 58.6°$	$B = 31.4°$	$a = .0409$
33	$a = 273$	$b = 418$	$A = 33.15°$	$B = 56.85°$	$c = 499.3$
34	$B = 23.15°$	$b = 75.48$	$A = 66.85°$	$a = 176.5$	$c = 191.9$
35	$A = 31.75°$	$a = 48.04$	$B = 58.25°$	$b = 77.64$	$c = 91.29$
36	$b = 512$	$c = 900$	$A = 55.32°$	$B = 34.68°$	$a = 740.2$
37	$a = 52$	$c = 60$	$A = 60.06°$	$B = 29.94°$	$b = 29.94$
38	$A = 2.49°$	$a = .83$	$B = 87.51°$	$b = 19.085$	$c = 19.107$
39	$A = 88.426°$	$b = 9$	$B = 1.574°$	$a = 327.5$	$c = 327.6$
40	$B = 4.963°$	$b = .07$	$A = 85.037°$	$a = .8062$	$c = .8092$
41	$B = 85.475°$	$c = 80$	$A = 4.525°$	$a = 6.313$	$b = 79.74$
42	$a = 100.87$	$b = 2$	$A = 88.844°$	$B = 1.136°$	$c = 100.9$

43. Find the areas of the first five of the above triangles.

Ans. (28) 5; (29) .0001677; (30) 283.6; (31) 35.83; (32) .00051.

44. The perimeter of a regular polygon of 11 sides is 23.47 ft. Find the radius of the circumscribed circle. *Ans.* 3.79 ft.

45. Two stations are 3 mi. apart on a plain. The angle of depression of one from a balloon directly over the other is observed to be $8° 15'$. How high is the balloon? *Ans.* .435 mi.

46. A rock on the bank of a river is 130 ft. above the water level. From a point just opposite the rock on the other bank of the river the angle of elevation of the rock is $14° 30' 21''$. Find the width of the river. *Ans.* 502.5 ft.

47. A rope 38 ft. long tied to the top of a tree 29 ft. high just reaches the level ground. Find the angle the rope makes with the tree. *Ans.* $40° 15'$.

48. A man 5 ft. 10 in. high stands at a distance of 4 ft. 7 in. from a lamp-post, and casts a shadow 18 ft. long. Find the height of the lamp-post. *Ans.* 7.32 ft.

49. The shadow of a vertical cliff 113 ft. high just reaches a boat on the sea 93 ft. from its base. Find the altitude of the sun. *Ans.* $50° 33'$.

50. The top of a tree broken by the wind strikes the ground 15 ft. from the foot of the tree and makes an angle of $42° 28'$ with the ground. Find the original height of the tree. *Ans.* 34.07 ft.

51. A building is 121 ft. high. From a point directly across the street its angle of elevation is $65° 3'$. Find the width of the street. *Ans.* 56.3 ft.

52. Given that the sun's distance from the earth is 92,000,000 mi., and the angle it subtends from the earth is $32'$. Find diameter of the sun.

Ans. About 856,400 mi.

53. Given that the radius of the earth is 3963 mi., and that it subtends an angle of $57'$ at the moon. Find the distance of the moon from the earth.

Ans. About 239,100 mi.

54. The radius of a circle is 12,732, and the length of a chord is 18,321. Find the angle the chord subtends at the center. *Ans.* 92° 2′.

55. If the radius of a circle is 10 in., what is the length of a chord which subtends an angle of 77° 17′ 40″ at the center? *Ans.* 12.488 in.

56. The angle between the legs of a pair of dividers is 43°, and the legs are 7 in. long. Find the distance between the points. *Ans.* 5.13 in.

82. Use of logarithms in the solution of oblique triangles. As has already been pointed out, formulas involving principally products, quotients, powers, and roots are well adapted to logarithmic computation; while in the case of formulas involving in the main sums and differences, the labor-saving advantages of logarithmic computation are not so marked. Thus, in solving oblique triangles, the law of sines

$$\frac{a}{\sin A} = \frac{b}{\sin B} = \frac{c}{\sin C},$$

and the law of tangents

$$\tan \tfrac{1}{2}(A - B) = \frac{a - b}{a + b} \tan \tfrac{1}{2}(A + B),$$

are well adapted to the use of logarithms, while this is not the case with the law of cosines, namely,

$$a^2 = b^2 + c^2 - 2\,bc \cos A.$$

In solving oblique triangles by logarithmic computation, it is convenient to classify the problems as follows:

CASE I. *When two angles and a side are given.*

CASE II. *When two sides and the angle opposite one of them are given (ambiguous case).*

CASE III. *When two sides and included angle are given.*

CASE IV. *When all three sides are given.*

CASE I. **When two angles and a side are given.**

First step. *To find the third angle, subtract the sum of the two given angles from 180°.*

Second step. *To find an unknown side, choose a pair of ratios from the law of sines*

$$\frac{a}{\sin A} = \frac{b}{\sin B} = \frac{c}{\sin C},$$

which involve only one unknown part, and solve for that part.

Check: See if the sides found satisfy the law of tangents.

Ex. 1. Having given $b = 20$, $A = 104°$, $B = 19°$; solve the triangle.

Solution. Drawing a figure of the triangle on which we indicate the known and unknown parts, we see that the problem comes under Case I.

First step. $C = 180° - (A + B) = 180° - 123° = 57°.$

Second step. Solving $\dfrac{a}{\sin A} = \dfrac{b}{\sin B}$ for a, we get

$$a = \frac{b \sin A}{\sin B},$$

or, $\qquad \log a = \log b + \log \sin A - \log \sin B.$

$$
\begin{aligned}
\log b =&\ \ 1.3010 \\
\log \sin A =&\ \ 9.9869 - 10 \ {*} \\
 &\ 11.2879 - 10 \\
\log \sin B =&\ \ 9.5126 - 10 \\
\log a =&\ \ 1.7753
\end{aligned}
$$

$$a = 59.61.$$

Solving $\dfrac{b}{\sin B} = \dfrac{c}{\sin C}$ for c, we get

$$c = \frac{b \sin C}{\sin B},$$

or, $\qquad \log c :: \log b + \log \sin C - \log \sin B.$

$$
\begin{aligned}
\log b =&\ \ 1.3010 \\
\log \sin C =&\ \ 9.9236 - 10 \\
 &\ 11.2246 - 10 \\
\log \sin B =&\ \ 9.5126 - 10 \\
\log c =&\ \ 1.7120
\end{aligned}
$$

$$c = 51.52.$$

Check: $\qquad a + c = 111.13, \qquad a - c = 8.09\,;$

$\qquad\qquad\qquad A + C = 161°, \qquad A - C = 47°\,;$

$\qquad\qquad\qquad \tfrac{1}{2}(A + C) = 80° \, 30', \quad \tfrac{1}{2}(A - C) = 23° \, 30'.$

Here, $\qquad \tan \tfrac{1}{2}(A - C) = \dfrac{a - c}{a + c} \tan \tfrac{1}{2}(A + C),$

or, $\qquad \log \tan \tfrac{1}{2}(A - C) = \log (a - c) + \log \tan \tfrac{1}{2}(A + C) - \log (a + c).$

$$
\begin{aligned}
\log (a - c) =&\ \ 0.9079 \\
\log \tan \tfrac{1}{2}(A + C) =&\ \ 10.7764 - 10 \\
 &\ 11.6843 - 10 \\
\log (a + c) =&\ \ 2.0458 \\
\log \tan \tfrac{1}{2}(A - C) &\ \ 9.6385 \cdot 10
\end{aligned}
$$

$$\therefore \ \tfrac{1}{2}(A - C) = 23° \, 31',$$

which substantially agrees with the above results.

---

* $\sin A = \sin 104° = \sin (180° - 104°) = \sin 76°$. Hence $\log \sin 104° = \log \sin 76° = 9.9869 - 10$.

EXAMPLES

Solve the following oblique triangles, using logarithmic Tables I and II.

No.	Given Parts			Required Parts		
1	$a=10$	$A=38°$	$B=77°10'$	$C=64°50'$	$b=15.837$	$c=14.703$
2	$a=795$	$A=79°59'$	$B=44°41'$	$C=55°20'$	$b=567.6$	$c=664$
3	$b=.8037$	$B=52°20'$	$C=101°40'$	$A=26°$	$a=.445$	$c=.9942$
4	$c=.032$	$A=36°8'$	$B=44°27'$	$C=99°25'$	$a=.01913$	$b=.02272$
5	$b=29.01$	$A=87°40'$	$C=33°15'$	$B=59°5'$	$a=33.78$	$c=18.54$
6	$a=804$	$A=99°55'$	$B=45°1'$	$C=35°4'$	$b=577.3$	$c=468.9$
7	$a=400$	$A=54°28'$	$C=60°$	$B=65°32'$	$b=447.4$	$c=425.7$
8	$c=161$	$A=35°15'$	$C=123°39'$	$B=21°6'$	$a=111.6$	$b=69.62$
9	$a=5.42$	$B=42°17.3'$	$C=82°28.4'$	$A=55°14.3'$	$b=4.439$	$c=6.542$
10	$b=2056$	$A=63°52.8'$	$B=70°$	$C=46°7.2'$	$a=1965.5$	$c=1578$
11	$a=7.86$	$B=32°2'52''$	$C=43°25'26''$	$A=104°31'42''$	$b=4.309$	$c=5.583$
12	$b=8$	$A=80°$	$B=2°15'46''$	$C=97°44'14''$	$a=199.53$	$c=200.73$

Solve the following oblique triangles, using logarithmic Tables I and III.

No.	Given Parts			Required Parts		
'13	$a=500$	$A=10.2°$	$B=46.6°$	$C=123.2°$	$b=2052$	$c=2363$
14	$a=45$	$A=36.8°$	$C=62°$	$B=81.2°$	$b=74.25$	$c=66.33$
15	$b=.085$	$B=95.6°$	$C=24.2°$	$A=60.2°$	$a=.0741$	$c=.035$
16	$b=5685$	$B=48.63°$	$C=83.26°$	$A=48.11°$	$a=5640$	$c=7523$
17	$c=7$	$A=59.58°$	$C=60°$	$B=60.42°$	$a=6.971$	$b=7.03$
18	$c=.0059$	$B=75°$	$C=36.87°$	$A=68.13°$	$a=.00916$	$b=.0095$
19	$a=76.08$	$B=126°$	$C=12.44°$	$A=41.56°$	$b=92.8$	$c=24.7$
20	$a=22$	$A=3.486°$	$B=73°$	$C=103.514°$	$b=346$	$c=351.8$
21	$b=8000$	$A=24.5°$	$B=86.495°$	$C=69.005°$	$a=3324$	$c=7483$
22	$b=129.38$	$A=19.42°$	$C=64°$	$B=96.58°$	$a=43.29$	$c=117.05$
23	$c=95$	$A=2.086°$	$B=112°$	$C=65.914°$	$a=3.788$	$b=96.5$
24	$b=132.6$	$A=1°$	$C=75°$	$B=104°$	$a=2.385$	$c=131.08$

25. A ship S can be seen from each of two points A and B on the shore. By measurement $AB = 800$ ft., angle $SAB = 67°43'$, and angle $SBA = 74°21'$. Find the distance of the ship from A. *Ans.* 1253 ft.

26. Two observers 5 mi. apart on a plain, and facing each other, find that the angles of elevation of a balloon in the same vertical plane with themselves are 55° and 58° respectively. Find the distances of the balloon from the observers. *Ans.* 4.607 mi.; 4.45 mi.

27. One diagonal of a parallelogram is 11.237, and it makes the angles 19°1' and 42°54' with the sides. Find the sides. *Ans.* 4.15 and 8.67.

28. To determine the distance of a hostile fort A from a place B, a line BC and the angles ABC and BCA were measured and found to be 322.6 yd., 60°34', 56°10' respectively. Find the distance AB. *Ans.* 300 yd.

29. From points A and B at the bow and stern of a ship respectively, the foremast, C, of another ship is observed. The points A and B are 300 ft. apart, and the angles ABC and BAC are found to be $65.46°$ and $112.85°$ respectively. What is the distance between the points A and C of the two ships ?

Ans. 92.54 ft.

30. A lighthouse was observed from a ship to bear N. $34°$ E.; after the ship sailed due south 3 mi. it bore N. $23°$ E. Find the distance from the lighthouse to the ship in each position.   *Ans.* 6.143 mi. and 8.792 mi.

31. In a trapezoid the parallel sides are 15 and 7, and the angles one of them makes with the nonparallel sides are $70°$ and $40°$. Find the nonparallel sides.

Ans. 8 and 5.47.

Case II. **When two sides and the angle opposite one of them are given, as a, b, A (ambiguous case *).**

First step. *Using the law of sines as in Case I, calculate log sin B.*

*If log sin $B = 0$, $\sin B = 1$, $B = 90°$; it is a **right triangle.***

*If log sin $B > 0$, $\sin B > 1$ (impossible); there is **no solution.***

*If log sin $B < 0$ and $b < a$, only the acute value of B found from the table can be used; there is **one solution.**†*

*If log sin $B < 0$ and $b > a$, the acute value of B found from the table and also its supplement, should be used; and there are **two solutions.**‡*

Second step. *Find C (one or two values according as we have one or two values of B) from*

$$C = 180° - (A + B).$$

Third step. *Find c (one or two values), using law of sines.*

Check: Use law of tangents.

Ex. 1. Having given $a = 36$, $b = 80$, $A = 28°$; solve the triangle.

Solution. In attempting to draw a figure of the triangle, the construction appears impossible. To verify this, let us find log sin B in order to apply our tests.

First step. Solving $\dfrac{a}{\sin A} = \dfrac{b}{\sin B}$ for sin B,

$$\sin B = \frac{b \sin A}{a},$$

or, $$\log \sin B = \log b + \log \sin A - \log a.$$

$$\log b = \ 1.9031$$
$$\log \sin A = \ 9.6716 - 10$$
$$11.5747 - 10$$
$$\log a = \ 1.5563$$
$$\log \sin B = 10.0184 - 10$$
$$= \ 0.0184.$$

Since log sin $B > 0$, $\sin B > 1$ (which is impossible), and there is no solution.

---

* In this connection the student should read over § 58, pp. 104, 105.

† For if $b < a$, B must be less than A, and hence B must be acute.

‡ Since $b > a$, A must be acute, and hence B may be either acute or obtuse.

Ex. 2. Having given $a = 7.42$, $b = 3.39$, $A = 105°$; solve the triangle.

Solution. Draw figure.

First step. From law of sines,

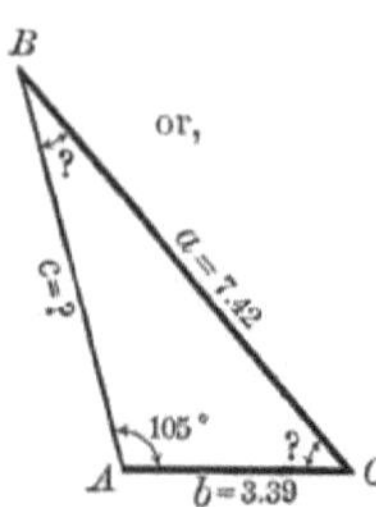

$$\sin B = \frac{b \sin A}{a},$$

or,

$$\log \sin B = \log b + \log \sin A - \log a.$$

$$
\begin{aligned}
\log b &= \ 0.5302 \\
\log \sin A &= \ 9.9849 - 10* \\
&\ \ \overline{10.5151 - 10} \\
\log a &= \ 0.8704 \\
\log \sin B &= \ 9.6447 - 10
\end{aligned}
$$

$$\therefore B = 26°11'. \qquad \text{Using Table II}$$

Since $\log \sin B < 0$ and $b < a$, there is only one solution.

Second step. $C = 180° - (A + B) = 180° - 131° \ 11' = 48° \ 49'$.

Third step. By law of sines,

$$c = \frac{a \sin C}{\sin A},$$

or,

$$\log c \quad \log a + \log \sin C - \log \sin A.$$

$$
\begin{aligned}
\log a &= \ 0.8704 \\
\log \sin C &= \ 9.8766 - 10 \\
&\ \ \overline{10.7470 - 10} \\
\log \sin A &= \ 9.9849 - 10 \\
\log c &= \ 0.7621
\end{aligned}
$$

$$\therefore c = 5.783.$$

Check: Use law of tangents.

$$\tan \tfrac{1}{2}(C - B) = \frac{c - b}{c + b} \tan \tfrac{1}{2}(C + B),$$

or, $\qquad \log \tan \tfrac{1}{2}(C - B) = \log(c - b) + \log \tan \tfrac{1}{2}(C + B) - \log(c + b).$

Substituting, we find that this equation is satisfied.

Ex. 3. Given $a = 732$, $b = 1015$, $A = 40°$; solve the triangle.

Solution. It appears from the construction of the triangle that there are two solutions.

First step. By law of sines,

$$\sin B = \frac{b \sin A}{a},$$

or, $\qquad \log \sin B = \log b + \log \sin A - \log a.$

$$
\begin{aligned}
\log b &= \ 3.0065 \\
\log \sin A &= \ 9.8081 - 10 \\
&= \ 12.8146 - 10 \\
\log a &= \ 2.8645 \\
\log \sin B &= \ 9.9501 - 10
\end{aligned}
$$

* $\sin A = \sin 105° = \sin (180° - 105°) = \sin 75°$. Hence $\log \sin A = \log \sin 75° = 9.9849 - 10$.

Since $\log \sin B < 0$ and $b > a$, we have two solutions, which test verifies our construction. From Table II we find the first value of B to be

$$B_1 = 63° 3'.$$

Hence the second value of B is

$$B_2 = 180° - B_1 = 116° 57'.$$

Second step. $\quad C_1 = 180° - (A + B_1) = 180° - 103° 3' = 76° 57';$

$$C_2 = 180° - (A + B_2) = 180° - 156° 57' = 23° 3'.$$

Third step. From law of sines,

$$c_1 = \frac{a \sin C_1}{\sin A},$$

or,

$$\log c_1 = \log a + \log \sin C_1 - \log \sin A.$$

$$\begin{aligned}
\log a &= 2.8645 \\
\log \sin C_1 &= \underline{9.9886 - 10} \\
&= 12.8531 - 10 \\
\log \sin A &= \underline{9.8081 - 10} \\
\log c_1 &= 3.0450
\end{aligned}$$

$$\therefore c_1 = 1109.3.$$

In the same manner, from $\quad c_2 = \dfrac{a \sin C_2}{\sin A}$

we get $\quad c_2 = 445.9.$

Check: Use $\tan \tfrac{1}{2}(C - B) = \dfrac{c - b}{c + b} \tan \tfrac{1}{2}(C + B)$ for both solutions.

EXAMPLES

Solve the following oblique triangles, using logarithmic Tables I and II.

No.	Given Parts			Required Parts		
1	$a=50$	$c=66$	$A=123° 11'$	Impossible		
2	$a=5.08$	$b=3.59$	$A=63° 50'$	$B=39° 21'$	$C=76° 49'$	$c=5.511$
3	$a=62.2$	$b=74.8$	$A=27° 18'$	$B_1=33° 28'$	$C_1=119° 14'$	$c_1=118.32$
				$B_2=146° 32'$	$C_2=6° 10'$	$c_2=14.567$
4	$b=.2337$	$c=.1982$	$B=109°$	$A=17° 41'$	$C=53° 19'$	$a=.07508$
5	$a=107$	$c=171$	$C=31° 53'$	$A=19° 18'$	$B=128° 49'$	$b=252.2$
6	$b=3069$	$c=1223$	$C=55° 52'$	Impossible		
7	$b=5.161$	$c=6.84$	$B=44° 3'$	$A_1=68° 47'$	$C_1=67° 10'$	$a_1=6.92$
				$A_2=23° 7'$	$C_2=112° 50'$	$a_2=2.913$
8	$a=8.656$	$c=10$	$A=59° 57'$	$B=30° 3'$	$C=90°$	$b=5.009$
9	$a=214.56$	$b=284.79$	$B=104° 20'$	$A=46° 53'$	$C=28° 47'$	$c=141.5$
10	$a=32.16$	$c=27.08$	$C=52° 24'$	$A_1=70° 12'$	$B_1=57° 24'$	$b_1=28.79$
				$A_2=109° 48'$	$B_2=17° 48'$	$b_2=10.45$
11	$b=811.3$	$c=606.4$	$B=126° 5' 20''$	$A=16° 44' 40''$	$C=37° 10'$	$a=289.2$

Solve the following oblique triangles, using logarithmic Tables I and III.

No.	GIVEN PARTS			REQUIRED PARTS		
12	$a = 840$	$b = 485$	$A = 21.5°$	$B = 12.21°$	$C = 146.29°$	$c = 1272$
13	$a = 72.63$	$b = 117.48$	$A = 80°$	Impossible		
14	$a = 177$	$b = 216$	$A = 35.6°$	$B_1 = 45.27°$	$C_1 = 99.13°$	$c_1 = 300.3$
				$B_2 = 134.73°$	$C_2 = 9.67°$	$c_2 = 51.09$
15	$b = 9.399$	$c = 9.197$	$B = 120.4°$	$A = 2.02°$	$C = 57.58°$	$a = .3841$
16	$b = .048$	$c = .0621$	$B = 57.62°$	Impossible		
17	$b = 19$	$c = 18$	$C = 15.8°$	$A_1 = 147.5°$	$B_1 = 16.7°$	$a_1 = 35.52$
				$A_2 = 0.9°$	$B_2 = 163.3°$	$a_2 = 1.0385$
18	$a = 55.55$	$c = 66.66$	$C = 77.7°$	$A = 54.5°$	$B = 47.8°$	$b = 50.54$
19	$a = 34$	$c = 22$	$C = 30.35°$	$A_1 = 51.37°$	$B_1 = 98.28°$	$b_1 = 43.07$
				$A_2 = 128.63°$	$B_2 = 21.02°$	$b_2 = 15.613$
20	$a = 528$	$b = 252$	$A = 124.6°$	$B = 23.14°$	$C = 32.26°$	$c = 342.3$
21	$b = 91.06$	$c = 77.04$	$B = 51.12°$	$A = 87.69°$	$C = 41.19°$	$a = 116.88$
22	$a = 17,060$	$b = 14,050$	$B = 40°$	$A_1 = 51.32°$	$C_1 = 88.68°$	$c_1 = 21,850$
				$A_2 = 128.68°$	$C_2 = 11.32°$	$c_2 = 4290$

23. One side of a parallelogram is 35, a diagonal is 63, and the angle between the diagonals is 21° 37′. Find the other diagonal. *Ans.* 124.62.

24. The distance from B to C is 145 ft., from A to C is 178 ft., and the angle ABC is 41° 10′. Find the distance from A to B. *Ans.* 259.4 ft.

25. Two buoys are 2789 ft. apart, and a boat is 4325 ft. from the nearer buoy. The angle between the lines from the buoys to the boat is 16° 13′. How far is the boat from the further buoy ? *Ans.* 6667 ft.

CASE III. **When two sides and the included angle are given, as a, b, C.**[*]
First step. *Calculate $a+b$, $a-b$; also $\frac{1}{2}(A+B)$ from $A+B=180°-C$.*
Second step. *From law of tangents,*

$$\tan \frac{1}{2}(A - B) = \frac{a - b}{a + b} \tan \frac{1}{2}(A + B),$$

we find $\frac{1}{2}(A - B)$. Adding this result to $\frac{1}{2}(A + B)$ gives A, and subtracting it gives B.

Third step. *To find side c use law of sines; for instance,*

$$c = \frac{a \sin C}{\sin A}.$$

Check: Check by law of sines,[†] that is, see if

$$\log a - \log \sin A = \log b - \log \sin B = \log c - \log \sin C.$$

[*] In case any other two sides and included angle are given, simply change the cyclic order of the letters throughout. Thus, if b, c, A are given, use

$$\tan \frac{1}{2}(B - C) = \frac{b - c}{b + c} \tan \frac{1}{2}(B + C), \text{ etc.}$$

[†] From law of sines, $\quad \dfrac{a}{\sin A} = \dfrac{b}{\sin B} = \dfrac{c}{\sin C}.$

Ex. 1. Having given $a = 540$, $b : : 420$, $C = 52°\,6'$; solve the triangle, using logarithms from Tables I and II.

Solution. Drawing a figure of the triangle on which we indicate the known and unknown parts, we see that the problem comes under Case II, since two sides and the included angle are given.

First step.

$$
\begin{array}{lll}
a = 540 & 540 & 180° \\
b = 420 & 420 & C = \underline{52°\,6'} \\
a + b = \overline{960} & a - b = \overline{120} & A + B = \overline{127°\,54'} \\
& & \therefore \tfrac{1}{2}(A + B) = 63°\,57'.
\end{array}
$$

Second step.
$$
\tan\frac{1}{2}(A - B) = \frac{a - b}{a + b}\tan\frac{1}{2}(A + B),
$$

or,
$$
\log \tan \tfrac{1}{2}(A - B) = \log(a - b) + \log \tan \tfrac{1}{2}(A + B) - \log(a + b).
$$

$$
\begin{array}{l}
\log(a - b) = 2.0792 \\
\log \tan \tfrac{1}{2}(A + B) = \underline{10.3108 - 10} \\
\phantom{\log \tan \tfrac{1}{2}(A + B) = } 12.3900 - 10 \\
\log(a + b) = \underline{2.9823} \\
\log \tan \tfrac{1}{2}(A - B) = 9.4077 - 10 \\
\therefore \tfrac{1}{2}(A - B) = 14°\,21'
\end{array}
$$

$$
\begin{array}{ll}
\tfrac{1}{2}(A + B) = 63°\,57' & 63°\,57' \\
\tfrac{1}{2}(A - B) = \underline{14°\,21'} & \underline{14°\,21'} \\
\end{array}
$$

Adding, $A = 78°\,18'.$ Subtracting, $B = 49°\,36'.$

Third step.
$$
c = \frac{a \sin C}{\sin A}. \qquad \text{From } \frac{c}{\sin C} = \frac{a}{\sin A}
$$

$$
\log c = \log a + \log \sin C - \log \sin A.
$$

$$
\begin{array}{l}
\log a = 2.7324 \\
\log \sin C = 9.8971 - 10 \\
 12.6295 - 10 \\
\log \sin A = \underline{9.9909 - 10} \\
\log c = 2.6386 \\
\therefore c = 435.1.
\end{array}
$$

Check: By law of sines,

$$
\begin{array}{lll}
\log a = 12.7324 - 10 & \log b = 12.6232 - 10 & \log c = 12.6386 - 10 \\
\log \sin A = \underline{9.9909 - 10} & \log \sin B = \underline{9.8817 - 10} & \log \sin C = \underline{9.8971 - 10} \\
 2.7415 & 2.7415 & 2.7415
\end{array}
$$

Ex. 2. Having given $a = 167$, $c -: 82$, $B = 98°$; solve the triangle, using logarithms from Tables I and III.

Solution. First step.

$$
\begin{array}{lll}
a = 167 & 167 & 180° \\
c = 82 & 82 & B = \underline{98°} \\
a + c = \overline{249} & a \cdot \cdot c = \overline{85} & A + C = \overline{82°} \\
& & \therefore \tfrac{1}{2}(A + C) = 41°.
\end{array}
$$

Second step.
$$\tan \tfrac{1}{2}(A - C) = \frac{a - c}{a + c} \tan \tfrac{1}{2}(A + C),$$

or,
$$\log \tan \tfrac{1}{2}(A - C) = \log (a - c) + \log \tan \tfrac{1}{2}(A + C) - \log (a + c).$$

$$\log (a - c) = 1.9294$$
$$\log \tan \tfrac{1}{2}(A + C) = 9.9392 - 10$$
$$11.8686 - 10$$
$$\log (a + c) = 2.3962$$
$$\log \tan \tfrac{1}{2}(A - C) = 9.4724 - 10$$

$$\therefore \tfrac{1}{2}(A - C) = 16.53°$$

$$\tfrac{1}{2}(A + C) = 41.00° \qquad\qquad 41.00°$$
$$\tfrac{1}{2}(A - C) = 16.53° \qquad\qquad 16.53°$$

Adding, $\qquad A = 57.53°.$ $\qquad$ Subtracting, $C = 24.47°.$

Third step. $\qquad b = \dfrac{a \sin B}{\sin A}.$ $\qquad$ from $\dfrac{b}{\sin B} = \dfrac{a}{\sin A}$

$$\log b = \log a + \log \sin B - \log \sin A.$$

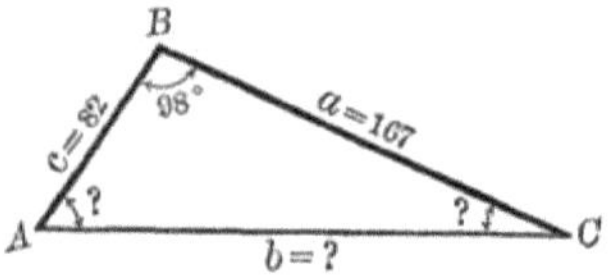

$$\log a = 2.2227$$
$$\log \sin B = 9.9958 - 10\;*$$
$$12.2185 - 10$$
$$\log \sin A = 9.9262 - 10$$
$$\log b = 2.2923$$
$$\therefore b = 196.$$

Check: By law of sines,

$$\log a = 12.2227 - 10 \qquad \log b = 12.2923 - 10 \qquad \log c = 11.9138 - 10$$
$$\log \sin A = 9.9262 - 10 \qquad \log \sin B = 9.9958 - 10 \qquad \log \sin C = 9.6172 - 10$$
$$2.2965 \qquad\qquad 2.2965 \qquad\qquad 2.2966$$

which substantially agree.

EXAMPLES

Solve the following oblique triangles, using logarithmic Tables I and II.

No.	Given Parts			Required Parts		
1	$a = 27$	$c = 15$	$B = 46°$	$A = 100° \, 57'$	$C = 33° \, 3'$	$b = 19.78$
2	$a = 486$	$b = 347$	$C = 51° \, 36'$	$A = 83° \, 15'$	$B = 45° \, 9'$	$c = 383.5$
3	$b = 2.302$	$c = 3.567$	$A = 62°$	$B = 39° \, 16'$	$C = 78° \, 44'$	$a = 3.211$
4	$a = 77.99$	$b = 83.39$	$C = 72° \, 16'$	$A = 51° \, 14'$	$B = 56° \, 30'$	$c = 95.24$
5	$a = 0.917$	$b = 0.312$	$C = 33° \, 7.2'$	$A = 132° \, 17'$	$B = 14° \, 35'$	$c = .6769$
6	$a = .3$	$b = .363$	$C = 124° \, 56'$	$A = 24° \, 41.8'$	$B = 30° \, 22.2'$	$c = .5886$
7	$b = 1192.1$	$c = 356.3$	$A = 26° \, 16'$	$B = 143° \, 29'$	$C = 10° \, 15'$	$a = 886.2$
8	$a = 7.4$	$c = 11.439$	$B = 82° \, 26'$	$A = 35° \, 2'$	$C = 62° \, 32'$	$b = 12.777$
9	$a = 53.27$	$b = 41.61$	$C = 78° \, 33'$	$A = 59° \, 16.5'$	$B = 42° \, 10.5'$	$c = 60.74$
10	$b = .02668$	$c = .05092$	$A = 115° \, 47'$	$B = 21° \, 1'$	$C = 43° \, 1.3'$	$a = .06697$
11	$a = 51.38$	$c = 67.94$	$B = 79° \, 12' \, 36''$	$A = 40° \, 52.7'$	$C = 59° \, 54.7'$	$b = 77.12$
12	$b = \sqrt{5}$	$c = \sqrt{3}$	$A = 35° \, 53'$	$B = 93° \, 28'$	$C = 50° \, 39'$	$a = 1.313$

---

* $\sin B = \sin 98° = \sin (180° - 98°) = \sin 82°.$ $\;\therefore\;$ $\log \sin 98° = \log \sin 82° = 9.9958 - 10.$

Solve the following oblique triangles, using logarithmic Tables I and III.

No.	GIVEN PARTS			REQUIRED PARTS		
13	$a = 17$	$b = 12$	$C = 59.3°$	$A = 77.2°$	$B = 43.5°$	$c = 14.99$
14	$a = 55.14$	$b = 33.09$	$C = 30.4°$	$A = 117.4°$	$B = 32.2°$	$c = 31.44$
15	$b = 101$	$c = 158$	$A = 37.38°$	$B = 38.26°$	$C = 104.37°$	$a = 99.04$
16	$a = 101$	$b = 29$	$C = 32.18°$	$A = 136.4°$	$B = 11.42°$	$c = 78$
17	$c = 45$	$b = 29$	$A = 42.8°$	$B = 39.72°$	$C = 97.48°$	$a = 30.84$
18	$a = .085$	$c = .0042$	$B = 56.5°$	$A = 121.07°$	$C = 2.43°$	$b = .08258$
19	$b = .9486$	$c = .8852$	$A = 84.6°$	$B = 49.88°$	$C = 45.52°$	$a = 1.235$
20	$b = 6$	$c = 9$	$A = 88.9°$	$B = 34.03°$	$C = 57.07°$	$a = 10.72$
21	$a = 12$	$b = 19$	$C = 5.24°$	$A = 8.84°$	$B = 78.54°$	$c = 7.132$
22	$a = 42,930$	$c = 73,480$	$B = 24.8°$	$A = 27.56°$	$C = 127.64°$	$b = 38,920$

23. In order to find the distance between two objects, A and B, separated by a swamp, a station C was chosen, and the distances $CA = 3825$ yd., $CB = 3476$ yd., together with the angle $ACB = 62° 31'$, were measured. What is the distance AB ? *Ans.* 3800 yd.

24. Two trains start at the same time from the same station and move along straight tracks that form an angle of $30°$, one train at the rate of 30 mi. an hour, the other at the rate of 40 mi. an hour. How far apart are the trains at the end of half an hour ? *Ans.* 10.27 mi.

25. In a parallelogram the two diagonals are 5 and 6 and form an angle of $49° 18'$. Find the sides. *Ans.* 5.002 and 2.338.

26. Two trees A and B are on opposite sides of a pond. The distance of A from a point C is 297.6 ft., the distance of B from C is 864.4 ft., and the angle ACB is $87.72°$. Find the distance AB. *Ans.* 903 ft.

27. Two stations A and B on opposite sides of a mountain are both visible from a third station C. The distances AC, BC, and the angle ACB were measured and found to be 11.5 mi., 9.4 mi., and $59° 31'$ respectively. Find the distance from A to B. *Ans.* 10.533 mi.

28. From a point 3 mi. from one end of an island and 7 mi. from the other end the island subtends an angle of $33° 55.8'$. Find the length of the island. *Ans.* 4.812 mi.

29. The sides of a parallelogram are 172.43 and 101.31, and the angle included by them is $61° 16'$. Find the two diagonals. *Ans.* 152.93 and 238.5.

30. Two yachts start at the same time from the same point, and sail, one due north at the rate of 10.44 mi. an hour, and the other due northeast at the rate of 7.71 mi. an hour. How far apart are they at the end of 40 minutes ? *Ans.* 4.936 mi.

CASE IV. **When all three sides a, b, c are given.**

First step. *Calculate* $s = \frac{1}{2}(a + b + c)$, $s - a$, $s - b$, $s - c$.

Second step. *Find $\log r$ from*

$$r = \sqrt{\frac{(s - a)(s - b)(s - c)}{s}}. \quad \textbf{(84)} \text{ to } \textbf{(87)}, \text{ p. } 115$$

Third step. *Find angles A, B, C from*

$$\tan \tfrac{1}{2} A = \frac{r}{s-a}, \quad \tan \tfrac{1}{2} B = \frac{r}{s-b}, \quad \tan \tfrac{1}{2} C = \frac{r}{s-c}.$$

Check : See if $A + B + C = 180°$.

Ex. 1. Having given $a = 51$, $b = 65$, $c = 20$; solve the triangle.

Solution. Drawing a figure of the triangle on which we indicate the known and unknown parts, we see that since the three sides are given, the problem comes under Case IV.

First step. $a = 51$ Hence

$$
\begin{aligned}
b &= 65 \\
c &= 20 \\
2s &= 136 \\
s &= 68.
\end{aligned}
\qquad
\begin{aligned}
s &= 68 \\
a &= 51 \\
\hline
s - a &= 17
\end{aligned}
\qquad
\begin{aligned}
s &= 68 \\
b &= 65 \\
\hline
s - b &= 3
\end{aligned}
\qquad
\begin{aligned}
s &= 68 \\
c &= 20 \\
\hline
s - c &= 48
\end{aligned}
$$

Second step.
$$r = \sqrt{\frac{(s-a)(s-b)(s-c)}{s}},$$

or,
$$\log r = \tfrac{1}{2}\left[\log(s-a) + \log(s-b) + \log(s-c) - \log s\right].$$

From the table of logarithms,

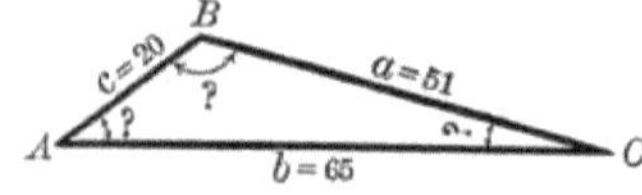

$$
\begin{aligned}
\log(s - a) &= 1.2304 \\
\log(s - b) &= 0.4771 \\
\log(s - c) &= 1.6812 \\
\hline
&\; 3.3887 \\
\log s &= 1.8325 \\
2\,|\; & 1.5562 \\
\log r &= 0.7781
\end{aligned}
$$

Third step. From the formula $\tan \tfrac{1}{2} A = \dfrac{r}{s-a}$,

$$\log \tan \tfrac{1}{2} A = \log r - \log(s - a).$$

$$
\begin{aligned}
\log r &= 10.7781 - 10 \\
\log(s - a) &= \;\;1.2304 \\
\log \tan \tfrac{1}{2} A &= \;\;9.5477 - 10 \qquad\text{using Table II *}
\end{aligned}
$$

$$\tfrac{1}{2} A = 19° 27',$$
or,
$$A = 38° 54'.$$

From the formula $\tan \tfrac{1}{2} B = \dfrac{r}{s-b}$,

$$\log \tan \tfrac{1}{2} A = \log r - \log(s - b).$$

$$
\begin{aligned}
\log r &= 10.7781 - 10 \\
\log(s - b) &= \;\;0.4771 \\
\log \tan \tfrac{1}{2} B &= 10.3010 - 10
\end{aligned}
$$

$$\tfrac{1}{2} B = 63° 26', \qquad\text{using Table II}$$
$$B = 126° 52'.$$

---

* If we use Table III instead, we get

$$
\begin{array}{lll}
\tfrac{1}{2} A = 19.44°, & \tfrac{1}{2} B = 63.43°, & \tfrac{1}{2} C = 7.12°, \\
A = 38.88°, & B = 126.86°, & C = 14.24°.
\end{array}
$$

and

Check : $A + B + C = 179.98°$.

From the formula $\tan \frac{1}{2} C = \dfrac{r}{s-c}$,

$$\log \tan \tfrac{1}{2} C = \log r - \log (s - c).$$
$$\log r = 10.7781 - 10$$
$$\log (s - c) = 1.6812$$
$$\log \tan \tfrac{1}{2} C = 9.0969 - 10$$
$$\tfrac{1}{2} C = 7^\circ 8'.$$
$$C = 14^\circ 16'.$$

Using Table II

Check :
$$A = 38^\circ 54'$$
$$B = 126^\circ 52'$$
$$C = 14^\circ 16'$$
$$A + B + C = 180^\circ 2'$$

EXAMPLES

Solve the following oblique triangles, using logarithmic Tables I and II.

No.	Given Parts			Required Parts		
1	$a = 2$	$b = 3$	$c = 4$	$A = 28^\circ 58'$	$B = 46^\circ 34'$	$C = 104^\circ 28'$
2	$a = 2.5$	$b = 2.79$	$c = 2.33$	$A = 57^\circ 36'$	$B = 70^\circ 28'$	$C = 51^\circ 54'$
3	$a = 5.6$	$b = 4.3$	$c = 4.9$	$A = 74^\circ 40'$	$B = 47^\circ 46'$	$C = 57^\circ 34'$
4	$a = 111$	$b = 145$	$c = 40$	$A = 27^\circ 20'$	$B = 143^\circ 8'$	$C = 9^\circ 32'$
5	$a = 79.3$	$b = 94.2$	$c = 66.9$	$A = 55^\circ 56'$	$B = 79^\circ 44'$	$C = 44^\circ 20'$
6	$a = 321$	$b = 361$	$c = 402$	$A = 49^\circ 24'$	$B = 58^\circ 38'$	$C = 71^\circ 58'$
7	$a = .641$	$b = .529$	$c = .702$	$A = 60^\circ 50'$	$B = 46^\circ 6'$	$C = 73^\circ 2'$
8	$a = 3.019$	$b = 6.731$	$c = 4.228$	$A = 18^\circ 12'$	$B = 135^\circ 50'$	$C = 25^\circ 56'$
9	$a = .8706$	$b = .0916$	$c = .7902$	$A = 149^\circ 50'$	$B = 3^\circ 2'$	$C = 27^\circ 10'$
10	$a = 73$	$b = 82$	$c = 91$	$A = 49^\circ 34'$	$B = 58^\circ 46'$	$C = 71^\circ 38'$
11	$a = 1.9$	$b = 3.4$	$c = 4.9$	$A = 16^\circ 26'$	$B = 30^\circ 26'$	$C = 133^\circ 10'$
12	$a = .21$	$b = .26$	$c = .31$	$A = 42^\circ 6'$	$B = 56^\circ 6'$	$C = 81^\circ 48'$
13	$a = 513.4$	$b = 726.8$	$c = 931.3$	$A = 33^\circ 16'$	$B = 50^\circ 56'$	$C = 95^\circ 48'$
14	$a = \sqrt{5}$	$b = \sqrt{6}$	$c = \sqrt{7}$	$A = 51^\circ 52'$	$B = 59^\circ 32'$	$C = 68^\circ 34'$

Solve the following oblique triangles, using logarithmic Tables I and III.

No.	Given Parts			Required Parts		
15	$a = 4$	$b = 7$	$c = 6$	$A = 34.78^\circ$	$B = 86.42^\circ$	$C = 58.82^\circ$
16	$a = 43$	$b = 50$	$c = 57$	$A = 46.82^\circ$	$B = 57.98^\circ$	$C = 75.18^\circ$
17	$a = .23$	$b = .26$	$c = .198$	$A = 58.44^\circ$	$B = 74.38^\circ$	$C = 47.18^\circ$
18	$a = 61.3$	$b = 84.7$	$c = 47.6$	$A = 45.2^\circ$	$B = 101.38^\circ$	$C = 33.44^\circ$
19	$a = .0291$	$b = .0184$	$c = .0358$	$A = 54.06^\circ$	$B = 30.8^\circ$	$C = 95.16^\circ$
20	$a = 705$	$b = 562$	$c = 639$	$A = 71.56^\circ$	$B = 49.14^\circ$	$C = 59.32^\circ$
21	$a = 56$	$b = 43$	$c = 49$	$A = 74.68^\circ$	$B = 47.78^\circ$	$C = 57.56^\circ$
22	$a = 301.9$	$b = 673.1$	$c = 422.8$	$A = 18.2^\circ$	$B = 135.86^\circ$	$C = 25.94^\circ$
23	$a = 2.51$	$b = 2.79$	$c = 2.33$	$A = 57.88^\circ$	$B = 70.3^\circ$	$C = 51.84^\circ$
24	$a = 80$	$b = 90$	$c = 100$	$A = 49.46^\circ$	$B = 58.76^\circ$	$C = 71.78^\circ$

25. The sides of a triangular field are 7 rd., 11 rd., and 9.6 rd. Find the angle opposite the longest side.   *Ans.* 81° 24′.

26. A pole 13 ft. long is placed 6 ft. from the base of an embankment, and reaches 8 ft. up its face. Find the slope of the embankment.   *Ans.* 44° 2′.

27. Under what visual angle is an object 7 ft. long seen when the eye of the observer is 5 ft. from one end of the object and 8 ft. from the other end ?

Ans. 60°.

28. The distances between three cities, A, B, and C, are as follows: $AB = 165$ mi., $AC = 72$ mi., and $BC = 185$ mi. B is due east from A. In what direction is C from A?   *Ans.* 7° 19.5′ N. of W.

29. Three towns, A, B, and C, are connected by straight roads. $AB = 4$ mi., $BC = 5$ mi., $AC = 7$ mi. Find the angle made by the roads AB and BC.

Ans. 101.54°.

30. The distances of two islands from a buoy are 3 and 4 mi. respectively. If the islands are 2 mi. apart, find the angle subtended by the islands at the buoy.

Ans. 28.96°.

31. A point P is 13,581 ft. from one end of a wall 12,342 ft. long, and 10,025 ft. from the other end. What angle does the wall subtend at the point P?

Ans. 60.86°.

83. Use of logarithms in finding the area of an oblique triangle. From § 62, p. 117, we have the following three cases.

CASE I. *When two sides and the included angle are given, use one of the formulas*

$$(88) \qquad S = \frac{ab \sin C}{2}, \quad S = \frac{bc \sin A}{2}, \quad S = \frac{ac \sin B}{2},$$

where $S = $ *area of the triangle.*

Ex. 1. Given $a = 25.6$, $b = 38.2$, $C = 41° 56′$; find the area of the triangle.

Solution. $S = \dfrac{ab \sin C}{2}.$

$$\log S = \log a + \log b + \log \sin C - \log 2.$$

$$
\begin{aligned}
\log a &= 1.4082 \\
\log b &= 1.5821 \\
\log \sin C &= \underline{9.8249 - 10} \\
&\quad\ 12.8152 - 10 \\
\log 2 &= \underline{0.3010} \\
\log S &= 12.5142 - 10 \\
&= 2.5142.
\end{aligned}
$$

$$\therefore S = 326.8. \quad Ans.$$

CASE II. *When the three sides are given, use formula*

$$(89) \qquad S = \sqrt{s(s - a)(s - b)(s - c)},$$

where $S = $ *area of the triangle,*

and $s = \tfrac{1}{2}(a + b + c).$

Ex. 2. Find the area of a triangle, having given $a = 12.53$, $b = 24.9$, $c = 18.91$.

Solution.

$$a = 12.53$$
$$b = 24.9$$
$$c = 18.91$$
$$2s = 56.34$$
$$s = 28.17.$$

Hence

$s = 28.17$	$s = 28.17$	$s = 28.17$
$a = 12.53$	$b = 24.9$	$c = 18.91$
$s - a = 15.64$	$s - b = 3.27$	$s - c = 9.26$

$$S = \sqrt{s(s-a)(s-b)(s-c)}.$$
$$\log S = \tfrac{1}{2}[\log s + \log(s-a) + \log(s-b) + \log(s-c)].$$

$$\log s = 1.4498$$
$$\log(s-a) = 1.1942$$
$$\log(s-b) = 0.5145$$
$$\log(s-c) = 0.9666$$
$$2\,\lfloor 4.1251$$
$$\log S = 2.0626$$

$$\therefore S = 115.5. \quad Ans.$$

CASE III. *Area problems which do not fall directly under Cases I or II may be solved by Case I if we first find an additional side or angle by the law of sines.*

Ex. 3. Given $A = 34° 22'$, $B = 66° 11'$, $c = 78.35$; find area of triangle.

Solution. This does not now come directly under either Case I or Case II. But

$$C = 180° - (A + B) = 180° - 100° 33' = 79° 27'.$$

And, by law of sines,

$$a = \frac{c \sin A}{\sin C}.$$
$$\log a = \log c + \log \sin A - \log \sin C.$$

$$\log c = 1.8941$$
$$\log \sin A = 9.7517 - 10$$
$$11.6458 - 10$$
$$\log \sin C = 9.9926 - 10$$
$$\log a = 1.6532$$

Now it comes under Case I.

$$S = \frac{ac \sin B}{2}.$$
$$\log S = \log a + \log c + \log \sin B - \log 2.$$

$$\log a = 1.6532$$
$$\log c = 1.8941$$
$$\log \sin B = 9.9614 - 10$$
$$13.5087 - 10$$
$$\log 2 = 0.3010$$
$$\log S = 13.2077 - 10$$
$$= 3.2077$$

$$\therefore S = 1613.3. \quad Ans.$$

EXAMPLES

Find the areas of the following oblique triangles, using Tables I and II for the first ten and Tables I and III for the rest.

No.	GIVEN PARTS			AREA
1	$a = 38$	$c = 61.2$	$B = 67° 56'$	1078
2	$b = 2.07$	$A = 70°$	$B = 36° 23'$	3.257
3	$b = 116.1$	$c = 100$	$A = 118° 16'$	5113
4	$a = 3.123$	$A = 53° 11'$	$B = 13° 57'$	1.344
5	$b = .439$	$A = 76° 38'$	$C = 40° 35'$	.0686
6	$a = .3228$	$c = .9082$	$B = 60° 16'$	.1273
7	$c = 80.25$	$B = 100° 5'$	$C = 31° 44'$	4506
8	$a = .010168$	$b = .018225$	$C = 11° 18.4'$	.000018155
9	$a = 18.063$	$A = 96° 30'$	$B = 35°$	70.55
10	$b = 142.8$	$c = 89.6$	$a = 95$	4174
11	$a = 100$	$B = 60.25°$	$C = 54.5°$	3891
12	$a = 145$	$b = 178$	$B = 41.17°$	12,383
13	$a = 886$	$b = 747$	$C = 71.9°$	314,600
14	$a = 266$	$b = 352$	$C = 73°$	44,770
15	$a = 960$	$b = 720$	$C = 25.67°$	149,730
16	$a = 79$	$b = 94$	$c = 67$	2604
17	$a = 23.1$	$b = 19.7$	$c = 25.2$	215.9
18	$a = 5.82$	$b = 6$	$c = 4.26$	11.733

19. The sides of a field $ABCD$ are $AB = 37$ rd., $BC = 63$ rd., and $DA = 20$ rd., and the diagonals AC and BD are 75 rd. and 42 rd., respectively. Required the area of the field.   1570 sq. rd.

20. In a field $ABCD$ the sides AB, BC, CD, and DA are 155 rd., 236 rd., 252 rd., and 105 rd., respectively, and the length from A to C is 311 rd. Find the area of the field.   29,800 sq. rd.

21. The area of a triangle is one acre ; two of its sides are 127 yd. and 150 yd. Find the angle between them.   30° 32′.

22. Given the area of a triangle $= 12$. Find the radius of the inscribed circle if $a = 60$ and $B = 40° 35.2'$.

84. Measurement of land areas. The following examples illustrate the nature of the measurements made by surveyors in determining land areas, and the usual method employed for calculating the area from the data found. The Gunter's chain is 4 rd., or 66 ft., in length. An acre equals 10 sq. chains, or 160 sq. rd.

EXAMPLES

1. A surveyor starting from a point A runs N. 27° E. 10 chains to B, thence N.E. by E. 8 chains to C, thence S. 5° W. 24 chains to D, thence N. 40° 44′ W. 13.94 chains to A. Calculate the area of the field $ABCD$.

Solution. Draw an accurate figure of the field. Through the extreme westerly point of the field draw a north-and-south line. From the figure, area

$ABCD =$ area trapezoid * $GCDE -$ (area trapezoid $GCBF +$ area triangle $FBA +$ area triangle $ADE) = 13.9$ acres. *Ans.*

2. A surveyor measures S. 50° 25′ E. 6.04 chains, thence S. 58° 10′ W. 4.15 chains, thence N. 28° 12′ W. 5.1 chains, thence to the starting point. Determine the direction and distance of the starting point from the last station, and find area of the field inclosed.

Ans. N. 39° 42′ E. 2 chains ; 1.66 acres.

3. One side of a field runs N. 83° 30′ W. 10.5 chains, the second side S. 22° 15′ W. 11.67 chains, the third side N. 71° 45′ E. 12.9 chains, the fourth side completes the circuit of the field. Find the direction and length of the fourth side, and calculate the area of the field.

Ans. N. 25° 1′ E. 6.15 chains ; 7.9 acres.

4. From station No. 1 to station No. 2 is S. 7° 20′ W. 4.57 chains, thence to station No. 3 S. 61° 55′ W. 7.06 chains, thence to station No. 4 N. 3° 10′ E. 5.06 chains, thence to station No. 5 N. 33° 50′ E. 325 chains, thence to station No. 1. Find the direction and distance of station No. 1 from No. 5, and calculate the area of the field inclosed.   *Ans.* E. 1° 15′ N. 4.7 chains ; 3.55 acres.

85. Parallel sailing. When a vessel sails due east or due west, that is, always travels on the same parallel of latitude, it is called *parallel sailing.* The distance sailed is the *departure,*† and it is expressed in

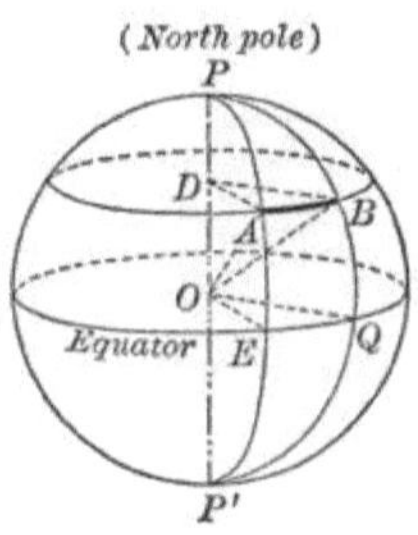

geographical ‡ miles. Thus, in the figure, arc AB is the departure between A and B. The latitudes of A and B are the same, i.e. arc $EA =$ angle $EOA =$ arc $QB =$ angle QOB. The difference in longitude

* From Geometry the area of a trapezoid equals one half the sum of the parallel sides times the altitude. Thus, area $GCDE = \frac{1}{2}(GC + ED) GE.$

† The *departure* between two meridians is the arc of a parallel of latitude comprehended between those meridians. It diminishes as the distance from the equator increases.

‡ A *geographical mile* or *knot* is the length of an arc of one minute on a great circle of the earth.

of A and $B = \text{arc } EQ$. The relation between *latitude, departure,* and *difference in longitude* may be found as follows: By Geometry,

$$\frac{\text{arc } AB}{\text{arc } EQ} = \frac{DA}{OE} = \frac{DA}{OA} = \cos OAD = \cos AOE = \cos \text{latitude}.$$

$$\therefore \text{ arc } AB = \text{arc } EQ \cos \text{latitude, or,}$$

(90) $$\text{Diff. long.} = \frac{\text{departure}}{\cos \text{latitude}}.$$

EXAMPLES

1. A ship whose position is lat. 25° 20′ N., long. 36° 10′ W. sails due west 140 knots. Find the longitude of the place reached.

Solution. Here departure $= 140$,
and latitude $= 25°\ 20'$ N.

Substituting in above formula (**90**),

$$\text{diff. long.} = \frac{140}{\cos 25°\ 20'}.$$

$$\log 140 = 12.1461 - 10$$
$$\log \cos 25°\ 20' = \ 9.9561 - 10$$
$$\log \text{diff. long.} = \ 2.1900$$
$$\text{diff. long.} = 154.9' = 2°\ 34.9'.$$

Hence longitude of place reached $= 36°\ 10' + 2°\ 34.9' = 38°\ 44.9'$ W. *Ans.*

2. A ship in lat. 42° 16′ N., long. 72° 16′ W., sails due east a distance of 149 geographical miles. What is the position of the point reached ?
Ans. Long. 68° 55′ W.

3. A vessel in lat. 44° 49′ S., long. 119° 42′ E., sails due west until it reaches long. 117° 16′ E. Find the departure. *Ans.* 103.6 knots.

4. A ship in lat. 36° 48′ N., long. 56° 15′ W., sails due east 226 mi. Find the longitude of the place reached. *Ans.* Long. 51° 33′ W.

5. A vessel in lat. 48° 54′ N., long. 10° 55′ W., sails due west until it is in long. 15° 12′ W. Find the number of knots sailed. *Ans.* 168.9 knots.

86. Plane sailing. When a ship sails in such a manner as to cross successive meridians at the same angle, it is said to sail on a *rhumb line.* This angle is called the *course,* and the *distance* between two places is measured on a rhumb line. Thus, in the figure, if a ship travels from A to B on a rhumb line,

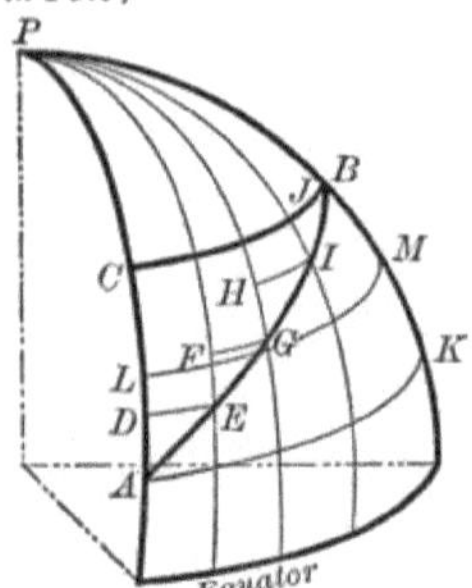

$$\text{arc } AB = \text{distance,}$$
$$\text{angle } CAB = \text{course,}$$
$$\text{arc } CB = \text{departure,}$$
$$\text{arc } AC = \text{difference in latitude}$$
$$\text{between } A \text{ and } B.$$

An approximate relation between the quantities involved is obtained by regarding the surface of the earth as a *plane* surface, that is, regarding ACB as a plane right triangle, the angle ACB being the right angle. This right triangle is called the *triangle of plane sailing.*

From this plane right triangle we get

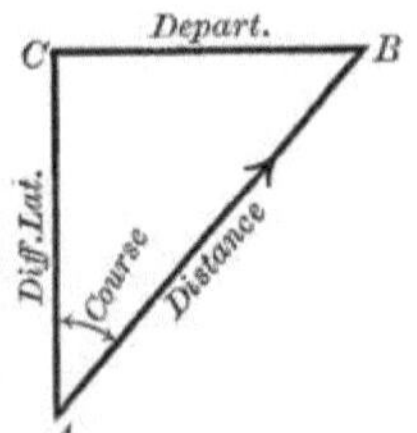

$$CB = AB \sin A, \text{ and}$$
$$AC = AB \cos A \; ; \text{ or,}$$

(91)   **Departure = distance × sin course, and**

(92)   **Diff. lat. = distance × cos course.**

If AB is long, the error caused by neglecting the curvature of the earth will be too great to make these results of any value. In that case AB may be divided into parts, such as AE, EG, GI, IB (figure on p. 174), which are so small that the curvature of the earth may be neglected.

EXAMPLES

1. A ship sails from lat. 8° 45′ S., on a course N. 36° E. 345 geographical mi. Find the latitude reached and the departure made.

Solution. Here distance = 345 and course = 36°.

∴ departure = 345 sin 36°.	diff. lat. = 345 cos 36°.
log 345 = 2.5378	log 345 = 2.5378
log sin 36° = 9.7692 − 10	log cos 36° = 9.9080 − 10
log departure = 2.3070	log diff. lat. = 2.4458
∴ departure = 202.8 mi. *Ans.*	diff. lat. = 279.1′ = 4° 39.1′.

As the ship is sailing in a northerly direction she will have reached latitude 8° 45′ − 4° 39.1′ = 4° 5.9′ S. *Ans.*

2. A ship sails from lat. 32° 18′ N., on a course between N. and W., a distance of 344 mi., and a departure of 103 mi. Find the course and the latitude reached.   *Ans.* Course N. 17° 25′ W., lat. 37° 46′ N.

3. A ship sails from lat. 43° 45′ S., on a course N. by E. 2345 mi. Find the latitude reached and the departure made.

Ans. Lat. 5° 25′ S., departure = 457.5 mi.

4. A ship sails on a course between S. and E. 244 mi., leaving lat. 2° 52′ S., and reaching lat. 5° 8′ S. Find the course and the departure.

Ans. Course S. 56° 8′ E., departure = 202.6 mi.

87. Middle latitude sailing. Here we take the departure between two places to be measured on that parallel of latitude which lies halfway between the parallels of the two places. Thus, in the figure on p. 174, the departure between A and B is LM, measured on a parallel of latitude midway between the parallels of A and B.

This will be sufficiently accurate for ordinary purposes if the run is not of great length nor too far away from the equator. The *middle latitude* is then the mean of the latitudes of A and B. The formula (**90**) on p. 174 will then become

$$(93) \qquad \text{Diff. long.} = \frac{\text{departure}}{\text{cos mid. lat.}}$$

EXAMPLES

1. A ship in lat. 42° 30′ N., long. 58° 51′ W., sails S. 33° 45′ E. 300 knots. Find the latitude and longitude of the position reached.

Solution. We know the latitude of the starting point A. To get the latitude of the final position B, we first find diff. in lat. from (**92**). This gives

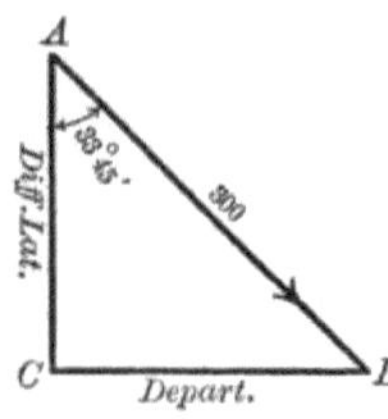

$$\text{diff. lat.} = 300 \cos 33° 45′.$$

$$\log 300 = 2.4771$$
$$\log \cos 33° 45′ = 9.9198 - 10$$
$$\overline{\log \text{diff. lat.} = 2.3969}$$

$$\text{diff. lat.} = 249.4′ = 4° 9.4′.$$

Since the ship sails in a southerly direction, she will have reached latitude $= 42° 30′ - 4° 9.4′ = 38° 20.6′$ N. *Ans.*

To get the longitude of B we must first calculate the departure and middle latitude for substitution in (**93**). From (**91**)

$$\text{departure} = 300 \sin 33° 45′.$$

$$\log 300 = 2.4771$$
$$\log \sin 33° 45′ = 9.7448 - 10$$
$$\overline{\log \text{departure} = 2.2219}$$

$$\text{departure} = 166.7′.$$
$$\text{Middle latitude} = \tfrac{1}{2} (42° 30′ + 38° 20.6′) = 40° 25.3′.$$

Substituting in (**93**), $\qquad \text{diff. long.} = \dfrac{166.7}{\cos 40° 25.3′}.$

$$\log 166.7 = 12.2219 - 10$$
$$\log \cos 40° 25.3′ = 9.8815 - 10$$
$$\overline{\log \text{diff. long.} = 2.3404}$$

$$\text{diff. long.} = 219′ = 3° 39′.$$

Since the ship sails in an easterly direction, she will have reached longitude $= 58° 51′ - 3° 39′ = 55° 12′$ W. *Ans.*

2. A vessel in lat. 26° 15′ N., long. 61° 43′ W., sails N.W. 253 knots. Find the latitude and longitude of the position reached.

$\qquad\qquad$ *Ans.* Lat. 29° 13.9′ N. ; long. 65° 5.1′ W.

3. A ship leaves lat. 31° 14′ N., long. 42° 19′ W., and sails E.N.E. 325 mi. Find the position reached. $\qquad$ *Ans.* Lat. 33° 18.4′ N. ; long. 36° 24′ W.

4. Leaving lat. 42° 30′ N., long. 58° 51′ W., a battleship sails S.E. by S. 300 mi. Find the place reached. $\qquad$ *Ans.* Lat. 38° 21′ N. ; long. 55° 12′ W.

5. A ship sails from a position lat. 49° 56′ N., long. 15° 16′ W., to another lat. 47° 18′ N., long. 20° 10′ W. Find the course and distance.

> *Ans.* Course, S. 50° 53′ W. ; distance = 250.5 mi.

Hint. The difference in latitude and the difference in longitude are known, also the middle latitude.

6. A torpedo boat in lat. 37° N., long. 32° 16′ W., steams N. 36° 56′ W., and reaches lat. 41° N. Find the distance steamed and the longitude of the position reached.   *Ans.* Distance = 300.3 mi.; long. 36° 8′ W.

7. A ship in lat. 42° 30′ N., long. 58° 51′ W., sails S.E. until her departure is 163 mi. and her latitude 38° 22′ N. Find her course and distance and the longitude of the position reached.

> *Ans.* Course, S. 33° 19′ E. ; distance = 296.7 mi.; long. 55° 17′ W.

8. A cruiser in lat. 47° 44′ N., long. 32° 44′ W., steams 171 mi. N.E. until her latitude is 50° 2′ N. Find her course and the longitude of the position reached.

> *Ans.* Course, N. 36° 11′ E.; long. 30° 10′ W.

9. A vessel in lat. 47° 15′ N., long. 20° 48′ W., sails S.W. 208 mi., the departure being 162 mi. Find the course and the latitude and longitude of the position reached.   *Ans.* Course, S. 51° 9′ W.; lat. 45° 4.5′ N.; long. 24° 42′ W.

CHAPTER IX

ACUTE ANGLES NEAR 0° OR 90°

88. When the angle x approaches the limit zero, each of the ratios $\dfrac{\sin x}{x}$, $\dfrac{\tan x}{x}$, approaches unity as a limit, x being the circular measure of the angle.

Proof. Let O be the center of a circle whose radius is unity. Let arc $AP = x$, and let arc $AP' = x$ in numerical value. Draw PP', and let PT and $P'T$ be the tangents drawn to the circle at P and P'. From Geometry

$$(A) \qquad PQP' < PAP' < PTP'.$$

But $\quad PQP' = PQ + QP' = 2 \sin x$ in numerical value,

$$PAP' = PA + AP' = 2\,x \text{ in numerical value,}$$

and $\quad PTP' = PT + TP' = 2 \tan x$ in numerical value.

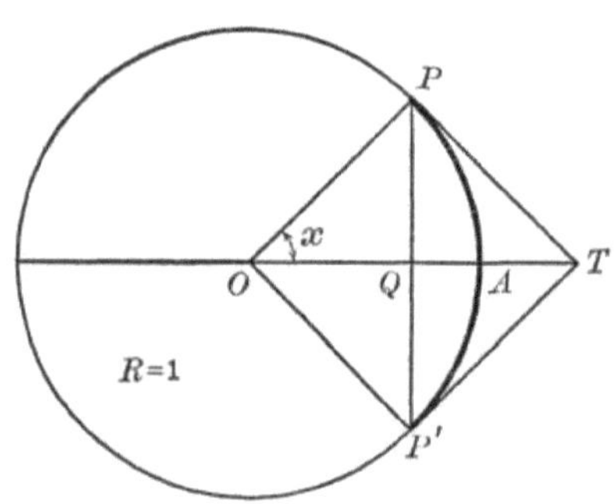

Substituting in (A),

$$2 \sin x < 2\,x < 2 \tan x.$$

Dividing through by 2, we have

$$(B) \quad \sin x < x < \tan x,$$

which proves that

If x be the circular measure of an acute angle, it will always lie between $\sin x$ and $\tan x$, being greater than $\sin x$ and less than $\tan x$.

Dividing (B) through by $\sin x$, we get

$$1 < \frac{x}{\sin x} < \frac{1}{\cos x}.$$

If we now let x approach the limit zero, it is seen that

$$\operatorname*{limit}_{x=0} \frac{x}{\sin x}$$

must lie between the constant 1 and $\operatorname*{limit}_{x=0} \dfrac{1}{\cos x}$, which is also 1.

Hence
$$\lim_{x=0} \frac{x}{\sin x} = 1, \text{ or,}$$

(C)
$$\lim_{x=0} \frac{\sin x}{x} = 1.$$

Similarly, if we divide (B) through by tan x, we get

$$\cos x < \frac{x}{\tan x} < 1.$$

As before, if x approaches zero as a limit,

$$\lim_{x=0} \frac{\tan x}{x}$$

must lie between the constant 1 and $\lim_{x=0} \cos x$, which is also 1.

Hence
$$\lim_{x=0} \frac{x}{\tan x} = 1, \text{ or,}$$

(D)
$$\lim_{x=0} \frac{\tan x}{x} = 1.$$

The limits (C) and (D) are of great importance both in pure and applied mathematics. These results may be stated as follows:

When x is the circular measure of a very small angle we may replace sin x and tan x in our calculations by x.

89. Functions of positive acute angles near 0° and 90°. So far we have assumed that the differences in the trigonometric functions are proportional to the differences in the corresponding angles. While this is not strictly true, it is in general sufficiently exact for most practical purposes unless the angles are very near 0° or 90°. In using logarithms we have also assumed that the differences in the logarithms of the trigonometric functions are proportional to the differences in the corresponding angles. This will give sufficiently accurate results for most purposes if we use Tables II or III in this book and confine ourselves to angles between .3° (= 18') and 89.7° (= 89° 42') inclusive. If, however, we have an angle between 0° and .3° (= 18') or one between 89.7° (= 89° 42') and 90°, and are looking for exact results, it is evident that the ordinary method will not do. For example, the tabular difference (Table II) between the logarithmic sine, tangent, or cotangent of 8' and the logarithm of the corresponding functions of 9' is 512, while between 9' and 10' it is 457. If we interpolate here in the usual way it is evident that our results will be only approximately correct. In case it is desired to obtain more accurate results we may use the principle established in the last section, namely :

We may replace sin x and tan x in our calculations by x when x is a very small angle and is expressed in circular measure.

From a table giving the natural functions of angles, we have

$$\sin 2.2° = 0.03839 = 0.0384,$$
$$\tan 2.2° = 0.03842 = 0.0384.$$

Also $\qquad 2.2° = 0.0384$ radians.

Hence it is seen that in any calculation we may replace the sine or tangent of any angle between 0° and 2.2° by the circular measure of the angle without changing the first four significant figures of the result. Also since

$$\cos 87.8° = \sin (90° - 87.8°) = \sin 2.2° = 0.0384,$$
$$\cot 87.8° = \tan (90° - 87.8°) = \tan 2.2° = 0.0384,$$

and $\qquad 2.2° = 90° - 87.8° = 0.0384$ radians,

we may replace the cosine or cotangent of any angle between 87.8° and 90° by the circular measure of the complement of that angle. We may then state the following rules :

90. Rule for finding the functions of acute angles near 0°.

$$\sin x = circular\ measure\ of\ x,^*$$
$$\tan x = circular\ measure\ of\ x,$$
$$\cot x = \frac{1}{circular\ measure\ of\ x},$$

$$cos\ x\ is\ found\ from\ the\ tables\ in\ the\ usual\ way.†$$

---

* The following equivalents may be used for reducing an angle to circular measure (radians), and in other computations.

$$1° = \frac{\pi}{180}\ \text{radians.}$$

$1° = 0.0174533$ radians,	$\log 0.0174533 = 8.2419 - 10.$
$1' = 0.0002909$ radians,	$\log 0.0002909 = 6.4637 - 10.$
$1'' = 0.0000048$ radians,	$\log 0.0000048 = 4.6856 - 10.$
$\dfrac{180°}{\pi} = 57.29578° = 1$ radian,	$\log 57.29578 = 1.7581.$
$\pi = 3.14159$	$\log \pi = 0.4971.$
$\quad = \frac{22}{7}$ approximately.	

† csc x and sec x are simply the reciprocals of sin x and cos x respectively.

91. Rule for finding the functions of acute angles near 90°.

$$\cos x = \text{circular measure of the complement of } x,^*$$
$$\cot x = \text{circular measure of the complement of } x,$$
$$\tan x = \frac{1}{\text{circular measure of the complement of } x},$$

sin x is found from the tables in the usual way.†

Since any function of an angle of any magnitude whatever, positive or negative, equals some function of a positive acute angle, it is evident that the above rules, together with those on p. 57, will suffice for finding the functions of angles near $\pm 90°$, $\pm 180°$, $\pm 270°$, $\pm 360°$.

Ex. 1. Find sine, tangent, and cotangent of 42′.

Solution. Reducing the angle to radians,

$$42' = 42 \times 0.0002909 \text{ radians} = 0.01222 \text{ radians.}$$

Therefore
$$\sin 42' = 0.01222,$$
$$\tan 42' = 0.01222,$$
$$\cot 42' = \frac{1}{0.01222} = 81.833. \quad \textit{Ans.}$$

Ex. 2. Find cosine, cotangent, and tangent of 89° 34.6′.

Solution. The complement of our angle is $90° - 89° \, 34.6' = 25.4'$. Reducing this remainder to radians,

$$25.4' = 25.4 \times 0.0002909 \text{ radians} = 0.00739 \text{ radians.}$$

Therefore
$$\cos 89° \, 34.6' = 0.00739,$$
$$\cot 89° \, 34.6' = 0.00739,$$
$$\tan 89° \, 34.6' = \frac{1}{0.00739} = 135.32. \quad \textit{Ans.}$$

When the function of a positive acute angle near 0° or 90° is given, to find the angle itself we reverse the process illustrated above. For instance:

Ex. 3. Find the angle subtended by a man 6 ft. tall at a distance of 1225 ft.

Solution.

From the figure $\tan x = \frac{6}{1225}$.

But, since the angle is very small, we may replace $\tan x$ by x, giving

$$x = \frac{6}{1225} \text{ radians} = 0.0049 \text{ radians.}$$

Or, reducing the angle to minutes of arc, we get

$$x = \frac{0.0049}{0.0002909} \text{ minutes of arc} = 16.8'. \quad \textit{Ans.}$$

---

* If the angle is given in degrees, subtract it from 90° and reduce the remainder to circular measure (radians). If the angle is given in circular measure (radians), simply subtract it from $\frac{\pi}{2}$ ($= 1.57079$).

† csc x and sec x are simply the reciprocals of sin x and cos x respectively.

92. Rules for finding the logarithms of the functions of angles near 0° and 90°.* For use in logarithmic computations the rules of the last two sections may be put in the following form:

If the angle is given in degrees, minutes, and seconds, it should first be reduced to degrees and the decimal part of a degree (see Conversion Table on p. 17 of Tables).

Rule I.   To find the logarithms of the functions of angles near 0°.

$$log\ sin\ x° = \overline{2}.2419 + log\ x.†$$
$$log\ tan\ x° = \overline{2}.2419 + log\ x.$$
$$log\ cot\ x° = 1.7581 - log\ x.‡$$
$$log\ cos\ x°\ is\ found\ from\ the\ tables\ in\ the\ usual\ way.$$

Rule II.   To find the logarithms of the functions of an angle near 90°.

$$log\ cos\ x° = \overline{2}.2419 + log\ (90 - x).$$
$$log\ cot\ x° = \overline{2}.2419 + log\ (90 - x).$$
$$log\ tan\ x° = 1.7581 - log\ (90 - x).$$
$$log\ sin\ x°\ is\ found\ from\ the\ tables\ in\ the\ usual\ way.$$

Ex. 1.   Find log tan 0.045°.

Solution. As is indicated in our logarithmic tables, ordinary interpolation will not give accurate results in this case. But from the above rule,

$$log\ tan\ 0.045° = \overline{2}.2419 + log\ 0.045$$
$$= \overline{2}.2419 + \overline{2}.6532.$$
$$\therefore\ log\ tan\ 0.045° = \overline{4}.8951.\ Ans.$$

On consulting a much larger table of logarithms, this result is found to be exact to four decimal places. Interpolating in the ordinary way, we get

$$log\ tan\ 0.045° = \overline{4}.8924,$$

which is correct to only two decimal places.

---

* These rules will give results accurate to four decimal places for all angles between 0° and 1.1° and between 88.9° and 90°.

† Since 1 degree = 0.017453 radians, the circular measure of

$$x\ degrees = 0.017453 \cdot x\ radians.$$

Hence, from p. 180, $sin\ x° = 0.017453 \cdot x,$
and $log\ sin\ x° = log\ 0.017453 + log\ x$
$$= 2.2419 + log\ x.$$

‡ From p. 180, $cot\ x° = \dfrac{1}{0.017453 \cdot x},$
and $log\ cot\ x° = -\ log\ 0.017453 - log\ x$
$$= 1.7581 - log\ x.$$

Ex. 2. Find log tan 89.935°.

Solution. From the above rule,

$$\log \tan 89.935° = 1.7581 - \log(90 - 89.935)$$
$$= 1.7581 - \log 0.065$$
$$= 1.7581 - 2.8129.$$

$$\therefore \ \log \tan 89.935° = 2.9452. \quad Ans.$$

If the tangent itself is desired, we look up the number in Table **I** corresponding to this logarithm. This gives

$$\tan 89.935° = 881.4.$$

93. Consistent measurements and calculations. In the examples given so far in this book it has generally been assumed that the given data were exact. That is, if two sides and the included angle of a triangle were given, as 135 ft., 217 ft., and 25.3° respectively, we have taken for granted that these numbers were not subject to errors made in measurement. This is in accordance with the plan followed in the problems that the student has solved in Arithmetic, Algebra, and Geometry. It should not be forgotten, however, that when we apply the principles of Trigonometry to the solution of practical problems, — engineering problems, for instance, — it is usually necessary to use data which have been found by actual measurement, and therefore are subject to error. In taking these measurements one should carefully see that they are made with about the same degree of accuracy. Thus, it would evidently be folly to measure one side of a triangle with much greater care than another, for, in combining these measurements in a calculation, the result would at best be no more accurate than the worst measurement. Similarly, the angles of a triangle should be measured with the same care as the sides.

The number of significant figures in a measurement is supposed to indicate the care that was intended when the measurement was made, and any two measurements showing the same number of significant figures will, in general, show about the same relative care in measurement. If the sides of a rectangle are about 936 ft. and 8 ft., the short side should be measured to at least two decimal places. A neglected 4 in the tenths place will alter the area by 374 sq. ft.

The following directions will help us to make consistent measurements and avoid unnecessary work in our calculations.

1. Let all measured lines and calculated lines show the same number of significant figures, as a rule.

2. When the lines show only one significant figure, let the angles read to the nearest 5°.

3. When the lines show two significant figures, let the angles read to the nearest half degree.

4. When the lines show three significant figures, let the angles read to the nearest 5'.

5. When the lines show four significant figures, let the angles read to the nearest minute.

EXAMPLES

1. The inclination of a railway to the horizontal is 40'. How many feet does it rise in a mile ? *Ans.* 61.43.

2. Given that the moon's distance from the earth is 238,885 mi. and subtends an angle of 31' 8" at the earth. Find the diameter of the moon in miles.

Ans. 2163.5.

3. Given that the sun's distance from the earth is 92,000,000 mi. and subtends an angle of 32' 4" at the earth. Find the sun's diameter. *Ans.* 858,200 mi.

4. Given that the earth's radius is 3963 mi. and subtends an angle of 57' 2" at the moon. Find the distance of the moon from the earth. *Ans.* 238,833 mi.

5. Given that the radius of the earth is 3963 mi. and subtends an angle of 9" at the sun. Find the distance of the sun from the earth. *Ans.* 90,840,000 mi.

6. Assuming that the sun subtends an angle of 32' 4" at the earth, how far from the eye must a dime be held so as to just hide the sun, the diameter of a dime being $\frac{9}{7}$ in. ? *Ans.* 69.83 in.

7. Find the angle subtended by a circular target 5 ft. in diameter at the distance of half a mile. *Ans.* 6' 30.6".

MISCELLANEOUS EXAMPLES

1. A balloon is at a height of 2500 ft. above a plain and its angle of elevation at a point in the plain is 40° 35'. How far is this point from the balloon ?

Ans. 3843 ft.

2. A tower standing on a horizontal plain subtends an angle of 37° 19.5' at a point in the plain distant 369.5 ft. from the foot of the tower. Find the height of the tower. *Ans.* 281.8 ft.

3. The shadow of a steeple on a horizontal plain is observed to be 176.23 ft. when the elevation of the sun is 33.2°. Find the height of the tower.

Ans. 115.3 ft.

4. From the top of a lighthouse 112.5 ft. high, the angles of depression of two ships, when the line joining the ships passes through the foot of the lighthouse, are 27.3° and 20.6° respectively. Find the distance between the ships.

Ans. 81 ft.

5. From the top of a cliff the angles of depression of the top and bottom of a lighthouse 97.25 ft. high are observed to be 23° 17' and 24° 19' respectively. How much higher is the cliff than the lighthouse ? *Ans.* 1947 ft.

6. The angle of elevation of a balloon from a station due south of it is 47° 18.5', and from another station due west of the former and 671.4 ft. from it the elevation is 41° 14'. Find the height of the balloon. *Ans.* 1000 ft.

7. A ladder placed at an angle of 75° with the street just reaches the sill of a window 27 ft. above the ground on one side of the street. On turning the ladder over without moving its foot, it is found that when it rests against a wall on the other side of the street it is at an angle of 15° with the street. Find the breadth of the street.   *Ans.* 34.24 ft.

8. A man traveling due west along a straight road observes that when he is due south of a certain windmill the straight line drawn to a distant church tower makes an angle of 30° with the direction of the road. A mile farther on the bearings of the windmill and church tower are N.E. and N.W. respectively. Find the distances of the tower from the windmill and from the nearest point on the road.   *Ans.* 2.39 mi., 1.37 mi.

9. Standing at a certain point, I observe the elevation of a house to be 45°, and the sill of one of its windows, known to be 20 ft. above the ground, subtends an angle of 20° at the same point. Find the height of the house.

Ans. 54.94 ft.

10. A hill is inclined 36° to the horizon. An observer walks 100 yd. away from the foot of the hill, and then finds that the elevation of a point halfway up the hill is 18°. Find the height of the hill.   *Ans.* 117.58 yd.

11. Two straight roads, inclined to one another at an angle of 60°, lead from a town A to two villages B and C; B on one road distant 30 mi. from A, and C on the other road distant 15 mi. from A. Find the distance from B to C.

Ans. 25.98 mi.

12. Two ships leave harbor together, one sailing N.E. at the rate of $7\frac{1}{2}$ mi. an hour and the other sailing north at the rate of 10 mi. an hour. Prove that the distance between the ships after an hour and a half is 10.6 mi.

13. A and B are two positions on opposite sides of a mountain; C is a point visible from A and B. From A to C and from B to C are 10 mi. and 8 mi. respectively, and the angle BCA is 60°. Prove that the distance between A and B is 9.165 mi.

14. A and B are two consecutive milestones on a straight road and C is a distant spire. The angles ABC and BAC are observed to be 120° and 45° respectively. Show that the distance of the spire from A is 3.346 mi.

15. If the spire C in the last example stands on a hill, and its angle of elevation at A is 15°, show that it is .866 mi. higher than A.

16. If in Example 14 there is another spire D such that the angles DBA and DAB are 45° and 90° respectively and the angle DAC is 45°, prove that the distance from C to D is very nearly $2\frac{3}{4}$ mi.

17. A and B are consecutive milestones on a straight road; C is the top of a distant mountain. At A the angle CAB is observed to be 38° 19′; at B the angle CBA is observed to be 132° 42′, and the angle of elevation of C at B is 10° 15′. Show that the top of the mountain is 1243.7 yd. higher than B.

18. A base line AB, 1000 ft. long, is measured along the straight bank of a river; C is an object on the opposite bank; the angles BAC and CBA are observed to be 65° 37′ and 53° 4′ respectively. Prove that the perpendicular breadth of the river at C is 829.8 ft.

19. The altitude of a certain rock is observed to be 47°, and after walking 1000 ft. towards the rock, up a slope inclined at an angle of 32° to the horizon, the observer finds that the altitude is 77°. Prove that the vertical height of the rock above the first point of observation is 1034 ft.

20. A privateer 10 mi. S.W. of a harbor sees a ship sail from it in a direction S. 80° E., at a rate of 9 mi. an hour. In what direction and at what rate must the privateer sail in order to come up with the ship in $1\frac{1}{2}$ hr.?

Ans. N. 76° 56′ E. 13.9 mi. per hour.

21. At the top of a chimney 150 ft. high, standing at one corner of a triangular yard, the angle subtended by the adjacent sides of the yard are 30° and 45° respectively, while that subtended by the opposite side is 30°. Show that the lengths of the sides are 150 ft., 86.6 ft., and 106.8 ft. respectively.

22. A person goes 70 yd. up a slope of 1 in $3\frac{1}{2}$ from the edge of a river, and observes the angle of depression of an object on the opposite bank to be $2\frac{1}{4}$°. Find the breadth of the river. *Ans.* 442.8 yd.

23. A flagstaff h ft. high stands on the top of a tower. From a point in the plain on which the tower stands the angles of elevation of the top and bottom of the flagstaff are observed to be α and β respectively. Prove that the height of the tower is $\dfrac{h\tan\beta}{\tan\alpha-\tan\beta}$ ft., i.e. $\dfrac{h\sin\beta\cdot\cos\alpha}{\sin(\alpha-\beta)}$ ft.

24. The length of a lake subtends at a certain point an angle of 46° 24′, and the distances from this point to the two extremities of the lake are 346 and 290 ft. Find the length of the lake. *Ans.* 255.8 ft.

25. From the top of a cliff h ft. high the angles of depression of two ships at sea in a line with the foot of the cliff are α and β respectively. Show that the distance between the ships is $h\,(\cot\beta-\cot\alpha)$ ft.

26. Two ships are a mile apart. The angular distance of the first ship from a fort on shore, as observed from the second ship, is 35° 14′ 10″; the angular distance of the second ship from the fort, observed from the first ship, is 42° 11′ 53″. Find the distance in feet from each ship to the fort.

Ans. 3121 ft., 3634 ft.

27. The angular elevation of a tower at a place due south of it is α, and at another place due west of the first and distant d from it, the elevation is β. Prove that the height of the tower is

$$\frac{d}{\sqrt{\cot^2\beta-\cot^2\alpha}}, \text{ i.e. } \frac{d\sin\alpha\cdot\sin\beta}{\sqrt{\sin(\alpha-\beta)\cdot\sin(\alpha+\beta)}}.$$

28. To find the distance of an inaccessible point C from either of two points A and B, having no instruments to measure angles. Prolong CA to a, and CB to b, and join AB, Ab, and Ba. Measure AB, 500; aA, 100; aB, 560; bB, 100; and Ab, 550. *Ans.* 500 and 536.

29. A man stands on the top of the wall of height h and observes the angular elevation α of the top of a telegraph post; he then descends from the wall and finds that the angular elevation is now β; prove that the height of the post exceeds the height of the man by $h\dfrac{\sin\beta\cdot\cos\alpha}{\sin(\beta-\alpha)}$.

30. Two inaccessible points A and B are visible from D, but no other point can be found whence both are visible. Take some point C, whence A and D can be seen, and measure CD, 200 ft.; ADC, 89°; ACD, 50° 30′. Then take some point E, whence D and B are visible, and measure DE, 200; BDE, 54° 30′; BED, 88° 30′. At D measure ADB, 72° 30′. Compute the distance AB.

Ans. 345.4 ft.

31. The angle of elevation of an inaccessible tower situated on a horizontal plane is 63° 26′; at a point 500 ft. farther from the base of the tower the elevation of its top is 32° 14′. Find the height of the tower. *Ans.* 460.5 ft.

32. To compute the horizontal distance between two inaccessible points A and B, when no point can be found whence both can be seen. Take two points C and D, distant 200 yd., so that A can be seen from C, and B from D. From C measure CF, 200 yd. to F, whence A can be seen; and from D measure DE, 200 yd. to E, whence B can be seen. Measure AFC, 83°; ACD, 53° 30′; ACF, 54° 31′; BDE, 54° 30′; BDC, 156° 25′; DEB, 88° 30′. *Ans.* 345.3 yd.

33. A tower is situated on the bank of a river. From the opposite bank the angle of elevation of the tower is 60° 13′, and from a point 40 ft. more distant the elevation is 50° 19′. Find the breadth of the river. *Ans.* 88.9 ft.

34. A ship sailing north sees two lighthouses 8 mi. apart, in a line due west; after an hour's sailing one lighthouse bears S.W. and the other S.S.W. Find the ship's rate. *Ans.* 13.6 mi. per hour.

35. A column in the north temperate zone is east-southeast of an observer, and at noon the extremity of its shadow is northeast of him. The shadow is 80 ft. in length, and the elevation of the column at the observer's station is 45°. Find the height of the column. *Ans.* 61.23 ft.

36. At a distance of 40 ft. from the foot of a tower on an inclined plane the tower subtends an angle of 41° 19′; at a point 60 ft. farther away the angle subtended by the tower is 23° 45′. Find the height of the tower. *Ans.* 56.5 ft.

37. A tower makes an angle of 113° 12′ with the inclined plane on which it stands; and at a distance of 89 ft. from its base, measured down the plane, the angle subtended by the tower is 23° 27′. Find the height of the tower.

Ans. 51.6 ft.

38. From the top of a hill the angles of depression of two objects situated in the horizontal plane of the base of the hill are 45° and 30°; and the horizontal angle between the two objects is 30°. Show that the height of the hill is equal to the distance between the objects.

39. I observe the angular elevation of the summits of two spires which appear in a straight line to be α, and the angular depressions of their reflections in still water to be β and γ. If the height of my eye above the level of the water be c, then the horizontal distance between the spires is

$$\frac{2\,c\,\cos^2\alpha\,\sin(\beta - \gamma)}{\sin(\beta - \alpha)\sin(\gamma - \alpha)}.$$

40. The angular elevation of a tower due south at a place A is 30°, and at a place B, due west of A and at a distance a from it, the elevation is 18°. Show that the height of the tower is $\dfrac{a}{\sqrt{2\sqrt{5} + 2}}$.

41. A boy standing c ft. behind and opposite the middle of a football goal sees that the angle of elevation of the nearer crossbar is A and the angle of elevation of the farther one is B. Show that the length of the field is $c(\tan A \cot B - 1)$.

42. A valley is crossed by a horizontal bridge whose length is l. The sides of the valley make angles A and B with the horizon. Show that the height of the bridge above the bottom of the valley is $\dfrac{l}{\cot A + \cot B}$.

43. A tower is situated on a horizontal plane at a distance a from the base of a hill whose inclination is α. A person on the hill, looking over the tower, can just see a pond, the distance of which from the tower is b. Show that, if the distance of the observer from the foot of the hill be c, the height of the tower is $\dfrac{bc \sin \alpha}{a + b + c \cos \alpha}$.

44. From a point on a hillside of constant inclination the angle of elevation of the top of an obelisk on its summit is observed to be α, and a ft. nearer to the top of the hill to be β; show that if h be the height of the obelisk, the inclination of the hill to the horizon will be

$$\cos^{-1}\left\{\frac{a}{h} \cdot \frac{\sin \alpha \sin \beta}{\sin(\beta - \alpha)}\right\}.$$

CHAPTER X

RECAPITULATION OF FORMULAS

Plane Trigonometry

Right triangles, pp. 2–11.

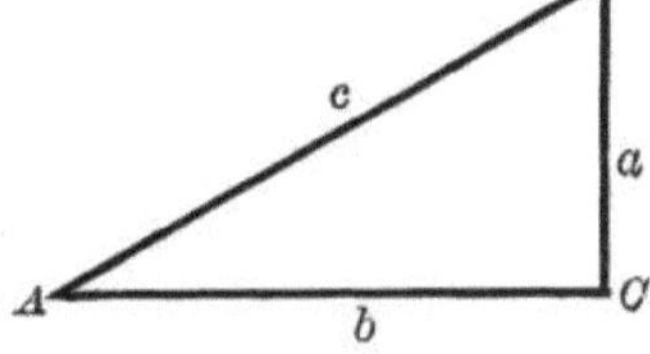

$$(1)\ \sin A = \frac{a}{c}. \qquad (4)\ \csc A = \frac{c}{a}.$$

$$(2)\ \cos A = \frac{b}{c}. \qquad (5)\ \sec A = \frac{c}{b}.$$

$$(3)\ \tan A = \frac{a}{b}. \qquad (6)\ \cot A = \frac{b}{a}.$$

(7) Side opposite an acute angle

$$= \text{hypotenuse} \times \text{sine of the angle.}$$

(8) Side adjacent an acute angle

$$= \text{hypotenuse} \times \text{cosine of the angle.}$$

(9) Side opposite an acute angle

$$= \text{adjacent side} \times \text{tangent of the angle.}$$

Fundamental relations between the functions, p. 59.

$$(19)\ \sin x = \frac{1}{\csc x}, \qquad \csc x = \frac{1}{\sin x}.$$

$$(20)\ \cos x = \frac{1}{\sec x}, \qquad \sec x = \frac{1}{\cos x}.$$

$$(21)\ \tan x = \frac{1}{\cot x}, \qquad \cot x = \frac{1}{\tan x}.$$

$$(22)\ \tan x = \frac{\sin x}{\cos x}, \qquad \cot x = \frac{\cos x}{\sin x}.$$

$$(23)\ \sin^2 x + \cos^2 x = 1.$$

$$(24)\ \sec^2 x = 1 + \tan^2 x. \qquad (25)\ \csc^2 x = 1 + \cot^2 x.$$

Functions of the sum and of the difference of two angles, pp. 63–69.

$$(40)\qquad \sin(x + y) = \sin x \cos y + \cos x \sin y.$$

$$(41)\qquad \sin(x - y) = \sin x \cos y - \cos x \sin y.$$

$$(42)\qquad \cos(x + y) = \cos x \cos y - \sin x \sin y.$$

$$(43)\qquad \cos(x - y) = \cos x \cos y + \sin x \sin y.$$

$$(44) \qquad \tan(x + y) = \frac{\tan x + \tan y}{1 - \tan x \tan y}.$$

$$(45) \qquad \tan(x - y) = \frac{\tan x - \tan y}{1 + \tan x \tan y}.$$

$$(46) \qquad \cot(x + y) = \frac{\cot x \cot y - 1}{\cot y + \cot x}.$$

$$(47) \qquad \cot(x - y) = \frac{\cot x \cot y + 1}{\cot y - \cot x}.$$

Functions of twice an angle, p. 70.

$$(48) \qquad \sin 2x = 2 \sin x \cos x.$$

$$(49) \qquad \cos 2x = \cos^2 x - \sin^2 x.$$

$$(50) \qquad \tan 2x = \frac{2 \tan x}{1 - \tan^2 x}.$$

Functions of an angle in terms of functions of half the angle, p. 72.

$$(51) \qquad \sin x = 2 \sin \frac{x}{2} \cos \frac{x}{2}.$$

$$(52) \qquad \cos x = \cos^2 \frac{x}{2} - \sin^2 \frac{x}{2}.$$

$$(53) \qquad \tan x = \frac{2 \tan \frac{x}{2}}{1 - \tan^2 \frac{x}{2}}.$$

Functions of half an angle, pp. 72–73.

$$(54) \quad \sin \frac{x}{2} = \pm \sqrt{\frac{1 - \cos x}{2}}. \qquad\qquad (58) \quad \tan \frac{x}{2} = \frac{1 - \cos x}{\sin x}.$$

$$(55) \quad \cos \frac{x}{2} = \pm \sqrt{\frac{1 + \cos x}{2}}. \qquad\qquad (59) \quad \cot \frac{x}{2} = \pm \sqrt{\frac{1 + \cos x}{1 - \cos x}}.$$

$$(56) \quad \tan \frac{x}{2} = \pm \sqrt{\frac{1 - \cos x}{1 + \cos x}}. \qquad (60) \quad \cot \frac{x}{2} = \frac{1 + \cos x}{\sin x}.$$

$$(57) \quad \tan \frac{x}{2} = \frac{\sin x}{1 + \cos x}. \qquad\qquad (61) \quad \cot \frac{x}{2} = \frac{\sin x}{1 - \cos x}.$$

Sums and differences of functions, p. 74.

$$(62) \qquad \sin A + \sin B = 2 \sin \tfrac{1}{2}(A + B) \cos \tfrac{1}{2}(A - B).$$

$$(63) \qquad \sin A - \sin B = 2 \cos \tfrac{1}{2}(A + B) \sin \tfrac{1}{2}(A - B).$$

$$(64) \qquad \cos A + \cos B = 2 \cos \tfrac{1}{2}(A + B) \cos \tfrac{1}{2}(A - B).$$

$$(65) \qquad \cos A - \cos B = -2 \sin \tfrac{1}{2}(A + B) \sin \tfrac{1}{2}(A - B).$$

$$(66) \qquad \frac{\sin A + \sin B}{\sin A - \sin B} = \frac{\tan \tfrac{1}{2}(A + B)}{\tan \tfrac{1}{2}(A - B)}.$$

Law of sines, p. 102.

$$(72) \qquad \frac{a}{\sin A} = \frac{b}{\sin B} = \frac{c}{\sin C}.$$

Law of cosines, p. 109.

$$(73) \qquad a^2 = b^2 + c^2 - 2\,bc\,\cos A.$$

Law of tangents, p. 112.

$$(77) \qquad \frac{a+b}{a-b} = \frac{\tan \frac{1}{2}(A+B)}{\tan \frac{1}{2}(A-B)}.$$

Functions of the half angles of a triangle in terms of the sides, pp. 113–115.

$$s = \tfrac{1}{2}(a+b+c).$$

$$(81) \qquad \sin \tfrac{1}{2} A = \sqrt{\frac{(s-b)(s-c)}{bc}}.$$

$$(82) \qquad \cos \tfrac{1}{2} A = \sqrt{\frac{s(s-a)}{bc}}.$$

$$(83) \qquad \tan \tfrac{1}{2} A = \sqrt{\frac{(s-b)(s-c)}{s(s-a)}}.$$

$$(84) \qquad r = \sqrt{\frac{(s-a)(s-b)(s-c)}{s}}.$$

$$(85) \qquad \tan \tfrac{1}{2} A = \frac{r}{s-a}.$$

$$(86) \qquad \tan \tfrac{1}{2} B = \frac{r}{s-b}.$$

$$(87) \qquad \tan \tfrac{1}{2} C = \frac{r}{s-c}.$$

Area of a triangle, p. 117.

$$(88) \qquad S = \tfrac{1}{2} bc \sin A.$$

$$(89) \qquad S = \sqrt{s(s-a)(s-b)(s-c)}.$$

SPHERICAL TRIGONOMETRY

CHAPTER I

RIGHT SPHERICAL TRIANGLES

1. Correspondence between the face angles and the diedral angles of a triedral angle on the one hand, and the sides and angles of a spherical triangle on the other. Take any triedral angle $O-A'B'C'$ and let a sphere of any radius, as OA, be described about the vertex O as a center. The intersections of this sphere with the faces of the

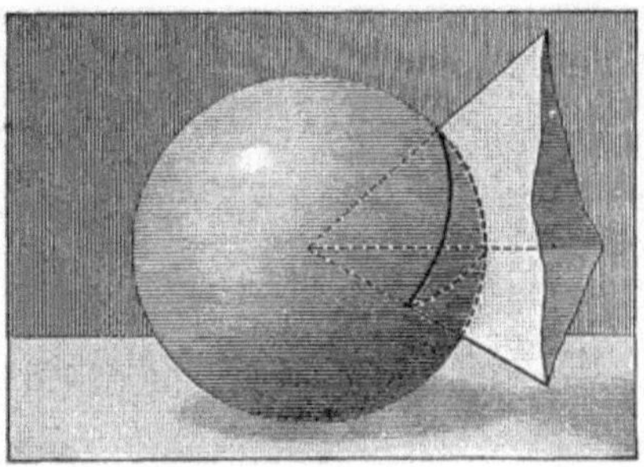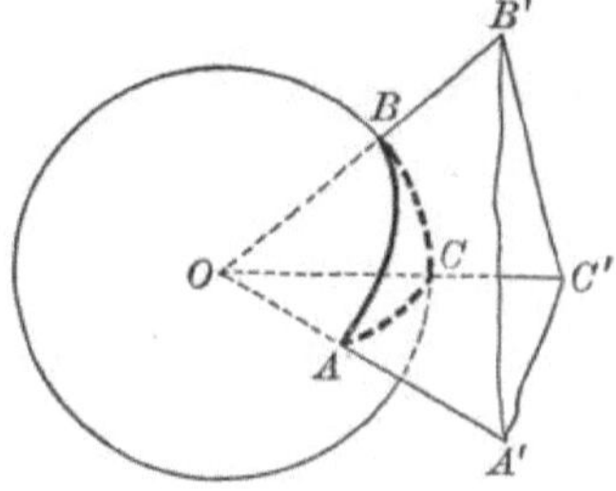

triedral angle will be three arcs of great circles of the sphere, forming a spherical triangle, as ABC. The sides (arcs) AB, BC, CA of this triangle measure the face angles $A'OB'$, $B'OC'$, $C'OA'$ of the triedral angle. The angles ABC, BCA, CAB, are measured by the plane angles which also measure the diedral angles of the triedral angle; for, by Geometry, each is measured by the angle between two straight lines drawn, one in each face, perpendicular to the edge at the same point.

Spherical Trigonometry treats of the trigonometric relations between the six elements (three sides and three angles) of a spherical triangle; or, what amounts to the same thing, between the face and diedral angles of the triedral angle which intercepts it, as shown in the figure. Hence we have the

Theorem. *From any property of triedral angles an analogous property of spherical triangles can be inferred, and vice versa.*

It is evident that the face and diedral angles of the triedral angle are not altered in magnitude by varying the radius of the sphere; hence the relations between the sides and angles of a spherical triangle are independent of the length of the radius.

The sides of a spherical triangle, being arcs, are usually expressed in degrees.* The length of a side (arc) may be found in terms of any linear unit from the proportion

circumference of great circle : length of arc :: 360° : degrees in arc.

A side or an angle of a spherical triangle may have any value from 0° to 360°, but any spherical triangle can always be made to depend on a spherical triangle having each element less than 180°.

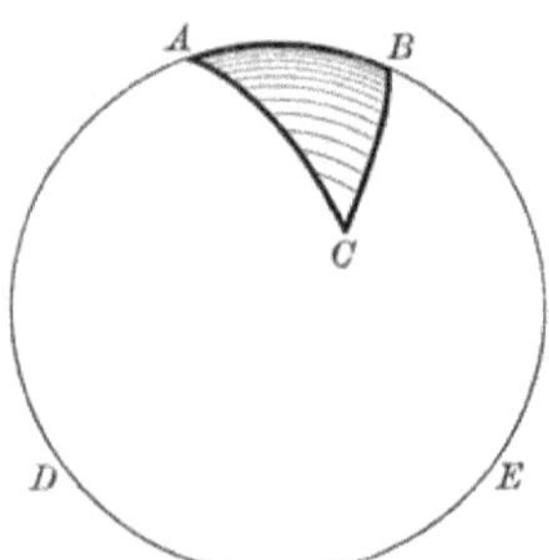

Thus, a triangle such as $ADEBC$ (unshaded portion of hemisphere in figure), which has a side $ADEB$ greater than 180°, need not be considered, for its parts can be immediately found from the parts of the triangle ABC, each of whose sides is less than 180°. For arc $ADEB = 360° -$ arc AB, angle $CAD = 180° -$ angle CAB, etc. Only triangles whose elements are less than 180° are considered in this book.

2. Properties of spherical triangles. The proofs of the following properties of spherical triangles may be found in any treatise on Spherical Geometry:

(*a*) Either side of a spherical triangle is less than the sum of the other two sides.

(*b*) If two sides of a spherical triangle are unequal, the angles opposite them are unequal, and the greater angle lies opposite the greater side, and conversely.

(*c*) The sum of the sides of a spherical triangle is less than 360°.†

(*d*) The sum of the angles of a spherical triangle is greater than 180° and less than 540°.‡

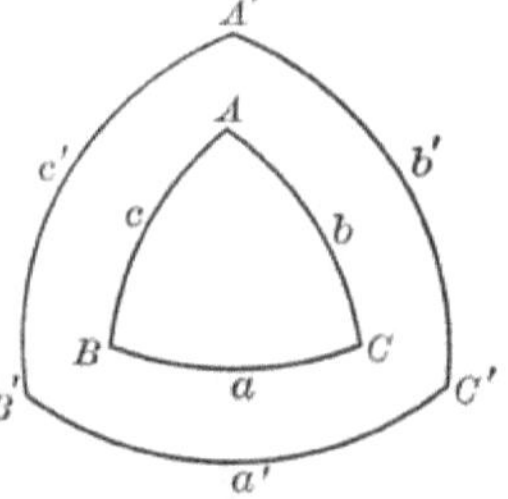

(*e*) If $A'B'C'$ is the polar triangle* of ABC, then, conversely, ABC is the polar triangle of $A'B'C'$.

(*f*) In two polar triangles each angle of one is the supplement of the side lying opposite to it in the other. Applying this to the last figure, we get

$$A = 180° - a', \qquad B = 180° - b', \qquad C = 180° - c',$$
$$A' = 180° - a, \qquad B' = 180° - b, \qquad C' = 180° - c.$$

A spherical triangle which has one or more right angles is called a *right spherical triangle*.

EXAMPLES

1. Find the sides of the polar triangles of the spherical triangles whose angles are as follows. Draw the figure in each case.

(a) $A = 70°$, $B = 80°$, $C = 100°$.   *Ans.* $a' = 110°$, $b' = 100°$, $c' = 80°$.

(b) $A = 56°$, $B = 97°$, $C = 112°$.

(c) $A = 68° \ 14'$, $B = 52° \ 10'$, $C = 98° \ 44'$.

(d) $A = 115.6°$, $B = 89.9°$, $C = 74.2°$.

2. Find the angles of the polar triangles of the spherical triangles whose sides are as follows:

(a) $a = 94°$, $b = 52°$, $c = 100°$.   *Ans.* $A' = 86°$, $B' = 128°$, $C' = 80°$.

(b) $a = 74° \ 42'$, $b = 95° \ 6'$, $c = 66° \ 25'$.

(c) $a = 106.4°$, $b = 64.3°$, $c = 51.7°$.

3. If a triangle has three right angles, show that the sides of the triangle are quadrants.

4. Show that if a triangle has two right angles, the sides opposite these angles are quadrants, and the third angle is measured by the opposite side.

5. Find the lengths of the sides of the triangles in Example 2 if the radius of the sphere is 4 ft.

3. Formulas relating to right spherical triangles. From the above Examples 3 and 4, it is evident that the only kind of right spherical triangle that requires further investigation is that which contains *only one* right angle.

In the figure shown on the next page let ABC be a right spherical triangle having only one right angle, the center of the sphere being at O. Let C be the right angle, and suppose first that each of the other elements is less than 90°, the radius of the sphere being unity.

---

* The *polar triangle* of any spherical triangle is constructed by describing arcs of great circles about the vertices of the original triangle as poles.

Pass an auxiliary plane through B perpendicular to OA, cutting OA at E and OC at D. Draw BE, BD, and DE. BE and DE are each perpendicular to OA;

[If a straight line is $\perp$ to a plane, it is $\perp$ to every line in the plane.]

therefore angle $BED =$ angle A. The plane BDE is perpendicular to the plane AOC;

$$\left[\begin{array}{l}\text{If a straight line is } \perp \text{ to a plane, every plane} \\ \text{passed through the line is } \perp \text{ to the first plane.}\end{array}\right]$$

hence BD, which is the intersection of the planes BDE and BOC, is perpendicular to the plane AOC,

$$\left[\begin{array}{l}\text{If two intersecting planes are each } \perp \text{ to a third} \\ \text{plane, their intersection is also } \perp \text{ to that plane.}\end{array}\right]$$

and therefore perpendicular to OC and DE.

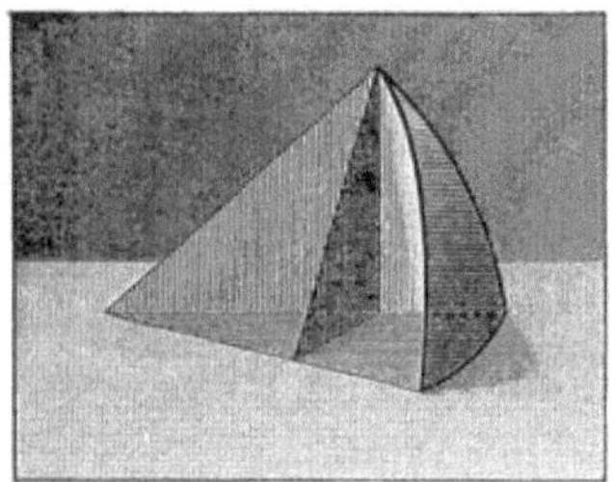
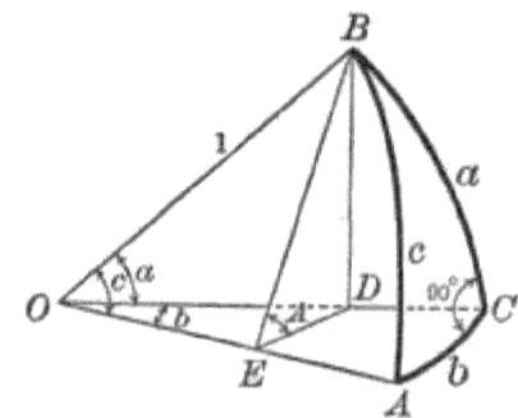

In triangle EOD, remembering that angle $EOD = b$, we have

$$\frac{OE}{OD} = \cos b,$$

or, clearing of fractions,

$(A) \qquad OE = OD \cdot \cos b.$

But $\qquad OE = \cos c \; (= \cos EOB)$,

and $\qquad OD = \cos a \; (= \cos DOB)$.

Substituting in (A), we get

(1) $\qquad \cos c = \cos a \cos b.$

In triangle BED, remembering that angle $BED =$ angle A, we have

$$\frac{BD}{BE} = \sin A,$$

or, clearing of fractions,

$(B) \qquad BD = BE \cdot \sin A.$

But $\qquad BD = \sin a \, (= \sin DOB)$,

and $\qquad BE = \sin c \, (= \sin EOB)$.

Substituting in (B), we get

(2) $\qquad\qquad \sin a = \sin c \sin A.$

Similarly, if we had passed the auxiliary plane through A perpendicular to OB,

(3) $\qquad\qquad \sin b = \sin c \sin B.$

Again, in the triangle BED,

(C) $\qquad\qquad \cos A = \dfrac{DE}{BE}.$

But $\qquad\qquad DE = OD \sin b, \qquad\qquad$ from $\sin b = \dfrac{DE}{OD}$

$\qquad\qquad\qquad\quad OD = \cos a \ (= \cos DOB),.$

and $\qquad\qquad\quad BE = \sin c \ (= \sin EOB).$

Substituting in (C),

(D) $\qquad\qquad \cos A = \dfrac{OD \sin b}{\sin c} = \cos a \cdot \dfrac{\sin b}{\sin c}.$

But from (3), $\qquad \dfrac{\sin b}{\sin c} = \sin B.$ Therefore

(4) $\qquad\qquad \cos A = \cos a \sin B.$

Similarly, if we had passed the auxiliary plane through A perpendicular to OB,

(5) $\qquad\qquad \cos B = \cos b \sin A.$

The above five formulas are fundamental; that is, from them we may derive all other relations expressing any one part of a right spherical triangle in terms of two others. For example, to find a relation between A, b, c, proceed thus:

From (4), $\qquad\qquad \cos A = \cos a \sin B$

$\qquad\qquad\qquad\qquad = \dfrac{\cos c \ \sin b}{\cos b \ \sin c}$

$\qquad\left[\text{Since } \cos a = \dfrac{\cos c}{\cos b} \text{ from (1), and } \sin B = \dfrac{\sin b}{\sin c} \text{ from (3).}\right]$

$\qquad\qquad\qquad\qquad = \dfrac{\sin b \ \cos c}{\cos b \ \sin c}.$

(6) $\qquad\qquad \therefore \cos A = \tan b \cot c.$

Similarly, we may get

(7) $\qquad\qquad \cos B = \tan a \cot c.$

(8) $\qquad\qquad \sin b = \tan a \cot A.$

(9) $\qquad\qquad \sin a = \tan b \cot B.$

(10) $\qquad\qquad \cos c = \cot A \cot B.$

These ten formulas are sufficient for the solution of right spherical triangles. In deriving these formulas we assumed all the elements except the right angle to be less than 90°. But the formulas hold when this assumption is not made. For instance, let us suppose that a is greater that 90°. In this case the auxiliary plane BDE will cut CO and AO produced beyond the center O, and we have, in triangle EOD,

$$(E) \qquad \cos DOE \; (= \cos b) = \frac{OE}{OD}.$$

But
$$OE = \cos EOB = -\cos AOB = -\cos c,$$
and
$$OD = \cos DOB = -\cos COB = -\cos a.$$

Substituting in (E), we get

$$\cos b = \frac{\cos c}{\cos a}, \text{ or } \cos c = \cos a \cos b,$$

which is the same as (1).

Likewise, the other formulas will hold true in this case. Similarly, they may be shown to hold true in all cases.

If the two sides including the right angle are either both less or

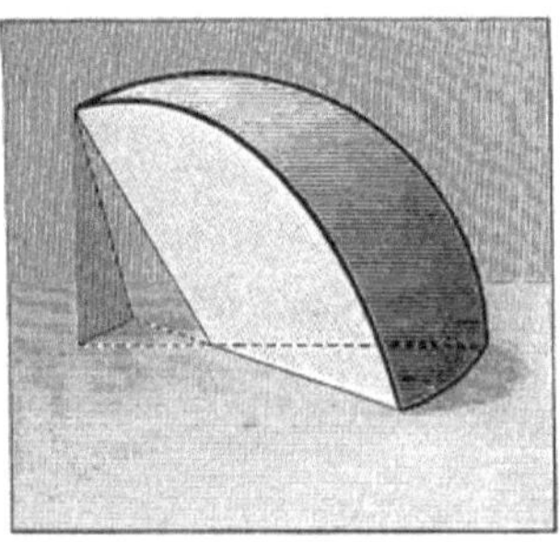
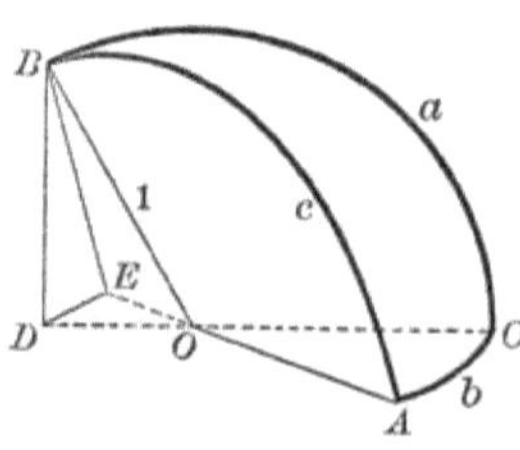

both greater than 90° (that is, $\cos a$ and $\cos b$ are either both positive or both negative), then the product

$$(F) \qquad\qquad \cos a \cos b$$

will always be positive, and therefore $\cos c$, from (1), will always be positive, that is, c will always be less than 90°. If, however, one of the sides including the right angle is less and the other is greater than 90°, the product (F), and therefore also $\cos c$, will be negative, and c will be greater than 90°.

Hence we have

Theorem I. *If the two sides including the right angle of a right spherical triangle are both less or both greater than 90°, the hypotenuse* *is less than 90°; if one side is less and the other is greater than 90°, the hypotenuse is greater than 90°.*

From (4) and (5), $\quad \sin B = \dfrac{\cos A}{\cos a}$, and $\sin A = \dfrac{\cos B}{\cos b}$.

Since A and B are less than 180°, $\sin A$ and $\sin B$ must always be positive. But then $\cos A$ and $\cos a$ must have the same sign, that is, A and a are either both less than 90° or both greater than 90°. Similarly, for B and b. Hence we have

Theorem II. *In a right spherical triangle an oblique angle and the side opposite are either both less or both greater than 90°.*

4. Napier's rules of circular parts. The ten formulas derived in the last section express the relations between the three sides and the two oblique angles of a right spherical triangle. All these relations may be shown to follow from two very useful rules discovered by Baron Napier, the inventor of logarithms.

For this purpose the right angle (not entering the formulas) is not taken into account, and we replace the hypotenuse and the two

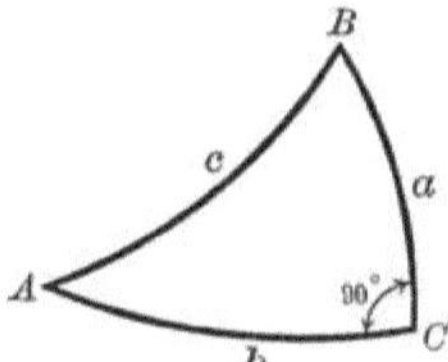 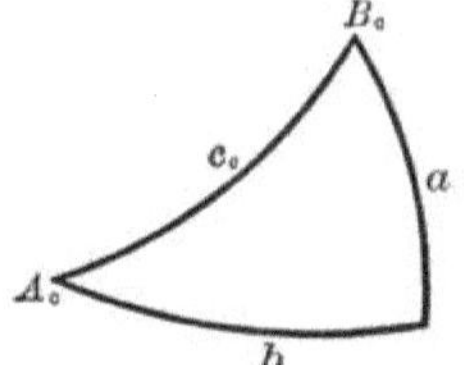

oblique angles by their respective *complements;* so that the five parts, called the *circular parts,* used in Napier's rules are a, b, A_c, c_c, B_c. The subscript c indicates that the complement is to be used. The first figure illustrates the ordinary method of representing a right spherical triangle. To emphasize the circular parts employed in Napier's rules, the same triangle might be represented as shown in the second figure. It is not necessary, however, to draw the triangle at all when using Napier's rules; in fact, it is found to be more convenient to simply write down the five parts in their proper order as on the circumference of a circle, as shown in the third figure (hence the name *circular parts*).

Any one of these parts may be called a *middle* part; then the two parts immediately adjacent to it are called *adjacent* parts, and the other two *opposite* parts. Thus, if a is taken as a middle part, A_c and b are the adjacent parts, while c_c and B_c are the opposite parts.

Napier's rules of circular parts.

Rule I. *The sine of any middle part is equal to the product of the tangents of the adjacent parts.*

Rule II. *The sine of any middle part is equal to the product of the cosines of the opposite parts.*

These rules are easily remembered if we associate the first one with the expression " **tan-adj.** " and the second one with " **cos-opp.** " *

Napier's rules may be easily verified by applying them in turn to each one of the five circular parts taken as a middle part, and comparing the results with (1) to (10).

For example, let c_c be taken as a middle part; then A_c and B_c are the adjacent parts, while a and b are the opposite parts.

Then, by Rule I, $\quad \sin c_c = \tan A_c \tan B_c,$

or, $\quad \cos c = \cot A \cot B;$

which agrees with (10), p. 197.

By Rule II, $\quad \sin c_c = \cos a \cos b,$

or, $\quad \cos c = \cos a \cos b;$

which agrees with (1), p. 196.

The student should verify Napier's rules in this manner by taking each one of the other four circular parts as the middle part.

Writers on Trigonometry differ as to the practical value of Napier's rules, but it is generally conceded that they are a great aid to the memory in applying formulas (1) to (10) to the solution of right spherical triangles, and we shall so employ them.

5. Solution of right spherical triangles. To solve a right spherical triangle, two elements (parts) must be given in addition to the right angle. For the sake of uniformity we shall continue to denote the right angle in a spherical triangle ABC by the letter C.

General directions for solving right spherical triangles.

First step. *Write down the five circular parts as in first figure.*

Second step. *Underline the two given parts and the required unknown part. Thus, if A_c and a are given to find b, we underline all three as is shown in the second figure.*

---

* Or by noting that a is the first vowel in the words "tangent" and "adjacent," while o is the first vowel in the words "cosine" and "opposite."

Third step. *Pick out the middle part (in this case b) and cross the line under it as indicated in the third figure.*

Fourth step. *Use Rule I if the other two parts are adjacent to the middle part (as in case illustrated), or Rule II if they are opposite, and solve for the unknown part.*

Check: *Check with that rule which involves the three required parts.**

Careful attention must be paid to the algebraic signs of the functions when solving spherical triangles; the cosines, tangents, and cotangents of angles or arcs greater than 90° being negative. When computing with logarithms we shall write (n) after the logarithms when the functions are negative. If the number of negative factors is even, the result will be positive; if it is odd, the result will be negative and (n) should be written after the resulting logarithm. In order to be able to show our computations in compact form, we shall write down all the logarithms of the trigonometric functions just as they are given in our table; that is, when a logarithm has a negative characteristic we will not write down -10 after it.†

Ex. 1. Solve the right spherical triangle, having given $B = 33°\ 50'$, $a = 108°$.
Solution. Follow the above general directions.

To find A	*To find b*	*To find c*
Using Rule II	Using Rule I	Using Rule I
$\sin A_c = \cos B_c \cos a$	$\sin a = \tan B_c \tan b$	$\sin B_c = \tan c_c \tan a$
$\cos A = \sin B \cos a$	$\tan b = \sin a \tan B$	$\cot c = \cos B \cot a$
$\log \sin B = 9.7457$	$\log \sin a = 9.9782$	$\log \cos B = 9.9194$
$\log \cos a = 9.4900\ (n)$	$\log \tan B = 9.8263$	$\log \cot a = 9.5118\ (n)$
$\log \cos A = 9.2357\ (n)$	$\log \tan b = 9.8045$	$\log \cot c = 9.4312\ (n)$
$\therefore 180° - A‡ = 80°\ 6'$	$\therefore b = 32°\ 31'.$	$\therefore 180° - c = 74°\ 54'$
and $\quad A = 99°\ 54'.$		and $\quad c = 105°\ 6'.$

The value of $\log \cos A$ found is the same as that found in our first computation. The student should observe that in checking our work in this example

---

* Thus, in above case, A_c and a are given; therefore we underline the three required parts and cross b as the middle part. Applying Rule II, c_c and B_c being opposite parts, we get $\sin b = \cos c_c \cos B_c$, or, $\sin b = \sin c \sin B$.

† For example, as in the table, we will write $\log \sin 24° = 9.6093$. To be exact, this should be written $\log \sin 24° = 9.6093 - 10$, or, $\log \sin 24° = 1.6093$.

‡ Since $\cos A$ is negative, we get the supplement of A from the table.

it was not necessary to look up any new logarithms. Hence the check in this case is only on the correctness of the logarithmic work.*

$$\underline{c_c}$$

$$\underset{\neq}{A_c} \qquad\qquad B_c$$

$$\underline{b} \qquad\qquad a$$

Check : Using Rule I

$$\sin A_c = \tan b \tan c_c$$
$$\cos A = \tan b \cot c$$
$$\log \tan b = 9.8045$$
$$\log \cot c = \underline{9.4312}\ (n)$$
$$\log \cos A = \overline{9.2357}\ (n)$$

In logarithmic computations the student should always write down an outline or skeleton of the computation before using his logarithmic table at all. In the last example this outline would be as follows:

$\log \sin B =$		$\log \sin a =$		$\log \cos B =$	
$\log \cos a = $ ____ (n)		$\log \tan B = $ ____		$\log \cot a = $ ____ (n)	
$\log \cos A = $ ____ (n)		$\log \tan b = $ ____		$\log \cot c = $ ____ (n)	
$\therefore\ 180° - A =$		$\therefore\ b =$		$\therefore\ 180° - c =$	
and $\quad A =$				and $\quad c =$	

It saves time to look up all the logarithms at once, and besides it reduces the liability of error to thus separate the theoretical part of the work from that which is purely mechanical. Students should be drilled in writing down forms like that given above before attempting to solve examples.

Ex. 2. Solve the right spherical triangle, having given $c = 70° 30'$, $A - 100°$.
Solution. Follow the general directions.

To find a	*To find b*	*To find B*
$\underline{c_c}$	$\underline{c_c}$	$\underset{\neq}{c_c}$
$\underline{A_c} \qquad B_c$	$\underset{\neq}{A_c} \qquad B_c$	$\underline{A_c} \qquad \underline{B_c}$
$b \qquad \underset{\neq}{a}$	$\underline{b} \qquad a$	$b \qquad a$
Using Rule II	Using Rule I	Using Rule I
$\sin a = \cos c_c \cos A_c$	$\sin A_c = \tan b \tan c_c$	$\sin c_c - \tan A_c \tan B_c$
$\sin a - \sin c \sin A$	$\tan b = \cos A \tan c$	$\cot B = \cos c \tan A$
$\log \sin c = 9.9743$	$\log \cos A = 9.2397\ (n)$	$\log \cos c = 9.5235$
$\log \sin A = 9.9934$	$\log \tan c = 0.4509$	$\log \tan A = 0.7537\ (n)$
$\log \sin a = 9.9677$	$\log \tan b = \overline{9.6906}\ (n)$	$\log \cot B = 0.2772\ (n)$
$\therefore\ 180° - a\dagger = 68° 10'$	$\therefore\ 180° - b = 26° 8'$	$\therefore\ 180° - B = 27° 51'$
and $\quad a = 111° 50'.$	and $\quad b = 153° 52'.$	and $\quad B = 152° 9'.$

The work of verifying the results is left to the student.

* In order to be sure that the angles and sides have been correctly taken from the tables, in such an example as this, we should use them together with some of the given data in relations not already employed.

† Since a is determined from its sine, it is evident that it may have the value 68° 10' found from the table, or the supplementary value 111° 50'. Since $A > 90°$, however, we know from Th. II, p. 199, that $a > 90°$; hence $a = 111° 50'$ is the only solution.

6. The ambiguous case. Two solutions. When the given parts of a right spherical triangle are an oblique angle and its opposite side, there are two triangles which satisfy the given conditions. For, in the triangle ABC, let $C = 90°$, and let A and CB ($= a$) be the given parts. If we extend AB and AC to A',

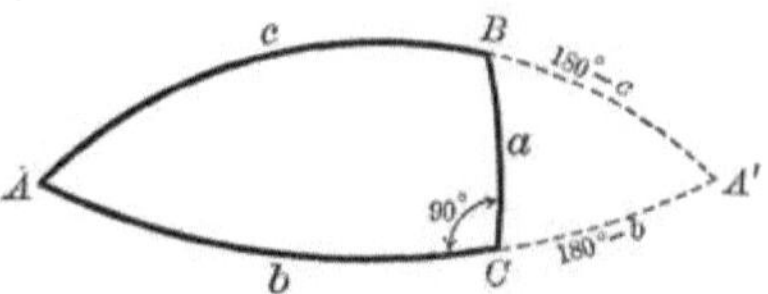

it is evident that the triangle $A'BC$ also satisfies the given conditions, since $BCA' = 90°$, $A' = A$, and $BC = a$. The remaining parts in $A'BC$ are supplementary to the respective remaining parts in ABC. Thus

$$A'B = 180° - c, \qquad A'C = 180° - b, \qquad A'BC = 180° - ABC.$$

This ambiguity also appears in the solution of the triangle, as is illustrated in the following example :

Ex. 3. Solve the right spherical triangle, having given $A = 105° 59'$, $a = 128° 33'$.

Solution. We proceed as in the previous examples.

To find b	To find B	To find c
c_c	c_c	c_c
$A_c \qquad B_c$	$A_c \qquad B_c$	$A_c \qquad B_c$
$b \qquad a$	$b \qquad a$	$b \qquad a$
$\sin b = \tan a \tan A_c$	$\sin A_c = \cos a \cos B_c$	$\sin a = \cos A_c \cos c_c$
$\sin b = \tan a \cot A$	$\sin B = \dfrac{\cos A}{\cos a}$	$\sin c = \dfrac{\sin a}{\sin A}$
$\log \tan a = 0.0986 \ (n)$	$\log \cos A = 9.4399 \ (n)$	$\log \sin a = 9.8932$
$\log \cot A = 9.4570 \ (n)$	$\log \cos a = 9.7946 \ (n)$	$\log \sin A = 9.9828$
$\log \sin B = 9.5556$	$\log \sin B = 9.6453$	$\log \sin c = 9.9104$
$\therefore b = 21° 4'$, or,	$\therefore B = 26° 14'$, or,	$\therefore c' = 54° 27'$, or,
$180° - b = 158° 56' = b'.*$	$180° - B = 153° 46' = B'.\dagger$	$180° - c' = 125° 33' = c.\ddagger$

Hence the two solutions are :

1. $b = 21° 4'$, $c = 125° 33'$, $B = 26° 14'$ (triangle ABC) ;
2. $b' = 158° 56'$, $c' = 54° 27'$, $B' = 153° 46'$ (triangle $A'BC$).

It is not necessary to check both solutions. We leave this to the student.

---

* Since $\sin B$ is positive and B is not known, we cannot remove the ambiguity. Hence both the acute angle taken from the table and its supplement must be retained.

† The two values of B must be retained, since b has two values which are supplementary.

‡ Since $a > 90°$ and b has two values, one $>$ and the other $< 90°$, it follows from Th. I, p. 198, that c will have two values, the first one $< 90°$ and the second $> 90°$.

EXAMPLES

Solve the following right spherical triangles:

No.	Given Parts		Required Parts		
1	$a = 132° 6'$	$b = 77° 51'$	$A = 131° 27'$	$B = 80° 55'$	$c = 98° 7'$
2	$a = 159°$	$c = 137° 20'$	$A = 148° 5'$	$B = 65° 23'$	$b = 38° 1'$
3	$A = 50° 20'$	$B = 122° 40'$	$a = 40° 42'$	$b = 134° 31'$	$c = 122° 7'$
4	$a = 160°$	$b = 38° 30'$	$A = 149° 41'$	$B = 66° 44'$	$c = 137° 20'$
5	$B = 80°$	$b = 67° 40'$	$A = 27° 12'$	$a = 25° 25'$	$c = 69° 54'$; or,
			$A' = 152° 48'$	$a' = 154° 35'$	$c = 110° 6'$
6	$B = 112°$	$c = 81° 50'$	$A = 109° 23'$	$a = 110° 58'$	$b = 113° 22'$
7	$a = 61°$	$B = 123° 40'$	$A = 66° 12'$	$b = 127° 17'$	$c = 107° 5'$
8	$a = 61° 40'$	$b = 144° 10'$	$A = 72° 29'$	$B = 140° 38'$	$c = 112° 38'$
9	$A = 99° 50'$	$a = 112°$	$B = 27° 7'$	$b = 25° 24'$	$c = 109° 46'$; or,
			$B' = 152° 53'$	$b' = 154° 36'$	$c' = 70° 14'$
10	$b = 15°$	$c = 152° 20'$	$A = 120° 44'$	$a = 156° 30'$	$B = 33° 53'$
11	$A = 62° 59'$	$B = 37° 4'$	$a = 41° 6'$	$b = 26° 25'$	$c = 47° 32'$
12	$A = 73° 7'$	$c = 114° 32'$	$a = 60° 31'$	$B = 143° 50'$	$b = 147° 32'$
13	$B = 144° 54'$	$b = 146° 32'$	$A = 78° 47'$	$a = 70° 10'$	$c = 106° 28'$; or,
			$A' = 101° 13'$	$a' = 109° 50'$	$c' = 73° 32'$
14	$B = 68° 18'$	$c = 47° 34'$	$A = 30° 32'$	$a = 22° 1'$	$b = 43° 18'$
15	$A = 161° 52'$	$b = 131° 8'$	$a = 166° 9'$	$B = 101° 49'$	$c = 50° 18'$
16	$a = 113° 25'$	$b = 110° 47'$	$A = 112° 3'$	$B = 109° 12'$	$c = 81° 54'$
17	$a = 137° 9'$	$B = 74° 51'$	$A = 135° 3'$	$b = 68° 17'$	$c = 105° 44'$
18	$A = 144° 54'$	$B = 101° 14'$	$a = 146° 33'$	$b = 109° 48'$	$c = 73° 35'$
19	$a = 69° 18'$	$c = 84° 27'$	$A = 70°$	$B = 75° 6'$	$b = 74° 7'$

20. For more examples take any two parts in the above triangles and solve for the other three.

7. Solution of isosceles and quadrantal triangles. Plane isosceles triangles were solved by dividing each one into two equal right triangles and then solving one of the right triangles. Similarly, we may solve an *isosceles spherical triangle* by dividing it into two symmetrical (equal) right spherical triangles by an arc drawn from the vertex perpendicular to the base, and then solving one of the right spherical triangles.

A *quadrantal triangle* is a spherical triangle one side of which is a quadrant ($= 90°$). By (f), p. 195, the polar triangle of a quadrantal triangle is a right triangle. Therefore, to solve a quadrantal triangle we have only to solve its polar triangle and take the *supplements* of the parts obtained by the calculation.

Ex. 1. Solve the triangle, having given $c = 90°$, $a = 67° 38'$, $b = 48° 50'$.

Solution. This is a quadrantal triangle since one side $c = 90°$. We then find the corresponding elements of its polar triangle by (f), p. 195. They are $C' = 90°$, $A' = 112° 22'$, $B' = 131° 10'$. We solve this right triangle in the usual way.

Construct the polar (right) triangle.
Given $A' = 112°\,22'$, $B' = 131°\,10'$:

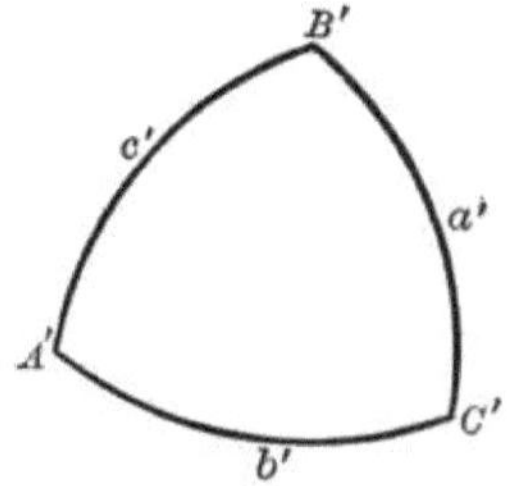

To find a'

$$c'_c$$

$$\underset{\neq}{A'_c} \qquad \underline{B'_c}$$

$$b' \qquad \underline{a'}$$

$$\sin A'_c = \cos a' \cos B'_c$$

$$\cos a' = \frac{\cos A'}{\sin B'}.$$

$\log \cos A' = 9.5804\ (n)$

$\log \sin B' = 9.8767$

$\log \cos a' = 9.7037\ (n)$

$180° - a' = 59°\,38'.$

$$a' = 120°\,22'.$$

Similarly, we get

$$b' = 135°\,23', \quad c' = 68°\,55'.$$

Hence in the given quadrantal triangle we have

$$A = 180° - a' = 59°\,38',$$
$$B = 180° - b' = 44°\,37',$$
$$C = 180° - c' = 111°\,5'.$$

EXAMPLES

Solve the following quadrantal triangles:

No.	Given Parts			Required Parts		
1	$A = 139°$	$b = 143°$	$c = 90°$	$a = 117°\,1'$	$B = 153°\,42'$	$C = 132°\,34'$
2	$A = 45°\,30'$	$B = 139°\,20'$	$c = 90°$	$a = 57°\,22'$	$b = 129°\,42'$	$C = 57°\,53'$
3	$a = 30°\,20'$	$C = 42°\,40'$	$c = 90°$	$A = 20°\,1'$	$B = 141°\,30'$	$b = 113°\,17'$
4	$B = 70°\,12'$	$C = 106°\,25'$	$c = 90°$	$A = 33°\,28'$	$a = 35°\,4'$	$b = 78°\,47'$
5	$A = 105°\,53'$	$a = 104°\,54'$	$c = 90°$	$B = 69°\,16'$	$b = 70°$	$C = 84°\,30'$; or
				$B = 110°\,44'$	$b = 110°$	$C = 95°\,30'$

Solve the following isosceles spherical triangles:

No.	Given Parts			Required Parts		
6	$a = 54°\,20'$	$c = 72°\,54'$	$A = B$	$b = 54°\,20'$ $A = B = 57°\,59'$		$C = 93°\,59'$
7	$a = 54°\,30'$	$C = 71°$	$A = B$	$b = 54°\,30'$ $A = B = 90°$		$c = 180°$
8	$a = 66°\,29'$	$A = B = 50°\,17'$		$b = 66°\,29'$	$c = 111°\,30'$	$C = 128°\,42'$
9	$c = 156°\,40'$	$C = 187°\,46'$	$A = B$	$a = b = 3°\,58'$ or $176°\,2'$		
				$A = B = 89°\,12'$ or $90°\,48'$		

Prove the following relations between the elements of a right spherical triangle ($C = 90°$):

10. $\cos^2 A \sin^2 c = \sin(c + a)\sin(c - a).$ 13. $\sin(b + c) = 2\cos^2 \tfrac{1}{2} A \cos b \sin c.$

11. $\tan a \cos c = \sin b \cot B.$ 14. $\sin(c - b) = 2\sin^2 \tfrac{1}{2} A \cos b \sin c.$

12. $\sin^2 A = \cos^2 B + \sin^2 a \sin^2 B.$

CHAPTER II

OBLIQUE SPHERICAL TRIANGLES

8. Fundamental formulas. In this chapter some relations between the sides and angles of any spherical triangle (whether right angled or oblique) will be derived.

9. Law of sines. *In a spherical triangle the sines of the sides are proportional to the sines of the opposite angles.*

Proof. Let ABC be any spherical triangle, and draw the arc CD perpendicular to AB. There will be two cases according as CD falls

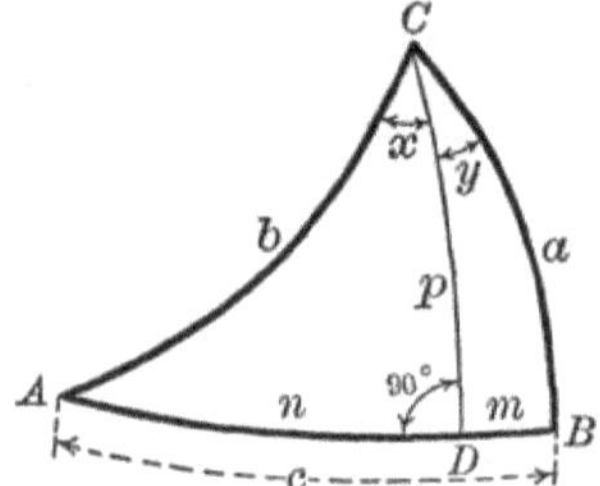
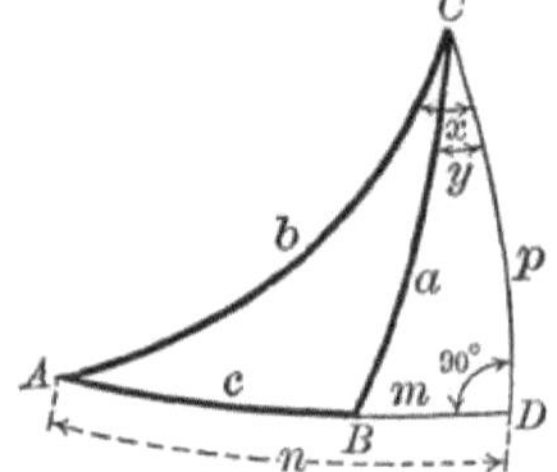

upon AB (first figure) or upon AB produced (second figure). For the sake of brevity let $CD = p$, $AD = n$, $BD = m$, angle $ACD = x$, angle $BCD = y$.

In the right triangle ADC (either figure)

(A) $\qquad\qquad\qquad \sin p = \sin b \sin A.$ $\qquad\qquad$ Rule II, p. 200

In the right triangle BCD (first figure)

(B) $\qquad\qquad\qquad \sin p = \sin a \sin B.$ $\qquad\qquad$ Rule II, p. 200

This also holds true in the second figure, for

$$\sin DBC = \sin (180° - B) = \sin B.$$

Equating the values of $\sin p$ from (A) and (B),

$$\sin a \sin B = \sin b \sin A,$$

or, dividing through by $\sin A \sin B$,

(C) $\qquad\qquad\qquad \dfrac{\sin a}{\sin A} = \dfrac{\sin b}{\sin B}.$

In like manner, by drawing perpendiculars from A and B, we get

(D)
$$\frac{\sin b}{\sin B} = \frac{\sin c}{\sin C}, \text{ and}$$

(E)
$$\frac{\sin c}{\sin C} = \frac{\sin a}{\sin A}, \text{ respectively.}$$

Writing (C), (D), (E) as a single statement, we get the **law of sines**,

$(\cdot\cdot)$
$$\frac{\sin a}{\sin A} = \frac{\sin b}{\sin B} = \frac{\sin c}{\sin C}. ^{*}$$

10. Law of cosines. *In a spherical triangle the cosine of any side is equal to the product of the cosines of the other two sides plus the product of the sines of these two sides and the cosine of their included angle.*

Proof. Using the same figures as in the last section, we have in the right triangle BDC,

$$\cos a = \cos p \cos m \qquad\qquad \text{Rule II, p. 200}$$
$$= \cos p \cos(c - n)$$
$$= \cos p \{\cos c \cos n + \sin c \sin n\}$$
(A)
$$= \cos p \cos c \cos n + \cos p \sin c \sin n.$$

In the right triangle ADC,

(B)
$$\cos p \cos n = \cos b.$$

Whence
$$\cos p = \frac{\cos b}{\cos n},$$

and, multiplying both sides by $\sin n$,

(C)
$$\cos p \sin n = \frac{\cos b}{\cos n} \cdot \sin n = \cos b \tan n.$$

But
$$\cos A = \tan n \cot b, \text{ or,} \qquad \text{Rule I, p. 200}$$

(D)
$$\tan n = \tan b \cos A.$$

Substituting value of $\tan n$ from (D) in (C), we have

(E)
$$\cos p \sin n = \cos b \tan b \cos A = \sin b \cos A.$$

Substituting the value of $\cos p \cos n$ from (B) and the value of $\cos p \sin n$ from (E) in (A), we get the law of cosines,

(F)
$$\cos a = \cos b \cos c + \sin b \sin c \cos A.$$

---

* Compare with the law of sines in Granville's *Plane Trigonometry*, p. 102.

Similarly, for the sides b and c we may obtain

$$(G) \qquad \cos b = \cos c \cos a + \sin c \sin a \cos B,$$

$$(H) \qquad \cos c = \cos a \cos b + \sin a \sin b \cos C.$$

11. Principle of Duality. Given any relation involving one or more of the sides a, b, c, and the angles A, B, C of any general spherical triangle. Now the polar triangle (whose sides are denoted by a', b', c', and angles by A', B', C') is also in this case a general spherical triangle, and the given relation must hold true for it also; that is, the given relation applies to the polar triangle if accents are placed upon the letters representing the sides and angles. Thus (F), (G), (H) of the last section give us the following law of cosines for the polar triangle:

$$(A) \qquad \cos a' = \cos b' \cos c' + \sin b' \sin c' \cos A'.$$

$$(B) \qquad \cos b' = \cos c' \cos a' + \sin c' \sin a' \cos B'.$$

$$(C) \qquad \cos c' = \cos a' \cos b' + \sin a' \sin b' \cos C'.$$

But by (f), p. 195,

$$a' = 180° - A, \qquad b' = 180° - B, \qquad c' = 180° - C,$$
$$A' = 180° - a, \qquad B' = 180° - b, \qquad C' = 180° - c.$$

Making these substitutions in (A), (B), (C), which refer to the polar triangle, we get

$$(D) \quad \cos(180° - A) = \cos(180° - B)\cos(180° - C)$$
$$+ \sin(180° - B)\sin(180° - C)\cos(180° - a),$$

$$(E) \quad \cos(180° - B) = \cos(180° - C)\cos(180° - A)$$
$$+ \sin(180° - C)\sin(180° - A)\cos(180° - b),$$

$$(F) \quad \cos(180° - C) = \cos(180° - A)\cos(180° - B)$$
$$+ \sin(180° - A)\sin(180° - B)\cos(180° - c),$$

which involve the sides and angles of the original triangle.

The result of the preceding discussion may then be stated in the following form:

Theorem. *In any relation between the parts of a general spherical triangle, each part may be replaced by the supplement of the opposite part, and the relation thus obtained will hold true.*

The Principle of Duality follows when the above theorem is applied to a relation involving one or more of the sides and the supplements of the angles (instead of the angles themselves).

Let the supplements of the angles of the triangle be denoted by α, β, γ^{*}, that is,

$$\alpha = 180° - A, \quad \beta = 180° - B, \quad \gamma = 180° - C,$$
$$\text{or,} \quad A = 180° - \alpha, \quad B = 180° - \beta, \quad C = 180° - \gamma.$$

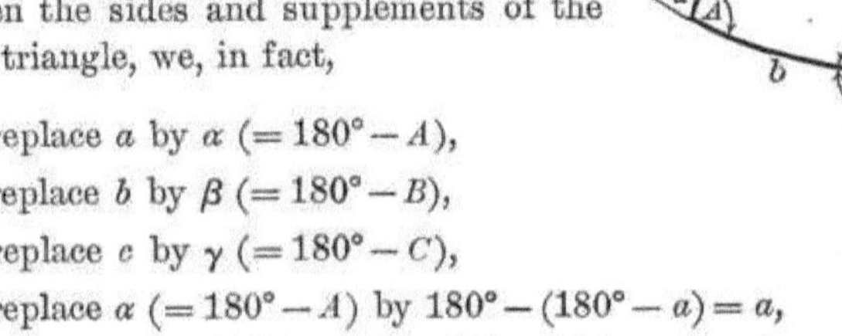

When we apply the above theorem to a relation between the sides and supplements of the angles of a triangle, we, in fact,

replace a by α $(= 180° - A)$,

replace b by β $(= 180° - B)$,

replace c by γ $(= 180° - C)$,

replace α $(= 180° - A)$ by $180° - (180° - a) = a$,

replace β $(= 180° - B)$ by $180° - (180° - b) = b$,

replace γ $(= 180° - C)$ by $180° - (180° - c) = c$,

or, what amounts to the same thing, *interchange the Greek and Roman letters.* For instance, substitute

$$A = 180° - \alpha, \qquad B = 180° - \beta, \qquad C = 180° - \gamma$$

in (F), (G), (H) of the last section. This gives the law of cosines for the sides in the new form

(12) $$\cos a = \cos b \cos c - \sin b \sin c \cos \alpha,$$

(13) $$\cos b = \cos c \cos a - \sin c \sin a \cos \beta,$$

(14) $$\cos c = \cos a \cos b - \sin a \sin b \cos \gamma.$$

[Since $\cos A = \cos(180° - \alpha) = -\cos \alpha$, etc.]

If we now apply the above theorem to these formulas, we get the law of cosines for the angles, namely,

(15) $$\cos \alpha = \cos \beta \cos \gamma - \sin \beta \sin \gamma \cos a,$$

(16) $$\cos \beta = \cos \gamma \cos \alpha - \sin \gamma \sin \alpha \cos b,$$

(17) $$\cos \gamma = \cos \alpha \cos \beta - \sin \alpha \sin \beta \cos c,$$

* α, β, γ are then the exterior angles of the triangle, as shown in the figure.

that is, we have derived three new relations between the sides and supplements of the angles of the triangle.* We may now state the

Principle of Duality. *If the sides of a general spherical triangle are denoted by the Roman letters a, b, c, and the supplements of the corresponding opposite angles by the Greek letters α, β, γ, then, from any given formula involving any of these six parts, we may write down a* **dual** *formula by simply interchanging the corresponding Greek and Roman letters.*

The immediate consequence of this principle is that formulas in Spherical Trigonometry occur in *pairs*, either one of a pair being the *dual* of the other.

Thus **(12)** and **(15)** are dual formulas; also **(13)** and **(16)**, or **(14)** and **(17)**.

If we substitute

$$A = 180° - \alpha, \qquad B = 180° - \beta, \qquad C = 180° - \gamma$$

in the law of sines (p. 207), we get

$$\frac{\sin a}{\sin \alpha} = \frac{\sin b}{\sin \beta} = \frac{\sin c}{\sin \gamma}.$$

[Since $\sin A = \sin(180° - \alpha) = \sin \alpha$, etc.]

Applying the Principle of Duality to this relation, we get

$$\frac{\sin \alpha}{\sin a} = \frac{\sin \beta}{\sin b} = \frac{\sin \gamma}{\sin c},$$

which is essentially the same as the previous form.

The forms of the law of cosines that we have derived involve algebraic sums. As these are not convenient for logarithmic calculations, we will reduce them to the form of products.

12. Trigonometric functions of half the supplements of the angles of a spherical triangle in terms of its sides. Denote half the sum of the sides of a triangle (i.e. half the perimeter) by s. Then

$$(A) \qquad 2s = a + b + c,$$

or,

$$s = \tfrac{1}{2}(a + b + c).$$

---

* If we had employed the interior angles of the triangle in our formulas (as has been the almost universal practice of writers on Spherical Trigonometry), the two sets of cosine formulas would not have been of the same form. That the method used here has many advantages will become more and more apparent as the reading of the text is continued. Not only are the resulting formulas much easier to memorize, but much labor is saved in that, when we have derived one set of formulas for the angles (or sides), the corresponding set of formulas for the sides (or angles) may be written down at once by mere inspection by applying this Principle of Duality. The great advantage of using this Principle of Duality was first pointed out by Mobius (1790–1868).

Subtracting $2\,c$ from both sides of (A),

$$2\,s - 2\,c = a + b + c - 2\,c, \text{ or,}$$

(B) $\qquad s - c = \tfrac{1}{2}(a + b - c).$

Similarly,

(C) $\qquad s - b = \tfrac{1}{2}(a - b + c), \text{ and}$

(D) $\qquad s - a = \tfrac{1}{2}(- a + b + c) = \tfrac{1}{2}(b + c - a).$

From Plane Trigonometry,

(E) $\qquad 2 \sin^2 \tfrac{1}{2}\alpha = 1 - \cos\alpha,$

(F) $\qquad 2 \cos^2 \tfrac{1}{2}\alpha = 1 + \cos\alpha.$

But from (12), p. 209, solving for $\cos\alpha$,

$$\cos\alpha = \frac{\cos b \cos c - \cos a}{\sin b \sin c};$$

hence (E) becomes

$$2 \sin^2 \tfrac{1}{2}\alpha = 1 - \frac{\cos b \cos c - \cos a}{\sin b \sin c}$$

$$= \frac{\sin b \sin c - \cos b \cos c + \cos a}{\sin b \sin c}$$

$$= \frac{\cos a - (\cos b \cos c - \sin b \sin c)}{\sin b \sin c}$$

$$= \frac{\cos a - \cos(b + c)}{\sin b \sin c}$$

$$= \frac{-2 \sin \tfrac{1}{2}(a + b + c) \sin \tfrac{1}{2}(a - b - c)}{\sin b \sin c}, * \text{ or,}$$

(G) $\qquad 2 \sin^2 \tfrac{1}{2}\alpha = \frac{2 \sin \tfrac{1}{2}(a + b + c) \sin \tfrac{1}{2}(b + c - a)}{\sin b \sin c}.$

[Since $\sin \tfrac{1}{2}(a - b - c) = - \sin \tfrac{1}{2}(- a + b + c) = - \sin \tfrac{1}{2}(b + c - a)$.]

Substituting from (A) and (D) in (G), we get

$$\sin^2 \tfrac{1}{2}\alpha = \frac{\sin s \sin(s - a)}{\sin b \sin c}, \text{ or,}$$

(18) $\qquad \sin \tfrac{1}{2}a = \sqrt{\dfrac{\sin s \sin(s - a)}{\sin b \sin c}}.$

---

* Let

$$\begin{array}{ll}
A = a & A = a \\
B = b + c & B = b + c \\
\overline{A + B = a + b + c} & \overline{A - B = a - b - c} \\
\tfrac{1}{2}(A + B) = \tfrac{1}{2}(a + b + c). & \tfrac{1}{2}(A - B) = \tfrac{1}{2}(a - b - c).
\end{array}$$

Hence, substituting in (65), p. 74, Granville's *Plane Trigonometry*, namely,

$$\cos A - \cos B = - 2 \sin \tfrac{1}{2}(A + B) \sin \tfrac{1}{2}(A - B),$$

we get

$$\cos a - \cos(b + c) = - 2 \sin \tfrac{1}{2}(a + b + c) \sin \tfrac{1}{2}(a - b - c).$$

Similarly, (F) becomes

$$2 \cos^2 \tfrac{1}{2}\, a = 1 + \frac{\cos b \cos c - \cos a}{\sin b \sin c}$$

$$= \frac{\sin b \sin c + \cos b \cos c - \cos a}{\sin b \sin c}$$

$$= \frac{\cos (b - c) - \cos a}{\sin b \sin c}$$

$$= \frac{- 2 \sin \tfrac{1}{2}(a + b - c) \sin \tfrac{1}{2}(b - c - a)}{\sin b \sin c},* \quad \text{or,}$$

$$(H) \qquad 2 \cos^2 \tfrac{1}{2}\, a = \frac{2 \sin \tfrac{1}{2}(a + b - c) \sin \tfrac{1}{2}(a - b + c)}{\sin b \sin c}.$$

[Since $\sin \tfrac{1}{2}(b - c - a) = - \sin \tfrac{1}{2}(- b + c + a) = - \sin \tfrac{1}{2}(a - b + c)$.]

Substituting from (B) and (C) in (H), we get

$$\cos^2 \tfrac{1}{2}\, a = \frac{\sin (s - c) \sin (s - b)}{\sin b \sin c}, \quad \text{or,}$$

$$(19) \qquad \cos \tfrac{1}{2}\, a = \sqrt{\frac{\sin (s - b) \sin (s - c)}{\sin b \sin c}}.$$

Since $\tan \tfrac{1}{2}\, a = \dfrac{\sin \tfrac{1}{2}\, a}{\cos \tfrac{1}{2}\, a}$, we get from this, by substitution from (18) and (19),

$$(20) \qquad \tan \tfrac{1}{2}\, a = \sqrt{\frac{\sin s \sin (s - a)}{\sin (s - b) \sin (s - c)}}.\dagger$$

---

* Let

$A = b - c$	$A = b - c$
$B = a$	$B = a$
$A + B = a + b - c$	$A - B = b - c - a$
$\tfrac{1}{2}(A + B) = \tfrac{1}{2}(a + b - c).$	$\tfrac{1}{2}(A - B) = \tfrac{1}{2}(b - c - a).$

Hence, substituting in formula (65), found on p. 74, Granville's *Plane Trigonometry*, namely,

$$\cos A - \cos B = - 2 \sin \tfrac{1}{2}(A + B) \sin \tfrac{1}{2}(A - B),$$

we get

$$\cos (b - c) - \cos a = - 2 \sin \tfrac{1}{2}(a + b - c) \sin \tfrac{1}{2}(b - c - a).$$

† In memorizing these formulas it will be found an aid to the memory to note the fact that under each radical

(a) only the sine function occurs.

(b) The denominators of the sine and cosine formulas involve those two sides of the triangle which are not opposite to the angle sought.

(c) When reading the numerator and denominator of the fraction in the tangent formula, s comes first and then the differences

$$s - a, \quad s - b, \quad s - c,$$

in cyclical order; s and the first difference occurring also in the numerator of the corresponding sine formula, while the last two differences occur in the numerator of the corresponding cosine formula.

In like manner, we may get

$$(21) \qquad \sin\tfrac{1}{2}\beta = \sqrt{\frac{\sin s \sin (s - b)}{\sin c \sin a}},$$

$$(22) \qquad \cos\tfrac{1}{2}\beta = \sqrt{\frac{\sin (s - c) \sin (s - a)}{\sin c \sin a}},$$

$$(23) \qquad \tan\tfrac{1}{2}\beta = \sqrt{\frac{\sin s \sin (s - b)}{\sin (s - c) \sin (s - a)}}.$$

Also

$$(24) \qquad \sin\tfrac{1}{2}\gamma = \sqrt{\frac{\sin s \sin (s - c)}{\sin a \sin b}},$$

$$(25) \qquad \cos\tfrac{1}{2}\gamma = \sqrt{\frac{\sin (s - a) \sin (s - b)}{\sin a \sin b}},$$

$$(26) \qquad \tan\tfrac{1}{2}\gamma = \sqrt{\frac{\sin s \sin (s - c)}{\sin (s - a) \sin (s - b)}}.$$

In solving triangles it is sometimes more convenient to use other forms of (20), (23), (26). Thus, in the right-hand member of (20), multiply both the numerator and denominator of the fraction under the radical by $\sin (s - a)$. This gives

$$\tan\tfrac{1}{2}\alpha = \sqrt{\frac{\sin s \sin^2 (s - a)}{\sin (s - a) \sin (s - b) \sin (s - c)}}$$

$$= \sin (s-a)\sqrt{\frac{\sin s}{\sin (s-a) \sin (s-b) \sin (s-c)}}.$$

Let

$$\tan\tfrac{1}{2} d^* = \sqrt{\frac{\sin (s - a) \sin (s - b) \sin (s - c)}{\sin s}},$$

then

$$\tan\tfrac{1}{2}\alpha = \frac{\sin (s - a)}{\tan\tfrac{1}{2} d}.$$

Similarly, for $\tan\tfrac{1}{2}\beta$ and $\tan\tfrac{1}{2}\gamma$. Hence

$$(27) \qquad \tan\tfrac{1}{2}d = \sqrt{\frac{\sin (s - a) \sin (s - b) \sin (s - c)}{\sin s}},$$

$$(28) \qquad \tan\tfrac{1}{2}\alpha = \frac{\sin (s - a)}{\tan\tfrac{1}{2} d},$$

$$(29) \qquad \tan\tfrac{1}{2}\beta = \frac{\sin (s - b)}{\tan\tfrac{1}{2} d},$$

$$(30) \qquad \tan\tfrac{1}{2}\gamma = \frac{\sin (s - c)}{\tan\tfrac{1}{2} d}.$$

* It may be shown that d = diameter of the circle inscribed in the spherical triangle.

13. Trigonometric functions of the half sides of a spherical triangle in terms of the supplements of the angles. By making use of the Principle of Duality on p. 208, we get at once from formulas (18) to (30), by replacing the supplement of an angle by the opposite side and each side by the supplement of the opposite angle, the following formulas:

$$(31) \qquad \sin \tfrac{1}{2} a = \sqrt{\frac{\sin \sigma \, \sin (\sigma - \alpha)}{\sin \beta \, \sin \gamma}},$$

$$(32) \qquad \cos \tfrac{1}{2} a = \sqrt{\frac{\sin (\sigma - \beta) \, \sin (\sigma - \gamma)}{\sin \beta \, \sin \gamma}},$$

$$(33) \qquad \tan \tfrac{1}{2} a = \sqrt{\frac{\sin \sigma \, \sin (\sigma - \alpha)}{\sin (\sigma - \beta) \, \sin (\sigma - \gamma)}},$$

$$(34) \qquad \sin \tfrac{1}{2} b = \sqrt{\frac{\sin \sigma \, \sin (\sigma - \beta)}{\sin \gamma \, \sin \alpha}},$$

$$(35) \qquad \cos \tfrac{1}{2} b = \sqrt{\frac{\sin (\sigma - \gamma) \, \sin (\sigma - \alpha)}{\sin \gamma \, \sin \alpha}},$$

$$(36) \qquad \tan \tfrac{1}{2} b = \sqrt{\frac{\sin \sigma \, \sin (\sigma - \beta)}{\sin (\sigma - \gamma) \, \sin (\sigma - \alpha)}},$$

$$(37) \qquad \sin \tfrac{1}{2} c = \sqrt{\frac{\sin \sigma \, \sin (\sigma - \gamma)}{\sin \alpha \, \sin \beta}},$$

$$(38) \qquad \cos \tfrac{1}{2} c = \sqrt{\frac{\sin (\sigma - \alpha) \, \sin (\sigma - \beta)}{\sin \alpha \, \sin \beta}}.$$

$$(39) \qquad \tan \tfrac{1}{2} c = \sqrt{\frac{\sin \sigma \, \sin (\sigma - \gamma)}{\sin (\sigma - \alpha) \, \sin (\sigma - \beta)}},$$

$$(40) \qquad \tan \tfrac{1}{2} \delta^* = \sqrt{\frac{\sin (\sigma - \alpha) \, \sin (\sigma - \beta) \, \sin (\sigma - \gamma)}{\sin \sigma}},$$

$$(41) \qquad \tan \tfrac{1}{2} a = \frac{\sin (\sigma - \alpha)}{\tan \tfrac{1}{2} \delta},$$

$$(42) \qquad \tan \tfrac{1}{2} b = \frac{\sin (\sigma - \beta)}{\tan \tfrac{1}{2} \delta},$$

$$(43) \qquad \tan \tfrac{1}{2} c = \frac{\sin (\sigma - \gamma)}{\tan \tfrac{1}{2} \delta},$$

where

$$\sigma = \tfrac{1}{2} (\alpha + \beta + \lambda)$$
$$= \tfrac{1}{2} (180° - A + 180° - B + 180° - C)$$
$$= 270° - \tfrac{1}{2} (A + B + C).$$

What we have done amounts to interchanging the corresponding Greek and Roman letters.

---

* It may be shown that δ is the supplement of the diameter of the circumscribed circle.

14. Napier's analogies. Dividing **(20)** by **(23)**, we get

$$\frac{\tan \tfrac{1}{2}\alpha}{\tan \tfrac{1}{2}\beta} = \sqrt{\frac{\sin s \sin (s-a)}{\sin (s-b)\sin (s-c)}} : \sqrt{\frac{\sin s \sin (s-b)}{\sin (s-c)\sin (s-a)}},$$

or,

$$\frac{\dfrac{\sin \tfrac{1}{2}\alpha}{\cos \tfrac{1}{2}\alpha}}{\dfrac{\sin \tfrac{1}{2}\beta}{\cos \tfrac{1}{2}\beta}} = \sqrt{\dfrac{\dfrac{\sin s \sin (s-a)}{\sin (s-b)\sin (s-c)}}{\dfrac{\sin s \sin (s-b)}{\sin (s-c)\sin (s-a)}}}.$$

Hence

$$\frac{\sin \tfrac{1}{2}\alpha \cos \tfrac{1}{2}\beta}{\cos \tfrac{1}{2}\alpha \sin \tfrac{1}{2}\beta} = \frac{\sin (s-a)}{\sin (s-b)}.$$

By composition and division, in proportion,

$$\frac{\sin \tfrac{1}{2}\alpha \cos \tfrac{1}{2}\beta + \cos \tfrac{1}{2}\alpha \sin \tfrac{1}{2}\beta}{\sin \tfrac{1}{2}\alpha \cos \tfrac{1}{2}\beta - \cos \tfrac{1}{2}\alpha \sin \tfrac{1}{2}\beta} = \frac{\sin (s-a)+\sin (s-b)}{\sin (s-a)-\sin (s-b)}.$$

From **(40)**, **(41)**, p. 63, and **(66)**, p. 74, Granville's *Plane Trigonometry*, the left-hand member equals

$$\frac{\sin (\tfrac{1}{2}\alpha + \tfrac{1}{2}\beta)}{\sin (\tfrac{1}{2}\alpha - \tfrac{1}{2}\beta)},$$

and the right-hand member

$$\frac{\sin (s-a)+\sin (s-b)}{\sin (s-a)-\sin (s-b)} = \frac{\tan \tfrac{1}{2}[s-a+(s-b)]}{\tan \tfrac{1}{2}[s-a-(s-b)]} = \frac{\tan \tfrac{1}{2}c}{\tan \tfrac{1}{2}(b-a)} \quad *$$

Equating these results, we get, noting that $\tan \tfrac{1}{2}(b-a)=-\tan \tfrac{1}{2}(a-b)$,

$$\frac{\sin \tfrac{1}{2}(\alpha + \beta)}{\sin \tfrac{1}{2}(\alpha - \beta)} = - \frac{\tan \tfrac{1}{2}c}{\tan \tfrac{1}{2}(a-b)}, \text{ or,}$$

$$(44)\qquad \tan \tfrac{1}{2}(a-b) = - \frac{\sin \tfrac{1}{2}(\alpha-\beta)}{\sin \tfrac{1}{2}(\alpha+\beta)}\tan \tfrac{1}{2}c.$$

In the same manner we may get the two similar formulas for $\tan \tfrac{1}{2}(b-c)$ and $\tan \tfrac{1}{2}(c-a)$.

Multiplying **(20)** and **(23)**, we get

$$\tan \tfrac{1}{2}\alpha \tan \tfrac{1}{2}\beta = \sqrt{\frac{\sin s \sin (s-a)}{\sin (s-b)\sin (s-c)}}\sqrt{\frac{\sin s \sin (s-b)}{\sin (s-c)\sin (s-a)}},$$

or,

$$\frac{\sin \tfrac{1}{2}\alpha \sin \tfrac{1}{2}\beta}{\cos \tfrac{1}{2}\alpha \cos \tfrac{1}{2}\beta} = \frac{\sin s}{\sin (s-c)}.$$

By composition and division, in proportion,

$$\frac{\cos \tfrac{1}{2}\alpha \cos \tfrac{1}{2}\beta - \sin \tfrac{1}{2}\alpha \sin \tfrac{1}{2}\beta}{\cos \tfrac{1}{2}\alpha \cos \tfrac{1}{2}\beta + \sin \tfrac{1}{2}\alpha \sin \tfrac{1}{2}\beta} = \frac{\sin (s-c)-\sin s}{\sin (s-c)+\sin s}.$$

* For $s-a+s-b=2s-a-b=a+b+c-a-b=c$, and $s-a-s+b=b-a$.

From (42), (43), p. 63, and (66), p. 74, Granville's *Plane Trigonometry*, the left-hand member equals

$$\frac{\cos\left(\tfrac{1}{2}\alpha + \tfrac{1}{2}\beta\right)}{\cos\left(\tfrac{1}{2}\alpha - \tfrac{1}{2}\beta\right)};$$

and the right-hand member

$$\frac{\sin(s-c)-\sin s}{\sin(s-c)+\sin s} = \frac{\tan\tfrac{1}{2}(s-c-s)}{\tan\tfrac{1}{2}(s-c+s)} = \frac{\tan\tfrac{1}{2}(-c)}{\tan\tfrac{1}{2}(a+b)}. \quad *$$

Equating these results, we get, noting that $\tan\tfrac{1}{2}(-c) = -\tan\tfrac{1}{2}c$,

$$\frac{\cos\tfrac{1}{2}(\alpha+\beta)}{\cos\tfrac{1}{2}(\alpha-\beta)} = -\frac{\tan\tfrac{1}{2}c}{\tan\tfrac{1}{2}(a+b)}, \text{ or,}$$

$$(45) \qquad \tan\tfrac{1}{2}(a+b) = -\frac{\cos\tfrac{1}{2}(\alpha-\beta)}{\cos\tfrac{1}{2}(\alpha+\beta)}\tan\tfrac{1}{2}c.$$

In the same manner we may get the two similar formulas for $\tan\tfrac{1}{2}(b+c)$ and $\tan\tfrac{1}{2}(c+a)$.

By making use of the Principle of Duality on p. 208, we get at once from formulas (44) and (45),

$$(46) \qquad \tan\tfrac{1}{2}(\alpha-\beta) = -\frac{\sin\tfrac{1}{2}(a-b)}{\sin\tfrac{1}{2}(a+b)}\tan\tfrac{1}{2}\gamma,$$

$$(47) \qquad \tan\tfrac{1}{2}(\alpha+\beta) = -\frac{\cos\tfrac{1}{2}(a-b)}{\cos\tfrac{1}{2}(a+b)}\tan\tfrac{1}{2}\gamma.$$

By changing the letters in cyclic order we may at once write down the corresponding formulas for $\tan\tfrac{1}{2}(\beta-\gamma)$, $\tan\tfrac{1}{2}(\gamma-\alpha)$, $\tan\tfrac{1}{2}(\beta+\gamma)$, and $\tan\tfrac{1}{2}(\gamma+\alpha)$.

The relations derived in this section are known as *Napier's analogies*.

Since $\cos\tfrac{1}{2}(a-b)$ and $\tan\tfrac{1}{2}\gamma = \tan\tfrac{1}{2}(180° - C) = \tan(90° - \tfrac{1}{2}C) = \cot\tfrac{1}{2}C$ are always positive, it follows from (47) that $\cos\tfrac{1}{2}(a+b)$ and $\tan\tfrac{1}{2}(\alpha+\beta)$ always have opposite signs; or, since $\tan\tfrac{1}{2}(\alpha+\beta) = \tan\tfrac{1}{2}(180° - A + 180° - B) = \tan\tfrac{1}{2}[360° - (A+B)] = \tan[180° - \tfrac{1}{2}(A+B)] = -\tan\tfrac{1}{2}(A+B)$, we may say that $\cos\tfrac{1}{2}(a+b)$ and $\tan\tfrac{1}{2}(A+B)$ always have the same sign. Hence we have the

Theorem. *In a spherical triangle the sum of any two sides is less than, greater than, or equal to 180°, according as the sum of the corresponding opposite angles is less than, greater than, or equal to 180°.*

15. Solution of oblique spherical triangles. We shall now take up the numerical solution of oblique spherical triangles. There are three cases to consider with two subdivisions under each case.

---

* For
$$s - c - s = -c,$$
and
$$s - c + s = 2s - c = a + b + c - c = a + b.$$

Case   I. (a) *Given the three sides.*

           (b) *Given the three angles.*

Case II. (a) *Given two sides and their included angle.*

           (b) *Given two angles and their included side.*

Case III. (a) *Given two sides and the angle opposite one of them.*

           (b) *Given two angles and the side opposite one of them.*

16. Case I. (a) Given the three sides. *Use formulas from p. 213,* namely,

$$(27) \qquad \tan \tfrac{1}{2} d = \sqrt{\frac{\sin(s-a)\sin(s-b)\sin(s-c)}{\sin s}},$$

$$(28) \qquad \tan \tfrac{1}{2} \alpha = \frac{\sin(s-a)}{\tan \tfrac{1}{2} d},$$

$$(29) \qquad \tan \tfrac{1}{2} \beta = \frac{\sin(s-b)}{\tan \tfrac{1}{2} d},$$

$$(30) \qquad \tan \tfrac{1}{2} \gamma = \frac{\sin(s-c)}{\tan \tfrac{1}{2} d},$$

to find α, β, γ, *and therefore* A, B, C, *and check by the law of sines,*

$$\frac{\sin a}{\sin A} = \frac{\sin b}{\sin B} = \frac{\sin c}{\sin C}.$$

Ex. 1. Given $a = 60°$, $b = 137° 20'$, $c = 116°$; find A, B, C.

Solution.

$a = 60°$	*To find* $\log \tan \tfrac{1}{2} d$ *use* (27)
$b = 137° 20'$	$\log \sin(s-a) = 9.9971$
$c = 116°$	$\log \sin(s-b) = 9.5199$
$2s = 313° 20'$	$\log \sin(s-c) = 9.8140$
$s = 156° 40'.$	$\ \overline{29.3310}$
$s - a = 96° 40'.$	$\log \sin s = 9.5978$
$s - b = 19° 20'.$	$2\,\lfloor 19.7332$
$s - c = 40° 40'.$	$\log \tan \tfrac{1}{2} d = 9.8666$

To find A use (28)	*To find B use* (29)	*To find C use* (30)
$\log \sin(s-a) = 9.9971$	$\log \sin(s-b) = 9.5199$	$\log \sin(s-c) = 9.8140$
$\log \tan \tfrac{1}{2} d = 9.8666$	$\log \tan \tfrac{1}{2} d = 9.8666$	$\log \tan \tfrac{1}{2} d = 9.8666$
$\log \tan \tfrac{1}{2} \alpha = 0.1305$	$\log \tan \tfrac{1}{2} \beta = 9.6533$	$\log \tan \tfrac{1}{2} \gamma = 9.9474$
$\tfrac{1}{2} \alpha = 53° 29'.$	$\tfrac{1}{2} \beta = 24° 14'.$	$\tfrac{1}{2} \gamma = 41° 32'.$
$\alpha = 106° 58'.$	$\beta = 48° 28'.$	$\gamma = 83° 4'.$

$\therefore A = 180° - 106°58' = 73°2'.\quad \therefore B = 180° - 48°28' = 131°32'.\quad \therefore C = 180° - 83°4' = 96°56'.$

Check: $\log \sin a = 9.9375$	$\log \sin b = 9.8311$	$\log \sin c = 9.9537$
$\log \sin A = 9.9807$	$\log \sin B = 9.8743$	$\log \sin C = 9.9969$
$\ 9.9568$	$\ 9.9568$	$\ 9.9568$

This checks up closer than is to be expected in general. There may be a variation of at most two units in the last figure when the work is accurate.

EXAMPLES

Solve the following oblique spherical triangles :

No.	Given Parts			Required Parts		
1	$a = 38°$	$b = 51°$	$c = 42°$	$A = 51° 58'$	$B = 83° 54'$	$C = 58° 53'$
2	$a = 101°$	$b = 49°$	$c = 60°$	$A = 142° 32'$	$B = 27° 52'$	$C = 32° 28'$
3	$a = 61°$	$b = 39°$	$c = 92°$	$A = 35° 32'$	$B = 24° 42'$	$C = 138° 24'$
4	$a = 62° 20'$	$b = 54° 10'$	$c = 97° 50'$	$A = 47° 22'$	$B = 42° 20'$	$C = 124° 38'$
5	$a = 58°$	$b = 80°$	$c = 96°$	$A = 55° 58'$	$B = 74° 14'$	$C = 103° 36'$
6	$a = 46° 30'$	$b = 62° 40'$	$c = 83° 20'$	$A = 43° 58'$	$B = 58° 14'$	$C = 108° 6'$
7	$a = 71° 15'$	$b = 39° 10'$	$c = 40° 35'$	$A = 130° 36'$	$B = 30° 26'$	$C = 31° 26'$
8	$a = 47° 30'$	$b = 55° 40'$	$c = 60° 10'$	$A = 56° 32'$	$B = 69° 7'$	$C = 78° 58'$
9	$a = 43° 30'$	$b = 72° 24'$	$c = 87° 50'$	$A = 41° 27'$	$B = 66° 26'$	$C = 106° 3'$
10	$a = 110° 40'$	$b = 45° 10'$	$c = 73° 30'$	$A = 144° 27'$	$B = 26° 9'$	$C = 36° 35'$

17. Case I. (*b*) **Given the three angles.** *Use formulas from p. 214, namely,*[*]

$$(40) \qquad \tan \tfrac{1}{2} \delta = \sqrt{\frac{\sin(\sigma - \alpha)\sin(\sigma - \beta)\sin(\sigma - \gamma)}{\sin \sigma}},$$

$$(41) \qquad \tan \tfrac{1}{2} a = \frac{\sin(\sigma - \alpha)}{\tan \tfrac{1}{2} \delta},$$

$$(42) \qquad \tan \tfrac{1}{2} b = \frac{\sin(\sigma - \beta)}{\tan \tfrac{1}{2} \delta},$$

$$(43) \qquad \tan \tfrac{1}{2} c = \frac{\sin(\sigma - \gamma)}{\tan \tfrac{1}{2} \delta},$$

to find a, b, c ; and check by the law of sines,

$$\frac{\sin a}{\sin A} = \frac{\sin b}{\sin B} = \frac{\sin c}{\sin C}.$$

Ex. 1. Given $A = 70°$, $B = 131° 10'$, $C = 94° 50'$; find a, b, c.

Solution. Here we use the supplements of the angles.

<table>
<tr><td>

$\alpha = 180° - A = 110°$
$\beta = 180° - B = \ 48° 50'$
$\gamma = 180° - C = \ 85° 10'$
$\quad 2\sigma = 244°$
$\quad\ \sigma = 122°.$
$\sigma - \alpha = \ 12°.$
$\sigma - \beta = \ 73° 10'.$
$\sigma - \gamma = \ 36° 50'.$

</td><td>

To find $\log \tan \tfrac{1}{2}\delta$ *use* (40)

$\log \sin(\sigma - \alpha) = \ 9.3179$
$\log \sin(\sigma - \beta) = \ 9.9810$
$\log \sin(\sigma - \gamma) = \ 9.7778$
$\qquad\qquad\qquad\quad 29.0767$
$\log \sin \sigma = \ 9.9284$
$\qquad\qquad 2\,|\,19.1483$
$\log \tan \tfrac{1}{2}\delta = \ 9.5742$

</td></tr>
</table>

[*] These formulas may be written down at once from those used in Case I, (*a*), p. 217, by simply interchanging the corresponding Greek and Roman letters.

To find a use (41)	*To find b use* (42)	*To find c use* (43)
$\log \sin (\sigma - \alpha) = 9.3179$	$\log \sin (\sigma - \beta) = 9.9810$	$\log \sin (\sigma - \gamma) = 9.7778$
$\log \tan \frac{1}{2}\delta = 9.5742$	$\log \tan \frac{1}{2}\delta = 9.5742$	$\log \tan \frac{1}{2}\delta = 9.5742$
$\log \tan \frac{1}{2} a = 9.7437$	$\log \tan \frac{1}{2} b = 0.4068$	$\log \tan \frac{1}{2} c = 0.2036$
$\frac{1}{2} a = 29°.$	$\frac{1}{2} b = 68° 36'.$	$\frac{1}{2} c = 57° 58'.$
$\therefore\ a = 58°.$	$b = 137° 12'.$	$c = 115° 56'.$

Check : $\log \sin a = 9.9284$	$\log \sin b = 9.8321$	$\log \sin c = 9.9539$
$\log \sin A = 9.9730$	$\log \sin B = 9.8767$	$\log \sin C = 9.9985$
9.9554	9.9554	9.9554

EXAMPLES

Solve the following oblique spherical triangles :

No.	Given Parts			Required Parts		
1	$A = 75°$	$B = 82°$	$C = 61°$	$a = 67° 52'$	$b = 71° 44'$	$c = 57°$
2	$A = 120°$	$B = 130°$	$C = 80°$	$a = 144° 10'$	$b = 148° 49'$	$c = 41° 44'$
3	$A = 91° 10'$	$B = 85° 40'$	$C = 72° 30'$	$a = 89° 51'$	$b = 85° 49'$	$c = 72° 32'$
4	$A = 138° 16'$	$B = 31° 11'$	$C = 35° 53'$	$a = 100° 5'$	$b = 49° 59'$	$c = 60° 6'$
5	$A = 78° 40'$	$B = 63° 50'$	$C = 46° 20'$	$a = 39° 30'$	$b = 35° 36'$	$c = 27° 59'$
6	$A = 121°$	$B = 102°$	$C = 68°$	$a = 130° 50'$	$b = 120° 18'$	$c = 54° 56'$
7	$A = 130°$	$B = 110°$	$C = 80°$	$a = 139° 21'$	$b = 126° 58'$	$c = 56° 52'$
8	$A = 28°$	$B = 92°$	$C = 85° 26'$	$a = 27° 56'$	$b = 85° 40'$	$c = 84° 2'$
9	$A = 59° 18'$	$B = 108°$	$C = 76° 22'$	$a = 61° 44'$	$b = 103° 4'$	$c = 84° 32'$
10	$A = 100°$	$B = 100°$	$C = 50°$	$a = 112° 14'$	$b = 112° 14'$	$c = 46° 4'$

**18. Case II. (a) Given two sides and their included angle, as a, b,
C.** *Use formulas on p. 216, namely,*

$$(46) \qquad \tan \tfrac{1}{2}(\alpha - \beta) = - \frac{\sin \frac{1}{2}(a - b)}{\sin \frac{1}{2}(a + b)} \tan \tfrac{1}{2}\gamma,$$

$$(47) \qquad \tan \tfrac{1}{2}(\alpha + \beta) = \cdot \frac{\cos \frac{1}{2}(a - b)}{\cos \frac{1}{2}(a + b)} \tan \tfrac{1}{2}\gamma,$$

to find α and β and therefore A and B; and from p. 215 use (44)
solved for $\tan \frac{1}{2} c$, *namely,*

$$(44) \qquad \tan \tfrac{1}{2} c = - \frac{\sin \frac{1}{2}(\alpha + \beta) \tan \frac{1}{2}(\alpha - \beta)}{\sin \frac{1}{2}(\alpha - \beta)},$$

to find c. Check by the law of sines.

Ex. 1. Given $a = 64° 24'$, $b = 42° 30'$, $C = 58° 40'$; find A, B, c.

Solution. $\gamma = 180° - C = 121° 20'$. $\therefore \frac{1}{2}\gamma = 60° 40'$.

$$
\begin{aligned}
a &= 64° 24' \\
b &= 42° 30' \\
\hline
a + b &= 106° 54' \\
\therefore \tfrac{1}{2}(a + b) &= 53° 27'.
\end{aligned}
\qquad\qquad
\begin{aligned}
a &= 64° 24' \\
b &= 42° 30' \\
\hline
a - b &= 21° 54' \\
\therefore \tfrac{1}{2}(a - b) &= 10° 57'.
\end{aligned}
$$

To find $\frac{1}{2}(\alpha - \beta)$ *use* (**46**)

$$\log \sin \tfrac{1}{2}(a - b) = 9.2786$$
$$\log \tan \tfrac{1}{2}\gamma = \underline{0.2503}$$
$$9.5289$$
$$\log \sin \tfrac{1}{2}(a + b) = 9.9049$$
$$\log \tan \tfrac{1}{2}(\alpha - \beta) = 9.6240\ (n)$$
$$\therefore\ \tfrac{1}{2}(\alpha - \beta) = -22°\ 49'.*$$

To find $\frac{1}{2}(\alpha + \beta)$ *use* (**47**)

$$\log \cos \tfrac{1}{2}(a - b) = 9.9920$$
$$\log \tan \tfrac{1}{2}\gamma = \underline{0.2503}$$
$$10.2423$$
$$\log \cos \tfrac{1}{2}(a + b) = 9.7749$$
$$\log \tan \tfrac{1}{2}(\alpha + \beta) = 0.4674\ (n)$$
$$180° - \tfrac{1}{2}(\alpha + \beta) = 71°\ 11'.\dagger$$
$$\therefore\ \tfrac{1}{2}(\alpha + \beta) = 108°\ 49'.$$

To find A and B

$$\tfrac{1}{2}(\alpha + \beta) = 108°\ 49'$$
$$\tfrac{1}{2}(\alpha - \beta) = -\ \underline{22°\ 49'}$$
Adding, $\qquad \alpha = \quad 86°$
Subtracting, $\quad \beta = 131°\ 38'.$
$$\therefore\ A = 180° - \alpha = 94°.$$
$$B = 180° - \beta = 48°\ 22'.$$

To find c use (**44**)

$$\log \sin \tfrac{1}{2}(\alpha + \beta) = 9.9761$$
$$\log \tan \tfrac{1}{2}(a - b) = \underline{9.2867}$$
$$19.2628$$
$$\log \sin \tfrac{1}{2}(\alpha - \beta) = 9.5886\ (n)$$
$$\log \tan \tfrac{1}{2}c = 9.6742\ \ddagger$$
$$\tfrac{1}{2}c = 25°\ 17'.$$
$$\therefore\ c = 50°\ 34'.$$

Check: $\ \log \sin a = 9.9551 \qquad \log \sin b :: 9.8297 \qquad \log \sin c = 9.8878$

$\qquad\qquad \log \sin A = \underline{9.9989} \qquad\ \log \sin B :: \underline{9.8735} \qquad \log \sin C = \underline{9.9315}$

$\qquad\qquad\qquad\quad 9.9562 \qquad\qquad\qquad 9.9562 \qquad\qquad\qquad 9.9563$

If c only is wanted, we may find it from the law of cosines, (**14**), p. 209, without previously determining A and B. But this formula is not well adapted to logarithmic calculations. Another method is

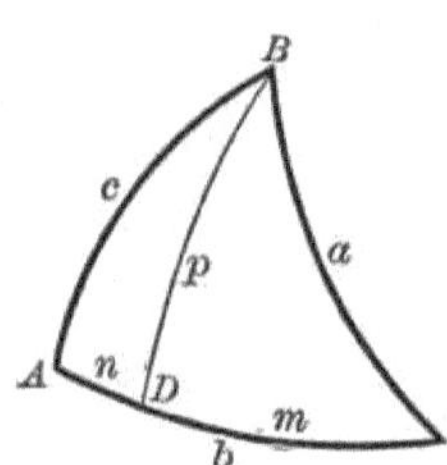

illustrated below, which depends on the solution of right spherical triangles, and hence requires only those formulas which follow from applying *Napier's rules of circular parts*, p. 200.

Through B draw an arc of a great circle perpendicular to $A C$, intersecting $A C$ at D. Let

$$BD = p, \qquad CD = m, \qquad AD = n.$$

Applying Rule I, p. 200, to the right spherical triangle BCD, we have
$$\cos C = \tan m \cot a, \text{ or,}$$
$$(A) \qquad \tan m = \tan a \cos C.$$

Applying Rule II, p. 200, to BCD,
$$\cos a = \cos m \cos p, \text{ or,}$$
$$(B) \qquad \cos p = \cos a \sec m.$$

---

* Since $\tan \frac{1}{2}(\alpha - \beta)$ is negative, $\frac{1}{2}(\alpha - \beta)$ may be an angle in the second or fourth quadrant. But $a > b$, therefore $A > B$ and $\alpha < \beta$, since α and β are the supplements of A and B. Hence $\frac{1}{2}(\alpha - \beta)$ must be a negative angle numerically less than 90°.

† Here $\frac{1}{2}(\alpha + \beta)$ must be a positive angle less than 180°. Since $\tan \frac{1}{2}(\alpha + \beta)$ is negative, $\frac{1}{2}(\alpha + \beta)$ must lie in the second quadrant, and we get its supplement from the table.

‡ $\tan \frac{1}{2}c$ is positive, since $\sin \frac{1}{2}(\alpha - \beta)$ is negative and there is a minus sign before the fraction.

Applying the same rule to ABD,

$$\cos c = \cos n \cos p, \text{ or,}$$

(C)
$$\cos p = \cos c \sec n.$$

Equating (B) and (C),

$$\cos c \sec n = \cos a \sec m, \text{ or,}$$

$$\cos c = \cos a \sec m \cos n.$$

But $n = b - m$; therefore

(D)
$$\cos c = \cos a \sec m \cos (b - m).$$

Now c may be computed from (A) and (D), namely,

(48)
$$\tan m = \tan a \cos C,$$

(49)
$$\cos c = \frac{\cos a \cos (b - m)}{\cos m}.$$

Ex. 2. Given $a = 98°$, $b = 80°$, $C = 110°$; find c.
Solution. Apply the method just explained.

<table>
<tr><td>

To find $b - m$ use **(48)**

log tan a = 0.8522 (n)
log cos C = 9.5341 (n)
log tan m = 0.3863
    $m = 67°\ 40'$.
∴ $b - m = 12°\ 20'$.

</td><td>

To find c use **(49)**

log cos a = 9.1436 (n)
log cos $(b - m)$ = 9.9899
           19.1335
log cos m = 9.5798
log cos c = 9.5537 (n)
$180° - c = 69°\ 2'$.
    $c = 110°\ 58'$.

</td></tr>
</table>

EXAMPLES

Solve the following oblique spherical triangles

No.	Given Parts			Required Parts		
1	$a = 137°\ 20'$	$c = 116°$	$A = 70°$	$B = 131°\ 17'$	$C = 94°\ 48'$	$a = 57°\ 57'$
2	$a = 72°$	$b = 47°$	$C = 33°$	$A = 121°\ 33'$	$B = 40°\ 57'$	$c = 37°\ 26'$
3	$a = 98°$	$c = 60°$	$B = 110°$	$A = 87°$	$C = 60°\ 51'$	$b = 111°\ 17'$
4	$b = 120°\ 20'$	$c = 70°\ 40'$	$A = 50°$	$B = 134°\ 57'$	$C = 50°\ 41'$	$a = 69°\ 9'$
5	$a = 125°\ 10'$	$b = 153°\ 50'$	$C = 140°\ 20'$	$A = 147°\ 29'$	$B = 163°\ 9'$	$c = 76°\ 8'$
6	$a = 93°\ 20'$	$b = 56°\ 30'$	$C = 74°\ 40'$	$A = 101°\ 24'$	$B = 54°\ 58'$	$c = 79°\ 10'$
7	$b = 76°\ 30'$	$c = 47°\ 20'$	$A = 92°\ 30'$	$B = 78°\ 21'$	$C = 47°\ 47'$	$a = 82°\ 42'$
8	$c = 40°\ 20'$	$a = 100°\ 30'$	$B = 46°\ 40'$	$A = 131°\ 29'$	$C = 29°\ 33'$	$b = 72°\ 40'$
9	$b = 76°\ 36'$	$c = 110°\ 26'$	$A = 46°\ 50'$	$B = 57°\ 43'$	$C = 125°\ 28'$	$a = 57°\ 13'$
10	$a = 84°\ 23'$	$b = 124°\ 48'$	$C = 62°$	$A = 68°\ 27'$	$B = 129°\ 51'$	$c = 70°\ 52'$

19. Case II. (*b*) **Given two angles and their included side, as** *A*, *B*, *c*.
Use formulas * *on pp. 215, 216, namely,*

$$(44) \qquad \tan \tfrac{1}{2}(a - b) = - \frac{\sin \tfrac{1}{2}(\alpha - \beta)}{\sin \tfrac{1}{2}(\alpha + \beta)} \tan \tfrac{1}{2} c,$$

$$(45) \qquad \tan \tfrac{1}{2}(a + b) = - \frac{\cos \tfrac{1}{2}(\alpha - \beta)}{\cos \tfrac{1}{2}(\alpha + \beta)} \tan \tfrac{1}{2} c,$$

to find a and b ; and from p. 216, use (**46**) *solved for tan* $\tfrac{1}{2}\gamma$*, namely,*

$$(46) \qquad \tan \tfrac{1}{2}\gamma = - \frac{\sin \tfrac{1}{2}(a + b) \tan \tfrac{1}{2}(\alpha - \beta)}{\sin \tfrac{1}{2}(a - b)},$$

to find γ *and therefore C.   Check by the law of sines.*

Ex. 1. Given $c = 116°$, $A = 70°$, $B = 131° 20'$; find a, b, C.
Solution. $\alpha = 180° - A = 110°$, and $\beta = 180° - B = 48° 40'$.

$\alpha = 110°$	$\alpha = 110°$	
$\beta = 48° 40'$	$\beta = 48° 40'$	
$\alpha + \beta = 158° 40'$	$\alpha - \beta = 61° 20'$	$c = 116°.$
$\therefore \tfrac{1}{2}(\alpha + \beta) = 79° 20'.$	$\therefore \tfrac{1}{2}(\alpha - \beta) = 30° 40'.$	$\therefore \tfrac{1}{2}c = 58°.$

To find $\tfrac{1}{2}(a - b)$ *use* (**44**)

$$\log \sin \tfrac{1}{2}(\alpha - \beta) = 9.7076$$
$$\log \tan \tfrac{1}{2} c = 0.2042$$
$$\overline{\phantom{\log \tan \tfrac{1}{2} c =} 9.9118}$$
$$\log \sin \tfrac{1}{2}(\alpha + \beta) = 9.9024$$
$$\log \tan \tfrac{1}{2}(a - b) = 9.9194 \; (n)$$
$$\therefore \tfrac{1}{2}(a - b) = - 39° 43'. \dagger$$

To find $\tfrac{1}{2}(a + b)$ *use* (**45**)

$$\log \cos \tfrac{1}{2}(\alpha - \beta) = 9.9346$$
$$\log \tan \tfrac{1}{2} c = 0.2042$$
$$\overline{\phantom{\log \tan \tfrac{1}{2} c =} 10.1388}$$
$$\log \cos \tfrac{1}{2}(\alpha + \beta) = 9.2674$$
$$\log \tan \tfrac{1}{2}(a + b) = 0.8714 \; (n)$$
$$180° - \tfrac{1}{2}(a + b) = 82° 21'.$$
$$\therefore \tfrac{1}{2}(a + b) = 97° 39'.$$

To find a and b

$$\tfrac{1}{2}(a + b) = 97° 39'$$
$$\tfrac{1}{2}(a - b) = - 39° 43'$$
Adding, $\qquad a = 57° 56'$
Subtracting, $\qquad b = 137° 22'.$

To find C use (**46**)

$$\log \sin \tfrac{1}{2}(a + b) = 9.9961$$
$$\log \tan \tfrac{1}{2}(\alpha - \beta) = 9.7730$$
$$\overline{\phantom{\log \tan \tfrac{1}{2}(\alpha - \beta) =} 19.7691}$$
$$\log \sin \tfrac{1}{2}(a - b) = 9.8055 \; (n)$$
$$\log \tan \tfrac{1}{2}\gamma = 9.9636$$
$$\tfrac{1}{2}\gamma = 42° 36'.$$
$$\gamma = 85° 12'.$$
$$\therefore C = 180° - \gamma = 94° 48'.$$

Check : $\log \sin a = 9.9281$ $\log \sin b = 9.8308$ $\log \sin c = 9.9537$
$ \log \sin A = 9.9730$ $\log \sin B = 9.8756$ $\log \sin C = 9.9985$
$ 9.9551$ $ 9.9552$ $ 9.9552$

* Same as those used in Case II, (*a*), p. 219, with Greek and Roman letters interchanged
† Since $A < B$ it follows that $a < b$, and $\tfrac{1}{2}(a - b)$ is negative.

If C only is wanted, we can calculate it without previously determining a and b, by dividing the given triangle into two right spherical triangles, as was illustrated on p. 220.

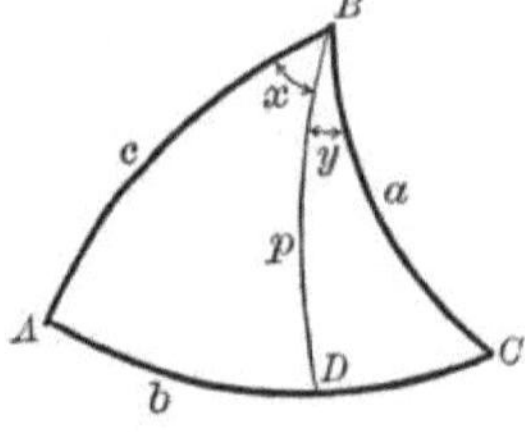

Through B draw an arc of a great circle perpendicular to AC, intersecting AC at D. Let $BD = p$, angle $ABD = x$, angle $CBD = y$. Applying Rule I of Napier's rules, p. 200, to the right spherical triangle ABD, we have

$$\cos c = \cot x \cot A, \text{ or,}$$

(A)
$$\cot x = \tan A \cos c.$$

Applying Rule II, p. 200, to ABD, we have

$$\cos A - \cos p \sin x, \text{ or,}$$

(B)
$$\cos p = \cos A \csc x.$$

Applying the same rule to CBD,

$$\cos C = \cos p \sin y, \text{ or,}$$

(C)
$$\cos p = \cos C \csc y.$$

Equating (B) and (C),

$$\cos C \csc y = \cos A \csc x, \text{ or,}$$

$$\cos C = \cos A \csc x \sin y.$$

But $y = B - x$; therefore

(D)
$$\cos C = \cos A \csc x \sin (B - x).$$

Now C may be computed from (A) and (D), namely,

(50)
$$\cot x = \tan A \cos c.$$

(51)
$$\cos C = \frac{\cos A \sin (B - x)}{\sin x}.$$

Ex. 2. Given $A = 35° 46'$, $B = 115° 9'$, $c = 51° 2'$; find C.

Solution. Apply the method just explained.

<table>
<tr><td>

To find $B - x$ use **(50)**

$\log \tan A = 9.8575$
$\log \cos c = 9.7986$
$\log \cot x = 9.6561$
$\qquad x = 65° 38'.$
$\therefore B - x = 49° 31'.$

</td><td>

To find C use **(51)**

$\log \cos A = 9.9093$
$\log \sin (B - x) = \underline{9.8811}$
$\qquad\qquad\qquad 19.7904$
$\log \sin x = 9.9595$
$\log \cos C = 9.8309$
$\qquad C = 47° 21'.$

</td></tr>
</table>

EXAMPLES

Solve the following oblique spherical triangles:

No.	Given Parts			Required Parts		
1	$A = 67° 30'$	$B = 45° 50'$	$c = 74° 20'$	$a = 63° 15'$	$b = 43° 54'$	$C = 95° 1'$
2	$B = 98° 30'$	$C = 67° 20'$	$a = 60° 40'$	$b = 86° 40'$	$c = 68° 40'$	$A = 59° 44'$
3	$C = 110°$	$A = 94°$	$b = 44°$	$a = 114° 10'$	$c = 120° 46'$	$B = 130° 34'$
4	$C = 70° 20'$	$B = 43° 50'$	$a = 50° 46'$	$b = 32° 59'$	$c = 47° 45'$	$A = 80° 14'$
5	$A = 78°$	$B = 41°$	$c = 108°$	$a = 95° 38'$	$b = 41° 52'$	$C = 110° 49'$
6	$B = 135°$	$C = 50°$	$a = 70° 20'$	$b = 120° 17'$	$c = 69° 20'$	$A = 50° 26'$
7	$A = 31° 40'$	$C = 122° 20'$	$b = 40° 40'$	$a = 34° 3'$	$c = 64° 19'$	$B = 37° 40'$
8	$A = 108° 12'$	$B = 145° 46'$	$c = 126° 32'$	$a = 69° 4'$	$b = 146° 26'$	$C = 125° 12'$
9	$A = 130° 36'$	$B = 30° 26'$	$c = 40° 35'$	$a = 71° 15'$	$b = 39° 10'$	$C = 31° 26'$
10	$A = 51° 58'$	$B = 83° 54'$	$c = 42°$	$a = 38°$	$b = 51°$	$C = 58° 53'$

20. Case III. (*a*) **Given two sides and the angle opposite one of them, as *a*, *b*, *B* (ambiguous case *).**

From the law of sines, p. 207, we get

$$(11) \qquad \sin A = \frac{\sin a \sin B}{\sin b},$$

which gives A†. To find C we use, from p. 216, formula (**46**), *solved for* $\tan \tfrac{1}{2}\gamma$, *namely,*

$$(46) \qquad \tan\tfrac{1}{2}\gamma = -\frac{\sin\tfrac{1}{2}(a+b)\tan\tfrac{1}{2}(\alpha-\beta)}{\sin\tfrac{1}{2}(a-b)}.$$

To find c, solve (**44**), *p. 215, for* $\tan\tfrac{1}{2}c$, *namely,*

$$(44) \qquad \tan\tfrac{1}{2}c = -\frac{\sin\tfrac{1}{2}(\alpha+\beta)\tan\tfrac{1}{2}(a-b)}{\sin\tfrac{1}{2}(\alpha-\beta)}.$$

Check by the law of sines.

Ex. 1. Given a 58°, $b = 137° 20'$, $B = 131° 20'$; find A, C, c.

Solution.

To find A use (**11**)

$$\log \sin a = \ 9.9284$$
$$\log \sin B = \ \underline{9.8756}$$
$$19.8040$$
$$\log \sin b = \ 9.8311$$
$$\log \sin A = \ 9.9729$$
$$\therefore A_1 = 69° 58',$$
or, $\quad A_2 = 180° - A_1 = 110° 2'.$

$a = 58°$ $\qquad a = 58°$
$b = 137° 20'$ $\qquad b = 137° 20'$
$a+b = 195° 20'$ $\qquad a-b = -79° 20'$
$\tfrac{1}{2}(a+b) = 97° 40'.$ $\quad \tfrac{1}{2}(a-b) = -39° 40'.$

$\beta = 180° - B = 48° 40'.$

Since $a < b$ and both A_1 and A_2 are $< B$, it follows that we have *two solutions.*

---

* Just as in the corresponding case in the solution of plane oblique triangles (Granville's *Plane Trigonometry*, pp. 105, 161), there may be *two solutions, one solution,* or *no solution,* depending on the given data.

† Since the angle A is here determined from its sine, it is necessary to consider both of the values found. If $a > b$ then $A > B$; and if $a < b$ then $A < B$. Hence [next page]

First solution. $\alpha_1 = 180° - A_1 = 110° \, 2'$.

$$\alpha_1 = 110° \, 2'$$
$$\beta = 48° \, 40'$$
$$\alpha_1 + \beta = 158° \, 42'$$
$$\tfrac{1}{2}(\alpha_1 + \beta) = 79° \, 21'.$$

$$\alpha_1 = 110° \, 2'$$
$$\beta = 48° \, 40'$$
$$\alpha_1 - \beta = 61° \, 22'$$
$$\tfrac{1}{2}(\alpha_1 - \beta) = 30° \, 41'.$$

To find C_1 use (**46**)

$$\log \sin \tfrac{1}{2}(a + b) = 9.9961$$
$$\log \tan \tfrac{1}{2}(\alpha_1 - \beta) = 9.7733$$
$$19.7694$$
$$\log \sin \tfrac{1}{2}(a - b) = 9.8050 \ (n)$$
$$\log \tan \tfrac{1}{2}\gamma_1 = 9.9644$$
$$\tfrac{1}{2}\gamma_1 = 42° \, 39'.$$
$$\gamma_1 = 85° \, 18'.$$
$$\therefore C_1 = 180° - \gamma_1 = 94° \, 42'.$$

To find c_1 use (**44**)

$$\log \sin \tfrac{1}{2}(\alpha_1 + \beta) = 9.9924$$
$$\log \tan \tfrac{1}{2}(a - b) = 9.9187 \ (n)$$
$$19.9111$$
$$\log \sin \tfrac{1}{2}(\alpha_1 - \beta) = 9.7078$$
$$\log \tan \tfrac{1}{2}c_1 = 10.2033$$
$$\tfrac{1}{2}c_1 = 57° \, 57'.$$
$$\therefore c_1 = 115° \, 54'.$$

Check:

$$\log \sin a = 9.9284 \qquad \log \sin b = 9.8311 \qquad \log \sin c_1 = 9.9541$$
$$\log \sin A_1 = 9.9729 \qquad \log \sin B = 9.8756 \qquad \log \sin C_1 = 9.9985$$
$$9.9555 \qquad\qquad 9.9555 \qquad\qquad 9.9556$$

Second solution. $\alpha_2 = 180° - A_2 = 69° \, 58'$.

$$\alpha_2 = 69° \, 58'$$
$$\beta = 48° \, 40'$$
$$\alpha_2 + \beta = 118° \, 38'$$
$$\tfrac{1}{2}(\alpha_2 + \beta) = 59° \, 19'.$$

$$\alpha_2 = 69° \, 58'$$
$$\beta = 48° \, 40'$$
$$\alpha_2 - \beta = 21° \, 18'$$
$$\tfrac{1}{2}(\alpha_2 - \beta) = 10° \, 39'.$$

To find C_2 use (**46**)

$$\log \sin \tfrac{1}{2}(a + b) = 9.9961$$
$$\log \tan \tfrac{1}{2}(\alpha_2 - \beta) = 9.2743$$
$$19.2704$$
$$\log \sin \tfrac{1}{2}(a - b) = 9.8050 \ (n)$$
$$\log \tan \tfrac{1}{2}\gamma_2 \quad 9.4654$$
$$\tfrac{1}{2}\gamma_2 : 16° \, 17'.$$
$$\gamma_2 = 32° \, 34'.$$
$$\therefore C_2 = 180° - \gamma_2 = 147° \, 26'.$$

To find c_1 use (**44**)

$$\log \sin \tfrac{1}{2}(\alpha_2 + \beta) = 9.9345$$
$$\log \tan \tfrac{1}{2}(a - b) = 9.9187 \ (n)$$
$$19.8532$$
$$\log \sin \tfrac{1}{2}(\alpha_2 - \beta) = 9.2667$$
$$\log \tan \tfrac{1}{2}c_2 = 10.5865$$
$$\tfrac{1}{2}c_2 = 75° \, 28'.$$
$$\therefore c_2 = 150° \, 56'.$$

Check:

$$\log \sin a = 9.9284 \qquad \log \sin b = 9.8311 \qquad \log \sin c_2 = 9.6865$$
$$\log \sin A_2 = 9.9729 \qquad \log \sin B = 9.8756 \qquad \log \sin C_2 = 9.7310$$
$$9.9555 \qquad\qquad 9.9555 \qquad\qquad 9.9555$$

If the side c or the angle C is wanted without first calculating the value of A, we may resolve the given triangle into two right triangles and then apply Napier's rules, as was illustrated under Cases II, (*a*), and II, (*b*), pp. 220, 223.

Theorem. *Only those values of A should be retained which are greater or less than B according as a is greater or less than b.*
If $\log \sin A = $ a positive number, there will be no solution.

EXAMPLES

Solve the following oblique spherical triangles :

No.	Given Parts			Required Parts		
1	$a=43° 20'$	$b=48° 30'$	$A=58° 40'$	$B_1=68° 47'$	$C_1=70° 40'$	$c_1=49° 18'$
				$B_2=111° 13'$	$C_2=14° 29'$	$c_2=11° 36'$
2	$a=56° 40'$	$b=30° 50'$	$A=103° 40'$	$B=36° 36'$	$C=52°$	$c=42° 39'$
3	$a=30° 20'$	$b=46° 30'$	$A=36° 40'$	$B_1=59° 4'$	$C_1=97° 39'$	$c_1=56° 57'$
				$B_2=120° 57'$	$C_2=28° 5'$	$c_2=23° 28'$
4	$b=99° 40'$	$c=64° 20'$	$B=95° 40'$	$C=65° 30'$	$A=97° 20'$	$a=100° 45'$
5	$a=40°$	$b=118° 20'$	$A=29° 40'$	$B_1=42° 40'$	$C_1=159° 54'$	$c_1=153° 30'$
				$B_2=137° 20'$	$C_2=50° 21'$	$c_2=90° 10'$
6	$a=115° 20'$	$c=146° 20'$	$C=141° 10'$	Impossible		
7	$a=109° 20'$	$c=82°$	$A=107° 40'$	$C=90°$	$B=113° 37'$	$b=114° 52'$
8	$b=108° 30'$	$c=40° 50'$	$C=39° 50'$	$B_1=68° 18'$	$A_1=132° 34'$	$a_1=131° 16'$
				$B_2=111° 42'$	$A_2=77° 5'$	$a_2=95° 50'$
9	$a=162° 20'$	$b=15° 40'$	$B=125°$	Impossible		
10	$a=55°$	$c=138° 10'$	$A=42° 30'$	$C=146° 38'$	$B=55° 1'$	$b=96° 34'$

21. **Case III.** (b) **Given two angles and the side opposite one of them, as A, B, b** (ambiguous case *).

From the law of sines, p. 207, we get

$$(11) \qquad \sin a = \frac{\sin A \sin b}{\sin B},$$

which gives a.† To find c we use, from p. 215, the formula ‡ (44), solved for tan ½ c, namely,

$$(44) \qquad \tan \tfrac{1}{2} c = - \frac{\sin \tfrac{1}{2}(\alpha + \beta) \tan \tfrac{1}{2}(a - b)}{\sin \tfrac{1}{2}(a - \beta)}.$$

To find C, solve (46), p. 216, for tan ½ γ, namely,

$$(46) \qquad \tan \tfrac{1}{2} \gamma = - \frac{\sin \tfrac{1}{2}(a + b) \tan \tfrac{1}{2}(a - \beta)}{\sin \tfrac{1}{2}(a - b)}.$$

Check by the law of sines.

---

* Just as in Case II, (b), we may have *two solutions, one solution*, or *no solution*, depending on the given data.

† Since the side is here determined from its sine, it is necessary to examine both of the values found. If $A > B$ then $a > b$; and if $A < B$ then $a < b$. Hence we have the

Theorem. *Only those values of a should be retained which are greater or less than b according as A is greater or less than B.*

If log sin a = a positive number, there will be no solution.

‡ Same as those used in Case III, (a), p. 224, when the Greek and Roman letters are interchanged.

Ex. 1. Given $A = 110°$, $B = 131° 20'$, $b = 137° 20'$; find a, b, C.

Solution. $\alpha = 180° - A = 70°$, and $\beta = 180° - B = 48° 40'$.

To find a use **(11)**

$$\log \sin A = 9.9730$$
$$\log \sin b = 9.8311$$
$$19.8041$$
$$\log \sin B = 9.8756$$
$$\log \sin a = 9.9285$$
$$\therefore a_1 = 58° 1',$$

or, $\quad a_2 = 180° - a_1 = 121° 59'.$

$\alpha = 70° \qquad \alpha = 70°$
$\beta = 48° 40' \qquad \beta = 48° 40'$
$\alpha + \beta = 118° 40' \qquad \alpha - \beta = 21° 20'$
$\frac{1}{2}(\alpha + \beta) = 59° 20'. \quad \frac{1}{2}(\alpha - \beta) = 10° 40'.$

Since $A < B$ and both a_1 and a_2 are $< b$, it follows that we have *two solutions.*

First solution.

$$a_1 = 58° 1'$$
$$b = 137° 20'$$
$$a_1 + b = 195° 21'$$
$$\tfrac{1}{2}(a_1 + b) = 97° 41'.$$

$$a_1 = 58° 1'$$
$$b = 137° 20'$$
$$a_1 - b = -79° 19'$$
$$\tfrac{1}{2}(a_1 - b) = -39° 40'.$$

To find c_1 use **(44)**

$$\log \sin \tfrac{1}{2}(\alpha + \beta) = 9.9346$$
$$\log \tan \tfrac{1}{2}(a_1 - b) = 9.9187 \ (n)$$
$$19.8533$$
$$\log \sin \tfrac{1}{2}(\alpha - \beta) = 9.2674$$
$$\log \tan \tfrac{1}{2} c_1 = 10.5859$$
$$\tfrac{1}{2} c_1 = 75° 27'.$$
$$\therefore c_1 = 150° 54'.$$

To find C_1 use **(46)**

$$\log \sin \tfrac{1}{2}(a_1 + b) = 9.9961$$
$$\log \tan \tfrac{1}{2}(\alpha - \beta) = 9.2750$$
$$19.2711$$
$$\log \sin \tfrac{1}{2}(a_1 - b) = 9.8050 \ (n)$$
$$\log \tan \tfrac{1}{2} \gamma_1 = 9.4661$$
$$\tfrac{1}{2} \gamma_1 = 16° 18'.$$
$$\gamma_1 = 32° 36'.$$
$$\therefore C_1 = 180° - \gamma_1 = 147° 24'.$$

Check: $\log \sin a_1 = 9.9285 \quad \log \sin b = 9.8311 \quad \log \sin c_1 = 9.6869$
$ \log \sin A = 9.9730 \quad \log \sin B = 9.8756 \quad \log \sin C_1 = 9.7314$
$ 9.9555 9.9555 9.9555$

Second solution. This gives $c_2 = 64° 8'$, and $C_2 = 85° 18'$.

Remembering that $a_2 = 121° 59'$, we may now check the second solution.

Check: $\log \sin a_2 = 9.9285 \quad \log \sin b = 9.8311 \quad \log \sin c_2 = 9.9542$
$ \log \sin A = 9.9730 \quad \log \sin B = 9.8756 \quad \log \sin C_2 = 9.9985$
$ 9.9555 9.9555 9.9557$

Hence the two solutions are

$$a_1 = 58° 1' \qquad c_1 = 150° 54' \quad C_1 = 147° 23',$$

and $\quad a_2 = 121° 59' \quad c_2 = 64° 8' \qquad C_2 = 85° 18'.$

If the angle C or the side c is wanted without first computing a, we may resolve the given triangle into two right triangles and then apply Napier's rules, as was illustrated under Cases II, (a), and II, (b), pp. 220, 223.

EXAMPLES

Solve the following oblique spherical triangles :

No.	Given Parts			Required Parts		
1	$A=108°\,40'$	$B=134°\,20'$	$a=145°\,36'$	$b=154°\,45'$	$c=34°\,9'$	$C=70°\,18'$
2	$B=116°$	$C=80°$	$c=84°$	$b=114°\,50'$	$A=79°\,20'$	$a=82°\,56'$
3	$A=132°$	$B=140°$	$b=127°$	$a_1=67°\,24'$	$C_1=164°\,6'$	$c_1=160°\,6'$
				$a_2=112°\,36'$	$C_2=128°\,21'$	$c_2=103°\,2'$
4	$A=62°$	$C=102°$	$a=64°\,30'$	$c=90°$	$B=63°\,43'$	$b=66°\,26'$
5	$A=133°\,50'$	$B=66°\,30'$	$a=81°\,10'$	Impossible		
6	$B=22°\,20'$	$C=146°\,40'$	$c=138°\,20'$	$b=27°\,22'$	$A=47°\,21'$	$a=117°\,9'$
7	$A=61°\,40'$	$C=140°\,20'$	$c=150°\,20'$	$a_1=43°\,3'$	$B_1=89°\,24'$	$b_1=129°\,8'$
				$a_2=136°\,57'$	$B_2=26°\,59'$	$b_2=20°\,36'$
8	$B=73°$	$C=81°\,20'$	$b=122°\,40'$	Impossible		
9	$B=36°\,20'$	$C=46°\,30'$	$b=42°\,12'$	$A_1=164°\,44'$	$a_1=162°\,38'$	$c_1=124°\,41'$
				$A_2=119°\,19'$	$a_2=81°\,19'$	$c_2=55°\,19'$
10	$A=110°\,10'$	$B=133°\,18'$	$a=147°\,6'$	$b=155°\,5'$	$c=33°\,2'$	$C=70°\,21'$

22. Length of an arc of a circle in linear units. From Geometry we know that *the length of an arc of a circle is to the circumference of the circle as the number of degrees in the arc is to 360.* That is

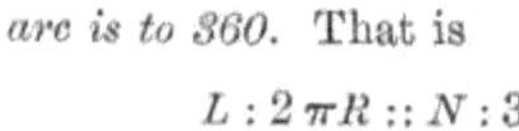

$$L : 2\,\pi R :: N : 360, \text{ or,}$$

$$(52) \qquad L = \frac{\pi R N}{180},$$

where

$L =$ length of arc,

$N =$ number of degrees in arc,

$R =$ length of radius.

In case the length of the arc is given to find the number of degrees in it, we instead solve for N, giving

$$(53) \qquad N = \frac{180\,L}{\pi R}.$$

Considering the earth as a sphere, *an arc of one minute* on a great circle is called *a geographical mile* or *a nautical mile.** Hence there are 60 nautical miles in an arc of 1 degree, and $360 \times 60 = 21,600$ nautical miles in the circumference of a great circle of the earth. If we assume the radius of the earth to be 3960 statute miles, the length

---

* In connection with a ship's rate of sailing a *nautical mile* is also called a *knot.*

of a nautical mile ($= 1$ min. $= \frac{1}{60}$ of a degree) in statute miles will be, from (52),

$$L = \frac{3.1416 \times 3960 \times \frac{1}{60}}{180} = 1.15 \text{ mi.}$$

Ex. 1. Find the length of an arc of $22° 30'$ in a circle of radius 4 in.

Solution. Here $N = 22° 30' = 22.5°$, and $R = 4$ in.

Substituting in (52), $L = \dfrac{3.1416 \times 4 \times 22.5}{180} = 1.57$ in. *Ans.*

Ex. 2. A ship has sailed on a great circle for $5\frac{1}{2}$ hr. at the rate of 12 statute miles an hour. How many degrees are there in the arc passed over?

Solution. Here $L = 5\frac{1}{2} \times 12 = 66$ mi., and $R = 3960$ mi.

Substituting in (53), $N = \dfrac{180 \times 66}{3.1416 \times 3960} = .955° = 57.3'$. *Ans.*

23. Area of a spherical triangle. From Spherical Geometry we know that *the area of a spherical triangle is to the area of the surface of the sphere as the number of degrees in its spherical excess* * *is to 720.* That is,

$$\text{Area of triangle} : 4\,\pi R^2 :: E : 720, \text{ or,}$$

(54) $$\text{Area of spherical triangle} = \frac{\pi R^2 E}{180}.$$

In case the three angles of the triangle are not given, we should first find them by solving the triangle. Or, if the three sides of the triangle are given, we may find E directly by Lhuilier's formula,† namely,

(55) $$\tan \tfrac{1}{4} E = \sqrt{\tan \tfrac{1}{2} s \tan \tfrac{1}{2}(s - a) \tan \tfrac{1}{2}(s - b) \tan \tfrac{1}{2}(s - c)},$$

where a, b, c denote the sides and $s = \tfrac{1}{2}(a + b + c)$.

The area of a spherical polygon will evidently be the sum of the areas of the spherical triangles formed by drawing arcs of great circles as diagonals of the polygon.

Ex. 1. The angles of a spherical triangle on a sphere of 25-in. radius are $A = 74° 40'$, $B = 67° 30'$, $C = 49° 50'$. Find the area of the triangle.

Solution. Here $E = (A + B + C) - 180° = 12°$.

Substituting in (54), Area $= \dfrac{3.1416 \times (25)^2 \times 12}{180} = 130.9$ sq. in. *Ans.*

---

* The spherical excess (usually denoted by E) of a spherical triangle is the excess of the sum of the angles of the triangle over 180°. Thus, if A, B, and C are the angles of a spherical triangle,

$$E = A + B + C - 180°.$$

† Derived in more advanced treatises.

EXAMPLES

1. Find the length of an arc of $5°\,12'$ in a circle whose radius is 2 ft. 6 in.

Ans. 2.72 in.

2. Find the length of an arc of $75°\,30'$ in a circle whose radius is 10 yd.

Ans. 13.17 yd.

3. How many degrees are there in a circular arc 15 in. long, if the radius is 6 in.?

Ans. $143°\,18'$.

4. A ship sailed over an arc of 4 degrees on a great circle of the earth each day. At what rate was the ship sailing?

Ans. 11.515 mi. per hour.

5. Find the perimeter in inches of a spherical triangle of sides $48°$, $126°$, $80°$, on a sphere of radius 25 in.

Ans. 110.78 in.

6. The course of the boats in a yacht race was in the form of a triangle having sides of length 24 mi., 20 mi., 18 mi. If we assume that they sailed on arcs of great circles, how many minutes of arc did they describe?

Ans. 53.85 min.

7. The angles of a spherical triangle are $A = 63°$, $B = 84°\,21'$, $C = 79°$; the radius of the sphere is 10 in. What is the area of the triangle?

Ans. 80.88 sq. in.

8. The sides of a spherical triangle are $a = 6.47$ in., $b = 8.39$ in., $c = 9.43$ in.; the radius of the sphere is 25 in. What is the area of the triangle?

Ans. 26.9 sq. in.

Hint. Find E by using formula **(55)**.

9. In a spherical triangle $A = 75°\,16'$, $B = 39°\,20'$, $c = 26$ ft.; the radius of the sphere is 14 ft. Find the area of the triangle.

Ans. 158.45 sq. ft.

10. Two ships leave Boston at the same time. One sails east 441 mi. and the other 287 mi. E. $38°\,21'$ N. the first day. If we assume that each ship sailed on an arc of a great circle, what is the area of the spherical triangle whose vertices are at Boston and at the positions of the ships at the end of the day?

Ans. 41,040 sq. mi.

11. A steamboat traveling at the rate of 15 knots per hour skirts the entire shore line of an island having the approximate shape of an equilateral triangle in 18 hr. What is the approximate area of the island?   *Ans.* 34,960 sq. mi.

12. Find the areas of the following spherical triangles, having given

(a) $a = 47°\,30'$, $b = 55°\,40'$, $c = 60°\,10'$; $R = 10$ ft.   *Ans.* 42.96 sq. ft.
(b) $a = 43°\,30'$, $b = 72°\,24'$, $c = 87°\,50'$; $R = 10$ in.   59.19 sq. in.
(c) $A = 74°\,40'$, $B = 67°\,30'$, $C = 49°\,50'$; $R = 100$ yd.   2094 sq. yd.
(d) $A = 112°\,30'$, $B = 83°\,40'$, $C = 70°\,10'$; $R = 25$ cm.   941.2 sq. cm.
(e) $a = 64°\,20'$, $b = 42°\,30'$, $C = 50°\,40'$; $R = 12$ ft.   46.73 sq. ft.
(f) $C = 110°$, $A = 94°$, $b = 44°$; $R = 40$ rd.   709.2 sq. rd.
(g) $a = 43°\,20'$, $b = 48°\,30'$, $A = 58°\,40'$; $R = 100$ rd.   24.88 acres.
(h) $A = 108°\,40'$, $B = 134°\,20'$, $a = 145°\,36'$; $R = 3960$ mi.   36,460,000 sq. mi.

CHAPTER III

APPLICATIONS OF SPHERICAL TRIGONOMETRY TO THE CELESTIAL AND TERRESTRIAL SPHERES

24. Geographical terms. In what follows we shall assume the earth to be a sphere of radius 3960 statute miles.

The **meridian** of a place on the earth is that great circle of the earth which passes through the place and the north and south poles.

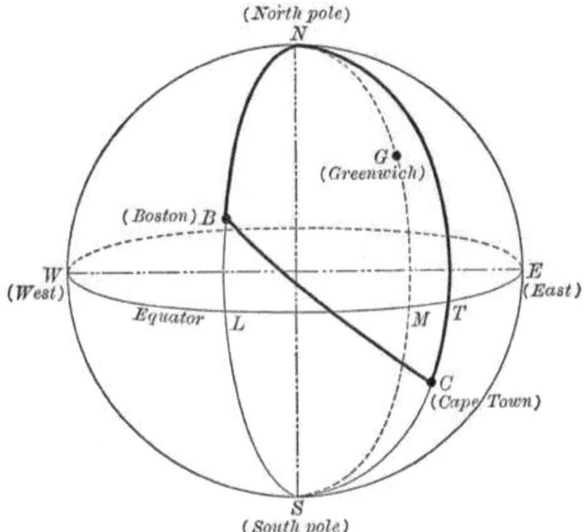

Thus, in the figure representing the earth, NGS is the meridian of Greenwich, NBS is the meridian of Boston, and NCS is the meridian of Cape Town.

The **latitude** of a place is the arc of the meridian of the place extending from the equator to the place. Latitude is measured north or south of the equator from 0° to 90°. Thus, in the figure, the arc LB measures the north latitude of Boston, and the arc TC measures the south latitude of Cape Town.

The **longitude** of a place is the arc of the equator extending from the zero meridian * to the meridian of the place. Longitude is

* As in this case, the zero meridian, or reference meridian, is usually the meridian passing through Greenwich, near London. The meridians of Washington and Paris are also used as reference meridians.

measured east or west from the Greenwich meridian from 0° to 180°.
Thus, in the figure, the arc MT measures the east longitude of Cape
Town, while the arc ML measures the west longitude of Boston.
Since the arcs MT and ML are the measures of the angles MNT and
MNL respectively, it is evident that we can also define the longitude
of a place as the angle between the reference meridian and the
meridian of the place. Thus, in the figure, the angle MNT is the
east longitude of Cape Town, while the angle MNL is the west
longitude of Boston.

The **bearing** of one place from a second place is the angle between
the arc of a great circle drawn from the second place to the first
place, and the meridian of the second place. Thus, in the figure, the
bearing of Cape Town from Boston is measured by the angle CBN
or the angle CBL, while the bearing of Boston from Cape Town is
measured by the angle NCB or the angle SCB.*

25. Distances between points on the surface of the earth. Since we
know from Geometry that *the shortest distance on the surface of a*

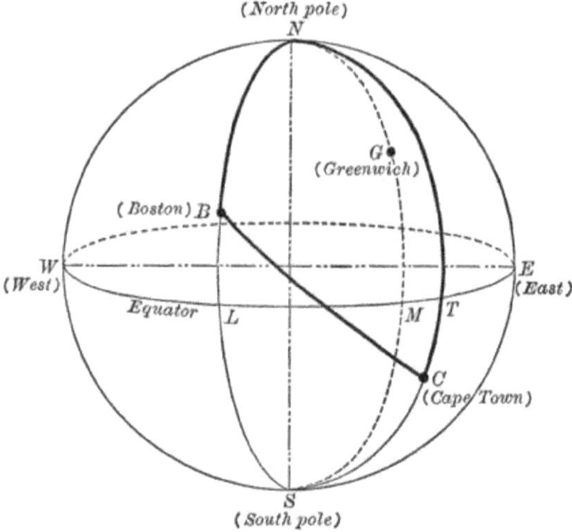

*sphere between any two points on that surface is the arc, not greater
than a semicircumference, of the great circle that joins them*, it is
evident that the shortest distance between two places on the earth
is measured in the same way. Thus, in the figure, the shortest

* The *bearing* or *course* of a ship at any point is the angle the path of the ship makes
with the meridian at that point.

distance between Boston and Cape Town is measured on the arc
BC of a great circle. We observe that this arc BC is one side of
a spherical triangle of which the two other sides are the arcs BN
and CN. Since

$$\text{arc } BN = 90° - \text{arc } LB = 90° - \text{north latitude of Boston,}$$

$$\text{arc } CN = 90° + \text{arc } TC = 90° + \text{south latitude of Cape Town,}$$

and   angle BNC = angle MNL + angle MNT

$$= \text{west longitude of Boston}$$
$$+ \text{ east longitude of Cape Town}$$
$$= \text{difference in longitude of Boston and Cape Town,}$$

it is evident that if we know the latitudes and longitudes of Boston
and Cape Town, we have all the data necessary for determining two
sides and the included angle of the triangle BNC. The third side
BC, which is the shortest distance between Boston and Cape Town,
may then be found as in Case II, (a), p. 219.

In what follows, north latitude will be given the sign + and south
latitude the sign —.

**Rule for finding the shortest distance between two points on the earth
and the bearing of each from the other, the latitude and longitude of each
point being given.**

First step. *Subtract the latitude of each place from 90°.* * *The
results will be the two sides of a spherical triangle.*

Second step. *Find the difference of longitude of the two places by
subtracting the lesser longitude from the greater if both are E. or both
are W., but add the two if one is E. and the other is W. This gives
the included angle of the triangle.*†

Third step. *Solving the triangle by Case II, (a), p. 219, the third
side gives the shortest distance between the two points in degrees of
arc,‡ and the angles give the bearings.*

---

* Note that this is *algebraic* subtraction. Thus, if the two latitudes were 25° N. and
42° S., we would get as the two sides of the triangle,

$$90° - 25° = 65° \quad \text{and} \quad 90° - (-42°) = 90° + 42° = 132°.$$

† If the difference of longitude found is greater than 180°, we should subtract it from
360° and use the remainder as the included angle.

‡ The number of minutes in this arc will be the distance between the two places in geographical (nautical) miles. The distance between the two places in statute miles is given
by the formula

$$L = \frac{3.1416 \times 3960 \times N}{180},$$

where N = the number of degrees in the arc.

Ex. 1. Find the shortest distance along the earth's surface between Boston (lat. 42° 21′ N., long. 71° 4′ W.) and Cape Town (lat. 33° 56′ S., long. 18° 26′ E.), and the bearing of each city from the other.

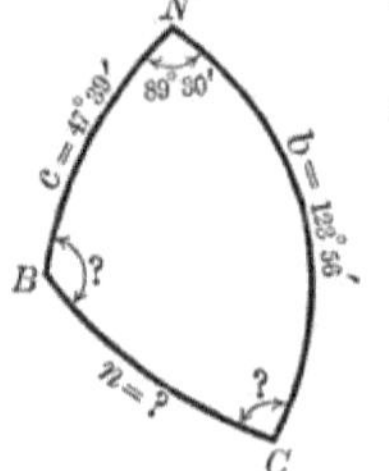

Solution. Draw a spherical triangle in agreement with the figure on p. 232.

First step.
$$c = 90° - 42° 21' = 47° 39',$$
$$b = 90° - (- 33° 56') = 123° 56'.$$

Second step.
$$N = 71° 4' + 18° 26' = 89° 30' = \text{difference in long.}$$

Third step. Solving the triangle by Case II, (a), p. 219, we get
$$n = 68° 14' = 68.23° = 4094 \text{ nautical miles,}$$
$$C = 52° 43' = \text{bearing of Boston from Cape Town,}$$
and
$$B = 116° 43' = \text{bearing of Cape Town from Boston.}$$

Hence a ship sailing from Boston to Cape Town on the arc of a great circle *sets out from Boston* on a course S. 63° 17′ E. and *approaches Cape Town* on a course S. 52° 43′ E.*

EXAMPLES

1. Find the shortest distance between Baltimore (lat. 39° 17′ N., long. 76° 37′ W.) and Cape Town (lat. 33° 56′ S., long. 18° 26′ E.), and the bearing of each from the other.	*Ans.* Distance = 65° 48′ = 3947 nautical miles,
S. 64° 58′ E. = bearing of Cape Town from Boston,
N. 57° 42′ W. = bearing of Boston from Cape Town.

2. What is the distance from New York (lat. 40° 43′ N., long. 74° W.) to Liverpool (lat. 53° 24′ N., 3° 4′ W.)? Find the bearing of each place from the other. In what latitude will a steamer sailing on a great circle from New York to Liverpool cross the meridian of 50° W., and what will be her course at that point?	*Ans.* Distance = 47° 50′ = 2870 nautical miles,
N. 75° 7′ W. = bearing of New York from Liverpool,
N. 49° 29′ E. = bearing of Liverpool from New York.
Lat. 51° 13′ N., with course N. 66° 54′ E.

3. Find the shortest distance between the following places in geographical miles :

(a) New York (lat. 40° 43′ N., long. 74° W.) and San Francisco (lat. 37° 48′ N., long. 122° 28′ W.).	*Ans.* 2230.

(b) Sandy Hook (lat. 40° 28′ N., long. 74° 1′ W.) and Madeira (lat. 32° 28′ N., long. 16° 55′ W.).	*Ans.* 2749.

(c) San Francisco (lat. 37° 48′ N., long. 122° 28′ W.) and Batavia (lat. 6° 9′ S., long. 106° 53′ E.).	*Ans.* 7516.

(d) San Francisco (lat. 37° 48′ N., long. 122° 28′ W.) and Valparaiso (lat. 33° 2′ S., long. 71° 41′ W.)	*Ans.* 5109.

---

* A ship that sails on a great circle (except on the equator or a meridian) must be continually changing her course. If the ship in the above example keeps constantly on the course S. 63° 17′ E., she will never reach Cape Town.

4. Find the shortest distance in statute miles (taking diameter of earth as 7912 mi.) between Boston (lat. 42° 21′ N., long. 71° 4′ W.) and Greenwich (lat. 51° 29′ N.), and the bearing of each place from the other.

> *Ans.* Distance = 3276 mi.,
>
> N. 53° 7′ E. = bearing of Greenwich from Boston,
> N. 71° 39′ W. = bearing of Boston from Greenwich.

5. As in last example, find the shortest distance between and bearings for Calcutta (lat. 22° 33′ N., long. 88° 19′ E.) and Valparaiso (lat. 33° 2′ S., long. 71° 42′ W.).   *Ans.* Distance = 10,860 mi.,

> S. 64° 20.5′ E. = bearing of Calcutta from Valparaiso,
> S. 54° 54.5′ W. = bearing of Valparaiso from Calcutta.

6. Find the shortest distance in statute miles from Oberlin (long. 82° 14′ W.) to New Haven (long. 72° 55′ W.), the latitude of each place being 41° 17′ N.

> *Ans.* 483.2 mi.

7. From a point whose latitude is 17° N. and longitude 130° W. a ship sailed an arc of a great circle over a distance of 4150 statute miles, starting S. 54° 20′ W. Find its latitude and longitude, if the length of 1° is 69⅛ statute miles.

> *Ans.* Lat. 19° 42′ S., long. 178° 21′ W.

26. Astronomical problems. One of the most important applications of Spherical Trigonometry is to Astronomy. In fact, Trigonometry was first developed by astronomers, and for centuries was studied only in connection with Astronomy. We shall take up the study of a few simple problems in Astronomy.

27. The celestial sphere. When there are no clouds to obstruct the view, the sky appears like a great hemispherical vault, with the observer at the center. The stars seem to glide upon the inner surface of this sphere from east to west,* their paths being parallel circles whose planes are perpendicular to the polar axis of the earth, and having their centers in that axis produced. Each star † makes a complete revolution, called its diurnal (daily) motion, in 23 hr. 56 min., ordinary clock time. We cannot estimate the distance of the surface of this sphere from us, further than to perceive that it must be very far away indeed, because it lies beyond even the remotest terrestrial objects. To an observer the stars all seem to be at the same enormous distance from him, since his eyes can judge their *directions only* and not their *distances*. It is therefore natural, and it is extremely convenient from a mathematical point of view, to regard this imaginary sphere on which all the heavenly bodies seem to be projected, as having a radius of unlimited length. This

---

* This apparent turning of the sky from east to west is in reality due to the rotation of the earth in the opposite direction, just as to a person on a swiftly moving train the objects outside seem to be speeding by, while the train appears to be at rest. The sky is really motionless, while the earth is rotating from west to east.

† By stars we shall mean fixed stars and nebulæ whose relative positions vary so slightly that it takes centuries to make the change perceptible.

sphere, called the **celestial sphere,** is conceived of as having such enormous proportions that the whole solar system (sun, earth, and planets) *lies at its center*, like a few particles of dust at the center of a great spherical balloon. The stars seem to retain the same relative positions with respect to each other, being in this respect like places on the earth's surface. As viewed from the earth, the sun, moon, planets, and comets are also projected on the celestial sphere, but they are changing their apparent positions with respect to the stars and with respect to each other. Thus, the sun appears to move eastward with respect to the stars about one degree each day, while the moon moves about thirteen times as far.

The following figure represents the celestial sphere, with the earth at the center showing as a mere dot.

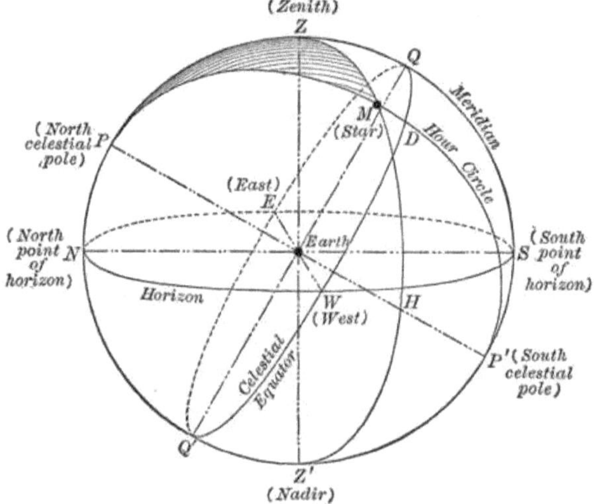

The **zenith** of an observer is the point on the celestial sphere directly overhead. A plumb line held by the observer and extended upwards will pierce the celestial sphere at his zenith (Z in figure).

The **nadir** is the point on the celestial sphere which is diametrically opposite to the zenith (Z' in the figure).

The **horizon** of an observer is the great circle on the celestial sphere having the observer's zenith for a pole; hence every point on the horizon ($SWNE$ in the figure) will be 90° from the zenith and from the nadir. A plane tangent* to a surface of still water

---

* On account of the great distance, a plane passed tangent to the earth at the place of the observer will cut the celestial sphere in a great circle which (as far as we are concerned) coincides with the observer's horizon.

at the place of the observer will cut the celestial sphere in his horizon.

All points on the earth's surface have different zeniths and horizons.

Every great circle passing through the zenith will be perpendicular to the horizon; such circles are called **vertical circles** (as $ZMHZ'$ and $ZQSP'Z'$ in figure).

The **celestial equator** or **equinoctial** is the great circle in which the plane of the earth's equator cuts the celestial sphere ($EQWQ'$ in the figure).

The **poles** of the celestial equator are the points (P and P' in the figure) where the earth's axis, if produced, would pierce the celestial sphere. The poles may also be defined as those two points on the sky where a star would have no diurnal (daily) motion. The Pole Star is near the north celestial pole, being about $1\tfrac{1}{2}°$ from it. Every point on the celestial equator is 90° from each of the celestial poles.

All points on the earth's surface have the same celestial equator and poles.

The geographical meridian of a place on the earth was defined as that great circle of the earth which passes through the place and the north and south poles. The **celestial meridian** of a point on the earth's surface is the great circle in which the plane of the point's geographical meridian cuts the celestial sphere ($ZQSP'Z'Q'NP$ in the figure). It is evidently that vertical circle of an observer which passes through the north and south points of his horizon. All points on the surface of the earth which do not lie on the same north-and-south line have different celestial meridians.

The **hour circle** of a heavenly body is that great circle of the celestial sphere which passes through the body * and through the north and south celestial poles. In the figure $PMDP'$ is the hour circle of the star M. The hour circles of all the heavenly bodies are continually changing with respect to any observer.

The spherical triangle PZM, having the north pole, the zenith, and a heavenly body at its three vertices, is a very important triangle in Astronomy. It is called the **astronomical triangle.**

28. Spherical coördinates.   When learning how to draw (or plot) the graph of a function, the student has been taught how to locate a point in a plane by measuring its distances from two fixed and mutually perpendicular lines called the axes of coordinates, the two distances being called the rectangular coordinates of the point.

---

* By this is meant that the hour circle passes through that point on the celestial sphere where we see the heavenly body projected.

If we now consider the surface to be spherical instead of plane, a similar system of locating points on it may be employed, two fixed and mutually perpendicular great circles being chosen as reference circles, and the *angular* distances of a point from these reference circles being used as the *spherical coordinates* of the point. Since the reference circles are perpendicular to each other, each one of them passes through the poles of the other.

In his study of Geography the student has already employed such a system for locating points on the earth's surface, for the latitude and longitude of a point on the earth are really the *spherical coordinates* of the point, the two reference circles being the equator and the zero meridian (usually the meridian of Greenwich). Thus, in the figure on p. 231, we may consider the spherical coordinates of Boston to be the arcs ML (west longitude) and LB (north latitude); and of Cape Town the spherical coordinates would be the arcs MT (east longitude) and TC (south latitude). Similarly, we have systems of spherical coordinates for determining the position of a point on the celestial sphere, and we shall now take up the study of the more important of these.

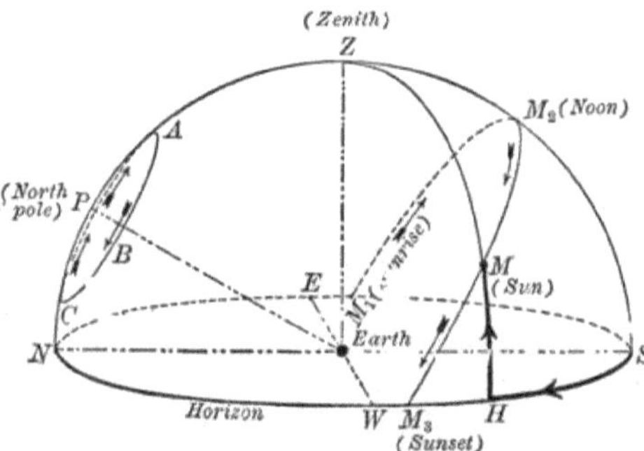

29. The horizon and meridian system. In this case the two fixed and mutually perpendicular great circles of reference are the *horizon* of the observer ($SHWNE$) and his meridian (SM_2ZPN), and the spherical coördinates of a heavenly body are its *altitude* and *azimuth*.

The **altitude** of a heavenly body is its angular distance above the horizon measured on a vertical circle from 0° to 90°.* Thus the altitude of the sun M is the arc HM. The distance of a heavenly body from the zenith is called its *zenith distance* (ZM in the figure), and it is evidently the complement of its altitude. The altitude of the zenith is 90°. The altitude of the sun at sunrise or sunset is zero.

The **azimuth** of a heavenly body is the angle between its vertical circle and the meridian of the observer. This angle is usually

---

* At sea the altitude is usually measured by the sextant, while on land a surveyor's transit is used.

measured along the horizon from the south point westward to the foot of the body's vertical circle.* Thus the azimuth of the sun M is the angle SZH, which is measured by the arc SH. The azimuth of the sun at noon is zero and at midnight 180°. The azimuth of a star directly west of an observer is 90°, of one north 180°, and of one east 270°.

Knowing the azimuth and altitude (spherical coördinates) of a heavenly body, we can locate it on the celestial sphere as follows. From the south point of the horizon, as S (which may be considered the origin of coordinates, since it is an intersection of the reference circles), lay off the azimuth, as SH. Then on the vertical circle passing through H lay off the altitude, as HM. The body is then located at M.

Ex. 1. In each of the following examples draw a figure of the celestial sphere and locate the body from the given spherical coordinates.

	Azimuth	*Altitude*			*Azimuth*	*Altitude*
(a)	45°	45°		(j)	0°	0°
(b)	60°	30°		(k)	180°	0°
(c)	90°	60°		(l)	0°	90°
(d)	120°	75°		(m)	90°	0°
(e)	180°	55°		(n)	270°	0°
(f)	225°	0°		(o)	360°	0°
(g)	300°	60°		(p)	330°	45°
(h)	315°	15°		(q)	75°	75°
(i)	178°	82°		(r)	90°	90°

Since any two places on the earth have, in general, different meridians and different horizons, it is evident that this system of spherical coordinates is purely local. The sun rises at M_1 on the eastern horizon (altitude zero), mounts higher and higher in the sky, on a circle ($M_1M_2M_3$) parallel to the celestial equator, until it reaches the observer's meridian M_2 (at noon, when its altitude is a maximum), then sinks downward to M_3 and sets on the western horizon.

Similarly, for any other heavenly body, so that all are continually changing their altitudes and azimuths. To an observer having the zenith shown in the figure, a star in the northern sky near the north pole will not set at all, and to the same observer a star near the south pole will not rise at all. If its path for one day were traced on the celestial sphere, it would be a circle (as ABC) with its center in the polar axis and lying in a plane parallel to the plane of the equator.

---

* That is, azimuth is measured from 0° to 360° clockwise.

30. The equator and meridian system. In this case the two fixed and mutually perpendicular great circles of reference are the *celestial equator* ($EQDWQ'$) and the *meridian* of the observer ($NPZQSP'Z'Q'$); and the spherical coördinates of a heavenly body are its *declination* and *hour angle*.

The **declination** of a heavenly body is its angular distance north or south of the celestial equator measured on the hour circle of the

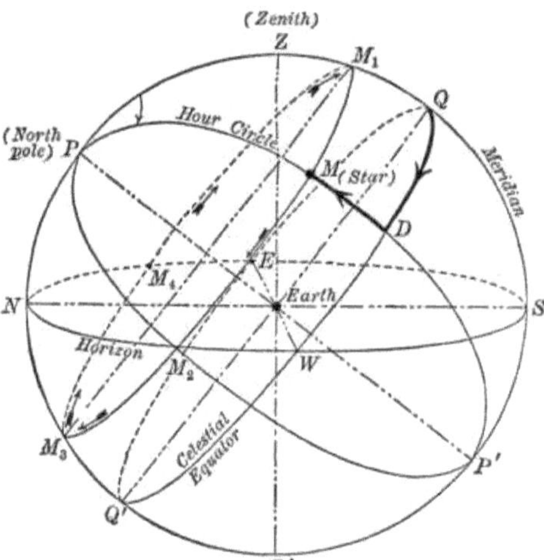

body from 0° to 90°.* Thus, in the figure, the arc DM is a measure of the north declination of the star M. *North declination* is always considered *positive* and *south declination nega-tive*. Hence the decli-nation of the north pole is + 90°, while that of the south pole is − 90°.

The declinations of the sun, moon, and planets are continually changing, but the dec-lination of a fixed star changes by an exceedingly small amount in the course of a year. The angular distance of a heavenly body from the north celestial pole, measured on the hour circle of the body, is called its *north polar distance* (PM in figure). The north polar distance of a star is evidently the complement of its declination.

The **hour angle** of a heavenly body is the angle between the merid-ian of the observer and the hour circle of the star measured *west-ward* from the meridian from 0° to 360°. Thus, in the figure, the hour angle of the star M is the angle QPD (measured by the arc QD). This angle is commonly used as a measure of time, hence the name *hour angle*. Thus the star M makes a complete circuit in 24 hours; that is, the hour angle QPD continually increases at the uniform rate of 360° in 24 hours, or 15° an hour. For this reason the hour angle of a heavenly body is usually reckoned in hours from

---

* The declinations of the sun, moon, planets, and some of the fixed stars, for any time of the year, are given in the *Nautical Almanac* or *American Ephemeris*, published by the United States government.

0 to 24, one hour being equal to 15°.* When the star is at M_1 (on the observer's meridian) its hour angle is zero. Then the hour angle increases until it becomes the angle M_1PM (when the star is at M). When the star sets on the western horizon its hour angle becomes M_1PM_2. Twelve hours after the star is at M_1 it will be at M_3, when its hour angle will be 180° ($= 12$ hours). Continuing on its circuit, the star rises at M_4 and finally reaches M_1, when its hour angle has become 360° ($= 24$ hours), or 0° again.

Knowing the hour angle and declination (spherical coordinates) of a heavenly body, we can locate it on the celestial sphere as follows. From the point, as Q, where the reference circles intersect, lay off the hour angle (or arc), as QD. Then on the hour circle passing through D lay off the declination, as DM. The body is then located at M.

Ex. 1. In each of the following examples draw a figure of the celestial sphere and locate the body from the given spherical coordinates.

	Hour angle	Declination		Hour angle	Declination
(a)	45°	N. 30°	(j)	60°	S. 45°
(b)	60°	N. 60°	(k)	0°	0°
(c)	90°	S. 45°	(l)	180°	0°
(d)	120°	S. 30°	(m)	90°	N. 90°
(e)	180°	N. 50°	(n)	270°	0°
(f)	5 hr.	N. 75°	(o)	12 hr.	S. 10°
(g)	15 hr.	− 25°	(p)	3 hr.	+ 80°
(h)	6 hr.	+ 79°	(q)	9 hr.	− 45°
(i)	0 hr.	− 90°	(r)	20 hr.	+ 60°

31. Practical applications. Among the practical applications of Astronomy the most important are:

(*a*) *To determine the position of an observer on the surface of the earth (i.e. his latitude and longitude).*

(*b*) *To determine the meridian of a place on the surface of the earth.*

(*c*) *To ascertain the exact time of day at the place of the observer.*

(*d*) *To determine the position of a heavenly body.*

The first of these, when applied to the determination of the place of a ship at sea, is the problem to which Astronomy mainly owes its economic importance. National astronomical observatories have been

---

* On account of the yearly revolution of the earth about the sun, it takes the sun about 4 minutes longer to make the circuit than is required by any particular fixed star. Hence the solar day is about 4 minutes longer than the sidereal (star) day, but each is divided into 24 hours; the first giving hours of ordinary clock time, while the second gives sidereal hours, which are used extensively in astronomical work. When speaking of the sun's hour angle it shall be understood that it is measured in hours of ordinary clock time, while the hour angle of a fixed star is measured in sidereal hours. In either case 1 hour = 15°.

established, and yearly nautical almanacs are being published by the principal nations controlling the commerce of the world, in order to supply the mariner with the data necessary to determine his position accurately and promptly.

32. Relation between the observer's latitude and the altitude of the celestial pole. To an observer on the earth's equator (latitude zero) the pole star is on the horizon; that is, the altitude of the star is zero. If the observer is traveling northward, the pole star will gradually rise; that is, the latitude of the observer and the altitude of the star are both increasing. Finally, when the observer reaches the north pole of the earth his latitude and the altitude of the star have both increased to 90°. The place of the pole in the sky then

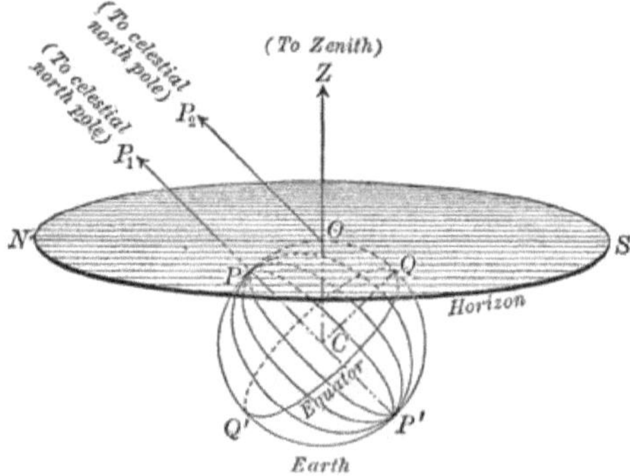

depends in some way on the observer's latitude, and we shall now prove that **the altitude of a celestial pole is equal to the latitude of the observer.**

Let O be the place of observation, say some place in the northern hemisphere; then the angle QCO (or arc QO) measures its north latitude. Produce the earth's axis CP until it pierces the celestial sphere at the celestial north pole. A line drawn from O in the direction (as OP_2) of the celestial north pole will be parallel to CP_1, since the celestial north pole is at an unlimited distance from the earth (see § 27, p. 235). The angle NOP_2 measures the altitude of the north pole. But CO is perpendicular to ON and CQ is perpendicular to OP_2 (since it is perpendicular to the parallel line CP_1); hence the angles NOP_2 and QCO are equal, and we find that *the altitude of the pole as observed at O is equal to the latitude of O.*

33. To determine the latitude of a place on the surface of the earth. If we project that part of the celestial sphere which lies above the horizon on the plane of the observer's celestial meridian, the horizon will be projected into a line (as *NS*), and the upper half of the celestial equator will also be projected into a line (as *OQ*). From the last section we know that the latitude of the observer equals the altitude of the elevated celestial pole (arc *NP* in figure), or, what amounts to the same thing, equals the angular distance between the zenith and the celestial equator (arc *ZQ* in figure). If then the elevated pole could be seen as a definitely marked point in the sky, the observer's latitude would be found by simply measuring the angular distance of that pole above the horizon. But there are no fixed stars visible at the exact points where the polar axis pierces the celestial

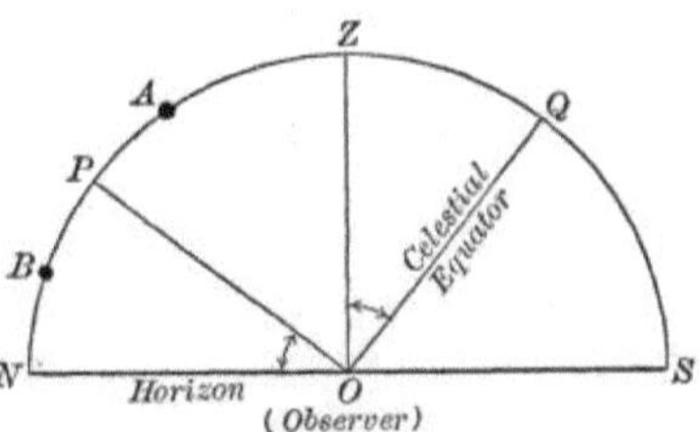

sphere, the so-called polar star being about $1\tfrac{1}{4}°$ from the celestial north pole. Following are some methods for determining the latitude of a place on the surface of the earth.

First method. *To determine latitude by observations on circumpolar stars.* The most obvious method is to observe with a suitable instrument the altitude of some star near the pole (so near the pole that it never sets; as, for instance, the star whose path in the sky is shown as the circle *ABC* in figure, p. 238) at the moment when it crosses the meridian above the pole, and again 12 hours later, when it is once more on the meridian but below the pole. In the first case its elevation will be the greatest possible; in the second, the least possible. *The mean of the two observed altitudes is evidently the latitude of the observer.* Thus, in the figure on this page, if *NA* is the maximum altitude and *NB* the minimum altitude of the star, then

$$\frac{NA + NB}{2} = NP = \text{altitude of pole}$$
$$= \text{latitude of place of observation.}$$

Ex. 1. The maximum altitude of a star near the pole star was observed to be 54° 16′, and 12 hours later its minimum altitude was observed to be 40° 24′. What is the latitude of the place of observation ?

Solution. 54° 16′ + 40° 24′ = 94° 40′.

Therefore $\dfrac{94° 40′}{2} = 47° 20′ = \text{altitude of north pole}$
$= \text{north latitude of place of observation.}$

Second method. *To determine latitude from the meridian altitude of a celestial body whose declination is known.* The altitude of a star

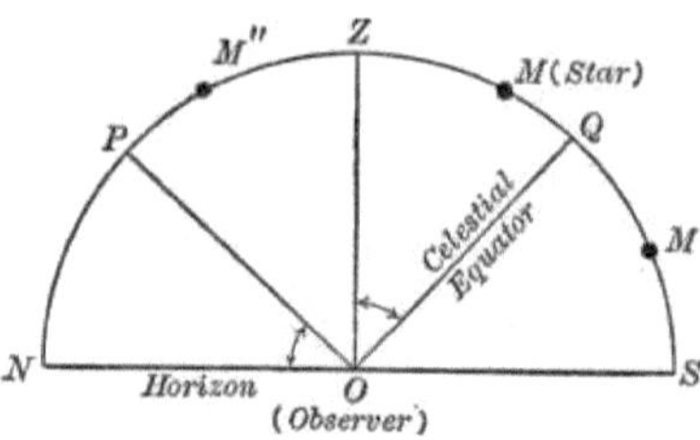

M is measured when it is on the observer's meridian. If we subtract this meridian altitude (arc SM in figure) from 90°, we get the star's zenith distance (ZM). In the Nautical Almanac we now look up the star's declination at the same instant; this gives us the arc QM. Adding the declination of the star to its zenith distance, we get

$$QM + MZ = QZ = NP = \text{altitude of pole} = \text{latitude of place.}$$

Therefore, when the observer is on the northern hemisphere and the star is on the meridian south of zenith,

*North latitude = zenith distance + declination.**

If the star is on the meridian between the zenith and the pole (as at M'' †), we will have

$$North\ latitude = NP = ZQ = QM'' - ZM''$$
$$= declination - zenith\ distance.$$

If the observer is on the southern hemisphere and the star M is on his meridian between the zenith and south pole, we would have

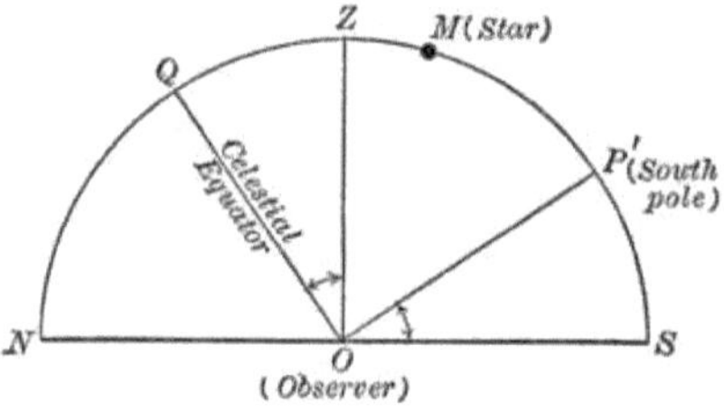

South latitude
$$= SP' = SM - MP'$$
$$= SM - (90° - QM)$$
$$= altitude - co\text{-}declination,$$

if we consider only the numerical value of the declination.

In working out examples the student should depend on the figure rather than try to memorize formulas to cover all possible cases.

Ex. 2. An observer in the northern hemisphere measured the altitude of a star at the instant it crossed his celestial meridian south of zenith, and found it to be 63° 40′. The declination of the star for the same instant was given by the Nautical Almanac as 21° 15′ N. What was the latitude of the observer ?

---

* If the star is south of the celestial equator (as at M'), the same rule will hold, for then the declination is negative (south), and the algebraic sum of the zenith distance and declination will still give the arc QZ.

† Maximum altitude, if a circumpolar star.

Solution. Draw the semicircle *NZSO.* Lay off the arc *SM* = altitude = 63° 40′,
which locates the star at *M.* Since the dec-
lination of the star is north, the celestial
equator may be located by laying off the
arc *MQ* = declination = 21° 15′ towards the
south. The line *QO* will then be the projection of the celestial equator, and *OP*,
drawn perpendicular to *QO*, will locate the
north pole *P.*

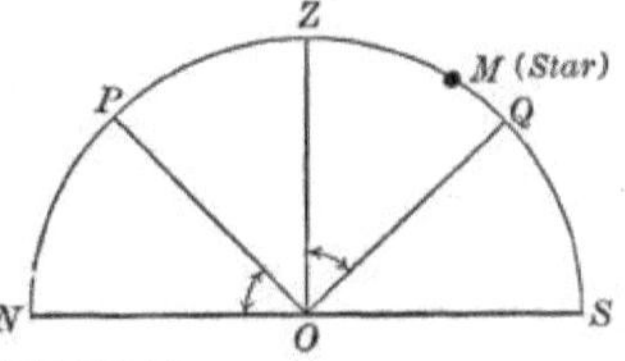

$$\text{Zenith distance} = ZM = 90° - SM \text{ (alt.)}$$
$$= 90° - 63° 40′ = 26° 20′.$$
$$\therefore \text{North latitude of observer} = NP = ZQ = ZM \text{ (zen. dist.)} + MQ \text{ (dec.)}$$
$$= 26° 20′ + 21° 15′ = 47° 35′.$$

Third method. *To determine latitude when the altitude, declination,
and hour angle of a celestial body are known.* Referring to the astronomical (spherical) triangle *PZM*, we see that

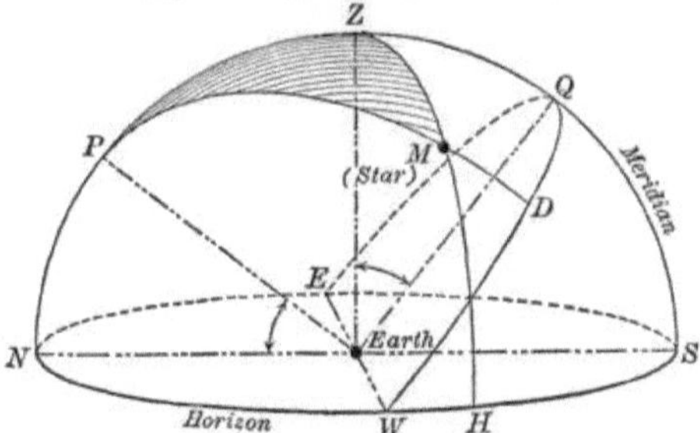

side *MZ*
$$= 90° - HM \text{ (alt.)}$$
$$= \text{co-altitude,}$$

the altitude of the star
being found by measure-
ment. Also

side *PM*
$$= 90° - DM \text{ (dec.)}$$
$$= \text{co-declination,}$$

the declination of the star being found from the Nautical Almanac.

Angle *ZPM* = hour angle, which is given. This hour angle will
be the local time when the observation is made on the sun. We then
have two sides and the angle opposite one of them given in the
spherical triangle *PZM.* Solving this for the side *PZ*, by Case
III, (*a*), p. 224, we get

$$\textit{Latitude of observer} = NP = 90° - PZ.$$

EX. 3. The declination of a star is 69° 42′ N. and its hour angle 60° 44′. What is
the north latitude of the place if the altitude of the star is observed to be 49° 40′?

Solution. Referring to the above figure, we have, in this example,

$$\text{side } MZ = \text{co-alt.} = 90° - 49° 40′ = 40° 20′,$$
$$\text{side } PM = \text{co-dec.} = 90° - 69° 42′ = 20° 18′,$$
$$\text{angle } ZPM = \text{hour angle} = 60° 44′.$$

Solving for the side *PZ* by Case III, (*a*), p. 224, we get side *PZ* = 47° 9′ = co-lat.

$$\therefore 90° - 47° 9′ = 42° 51′ = \text{north latitude of place.}$$

The angle *MZP* is found to be 27° 53′; hence the azimuth of the star
(angle *SZH*) is 180° − 27° 53′ = 152° 7′.

EXAMPLES

1. The following observations for altitude have been made on some north circumpolar star. What is the latitude of each place ?

		Maximum altitude	Minimum altitude	North latitude
(a)	New York	50° 46′	30° 40′	*Ans.* 40° 43′
(b)	Boston	44° 22′	40° 20′	42° 21′
(c)	New Haven	58° 24′	24° 10′	41° 17′
(d)	Greenwich	64° 36′	38° 22′	51° 29′
(e)	San Francisco	55° 6′	20° 30′	37° 48′
(f)	Calcutta	24° 18′	20° 48′	22° 33′

2. In the following examples the altitude of some heavenly body has been measured at the instant when it crossed the observer's celestial meridian. What is the latitude of the observer in each case, the declination being found from the Nautical Almanac ?

	Hemisphere	Meridian altitude	Declination	Body is	Latitude
(a)	Northern	60°	N. 20°	S. of zenith	*Ans.* 50° N.
(b)	Northern	75° 40′	N. 32° 13′	S. of zenith	46° 33′ N.
(c)	Northern	43° 27′	S. 10° 52′	S. of zenith	35° 41′ N.
(d)	Northern	38° 6′	S. 44° 26′	S. of zenith	7° 28′ N.
(e)	Northern	50°	N. 62°	N. of zenith	22° N.
(f)	Northern	28° 46′	N. 73° 16′	N. of zenith	12° 2′ N.
(g)	Southern	67°	S. 59°	S. of zenith	36° S.
(h)	Southern	45° 26′	S. 81° 48′	S. of zenith	37° 14′ S.
(i)	Southern	72°	S. 8°	N. of zenith	26° S.
(j)	Southern	22° 18′	N. 46° 25′	N. of zenith	21° 17′ S.

3. In the following examples the altitude of some heavenly body not on the observer's celestial meridian has been measured. The hour angle and declination is known for the same instant. Find the latitude of the observer in each case.

	Hemisphere	Altitude	Declination	Hour angle	Latitude
(a)	Northern	40°	N. 10°	50°	*Ans.* 27° 2′ N.
(b)	Northern	15°	S. 8°	65°	35° 38′ N.
(c)	Northern	52°	N. 19°	2 hr.	48° 16′ N.
(d)	Northern	64° 42′	N. 24° 20′	345°	3° 34′ N. or 46° 36′ N.
(e)	Northern	0°	S. 5°	5 hr.	77° 37′ N.
(f)	Northern	25°	0°	21 hr.	53° 18′ N.
(g)	Northern	0°	N. 11° 14′	68° 54′	No solution
(h)	Northern	9° 26′	0°	72° 22′	57° 14′ N.
(i)	Southern	38°	S. 12°	52°	33° 56′ S. or 4° 8′ S.
(j)	Southern	19°	N. 7°	3 hr.	52° 56′ S.
(k)	Southern	46° 18′	S. 15° 23′	326°	49° 14′ S.
(l)	Southern	0°	N. 14°	38°	72° 26′ S.
(m)	Southern	57° 36′	0°	2 hr.	12° 50′ S.

34. To determine the time of day. A very simple relation exists between the hour angle of the sun and the time of day at any place. The sun appears to move from east to west at the uniform rate of 15° per hour, and when the sun is on the meridian of a place it is apparent noon at that place. Comparing,

Hour angle of sun	*Time of day*
0°	Noon
15°	1 P.M.
30°	2 P.M.
45°	3 P.M.
90°	6 P.M.
180°	Midnight
195°	1 A.M.
210°	2 A.M.
270°	6 A.M.
300°	8 A.M.
360°	Noon

The hour angle of the sun M is the angle at P in the astronomical (spherical) triangle PZM. We may find this hour angle (time of

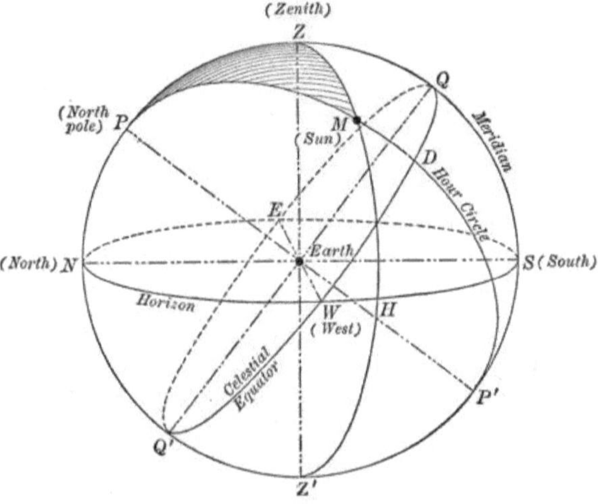

day) by solving the astronomical triangle for the angle at P, provided we know three other elements of the triangle.

DM = declination of sun, and is found from the Nautical Almanac.

$$\therefore \textit{ Side } PM = 90° - DM = \textit{co-declination of sun.}$$

HM = altitude of sun, and is found by measuring the angular distance of the sun above the horizon with a sextant or transit.

$$\therefore \textit{ side } MZ = 90° - HM = \textit{co-altitude of sun.}$$

$$NP = \text{altitude of the celestial pole}$$
$$= \text{latitude of the observer (p. 243).}$$

$$\therefore \textit{ Side } PZ = 90° - NP = \textit{co-latitude of observer.}$$

Hence we have

Rule for determining the time of day at a place whose latitude is known, when the declination and altitude of the sun at that time and place are known.

First step. *Take for the three sides of a spherical triangle*

the co-altitude of the sun,
the co-declination of the sun,
the co-latitude of the place.

Second step. *Solve this spherical triangle for the angle opposite the first-mentioned side. This will give the* **hour angle in degrees** *of the sun, if the observation is made in the afternoon. If the observation is made in the forenoon, the hour angle will be* **360° — the angle found.**

Third step. *When the observation is made in the afternoon the time of day will be*

$$\frac{\textit{hour angle}}{15} \text{ P.M.}$$

When the observation is made in the forenoon the time of day will be

$$\left(\frac{\textit{hour angle}}{15} - 12 \right) \text{A.M.}$$

Ex. 1. In New York (lat. 40° 43′ N.) the sun's altitude is observed to be 30° 40′. Having given that the sun's declination is 10° N. and that the observation is made in the afternoon, what is the time of day?

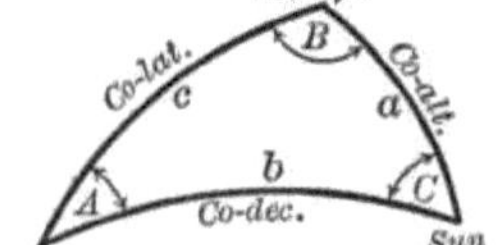

Solution. First step. Draw the triangle.

Side a = co-alt. = 90° − 30° 40′ = 59° 20′.
Side b = co-dec. = 90° − 10° = 80°.
Side c = co-lat. = 90° − 40° 43′ = 49° 17′.

Second step. As we have three sides given, the solution of this triangle comes under Case I, (*a*), p. 217. But as we only want the angle A (hour angle), some

labor may be saved by using one of the formulas (18), (19), (20), pp. 211, 212.
Let us use (18),

$$a = 59° 20'$$
$$b = 80°$$
$$c = 49° 17'$$
$$2s = 188° 37'$$
$$s = 94° 19'.$$
$$s - a = 34° 59'.$$

$$\sin \tfrac{1}{2}\alpha = \sqrt{\frac{\sin s \sin(s-a)}{\sin b \sin c}},$$

$$\log \sin \tfrac{1}{2}\alpha = \tfrac{1}{2}[\log \sin s + \log \sin(s-a) - \{\log \sin b + \log \sin c\}].$$

$$\log \sin s = 9.9988 \qquad\qquad \log \sin b = 9.9934$$
$$\log \sin(s-a) = 9.7584 \qquad\qquad \log \sin c = 9.8797$$
$$\log \text{ numerator} = 19.7572 \qquad\qquad \log \text{ denominator} = 19.8731$$
$$\log \text{ denominator} = 19.8731$$
$$\overline{9.8841}$$
$$2|19.8841$$
$$\log \sin \tfrac{1}{2}\alpha = 9.9421$$
$$\tfrac{1}{2}\alpha = 61° 4'.$$
$$\alpha = 122° 8'.$$

$$\therefore A = 180° - \alpha = 57° 52' = \text{ hour angle of sun.}$$

Third step. Time of day $= \dfrac{\text{hour angle}}{15}$ P.M. $= 3$ hr. 51 min. P.M. *Ans.*

EXAMPLES

1. In Milan (lat. 45° 30′ N.) the sun's altitude at an afternoon observation is 26° 30′. The sun's declination being 8° S., what is the time of day ?
Ans. 2 hr. 33 min. P.M.

2. In New York (lat. 40° 43′ N.) a forenoon observation on the sun gives 30° 40′ as the altitude. What is the time of day, the sun's declination being 10° S.? *Ans.* 9 hr. 46 min. A.M.

3. A mariner observes the altitude of the sun to be 60°, its declination at the time of observation being 6° N. If the latitude of the vessel is 12° S., and the observation is made in the morning, find the time of day. *Ans.* 10 hr. 24 min. A.M.

4. A navigator observes the altitude of the sun to be 35° 23′, its declination being 10° 48′ S. If the latitude of the ship is 26° 13′ N., and the observation is made in the afternoon, find the time of day. *Ans.* 2 hr. 45 min. P.M.

5. At a certain place in latitude 40° N. the altitude of the sun was found to be 41°. If its declination at the time of observation was 20° N., and the observation was made in the morning, how long did it take the sun to reach the meridian ? *Ans.* 3 hr. 31 min.

6. In London (lat. 51° 31′ N.) at an afternoon observation the sun's altitude is 15° 40′. Find the time of day, given that the sun's declination is 12° S.
Ans. 2 hr. 59 min. P.M.

7. A government surveyor observes the sun's altitude to be 21°. If the latitude of his station is 27° N. and the declination of the sun 16° N., what is the time of day if the observation was made in the afternoon ? *Ans.* 4 hr. 57 min. P.M.

8. The captain of a steamship observes that the altitude of the sun is 26° 30′. If he is in latitude 45° 30′ N. and the declination of the sun is 18° N., what is the time of day if the observation was made in the afternoon ? *Ans.* 4 hr. 41 min. P.M.

35. To find the time of sunrise or sunset. If the latitude of the place and the declination of the sun is known, we have a special case of the preceding problem; for at sunrise or sunset the sun is on the horizon and its altitude is zero. Hence the co-altitude, which is one side of the astronomical triangle, will be 90°, and the triangle will be a quadrantal triangle (p. 204). The triangle may then be solved by the method of the last section or as a quadrantal triangle.

EXAMPLES

1. At what hour will the sun set in Montreal (lat. 45° 30′ N.), if its declination at sunset is 18° N.?　　　　　*Ans.* 7 hr. 17 min. p.m.

2. At what hour will the sun rise in Panama (lat. 8° 57′ N.), if its declination at sunrise is 23° 2′ S.?　　　　　*Ans.* 6 hr. 15 min. a.m.

3. About the first of April of each year the declination of the sun is 4° 30′ N. Find the time of sunrise on that date at the following places:

    (a) New York (lat. 40° 43′ N.).　　　　　*Ans.* 5 hr. 45 min. a.m.
    (b) London (lat. 51° 31′ N.).　　　　　5 hr. 37 min. a.m.
    (c) St. Petersburg (lat. 60° N.).　　　　　5 hr. 29 min. a.m.
    (d) New Orleans (lat. 29° 58′ N.).　　　　　5 hr. 50 min. a.m.
    (e) Sydney (lat. 33° 52′ S.).　　　　　5 hr. 48 min. a.m.

36. To determine the longitude of a place on the earth. From the definition of terrestrial longitude given on p. 231 it is evident that the meridians on the earth are projected into hour circles on the celestial sphere. Hence the same angle (or arc) which measures the angle between the celestial meridians (hour circles) of the place of observation and of Greenwich may be taken as a measure of the longitude of the place. Thus, in the figure, if PQP' is the meridian (hour circle) of

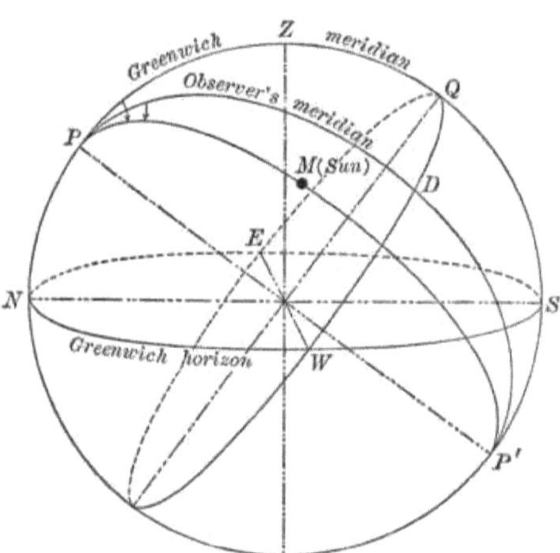

Greenwich and PDP' the meridian (hour circle) of the place of observation, then the angle QPD (or arc QD) measures the west longitude of the place. If PMP' is the hour circle of the sun, it is evident that

$$\text{angle } QPM = \text{hour angle of sun for Greenwich}$$
$$= \text{local time at Greenwich;}$$

$$\text{angle } DPM = \text{hour angle of sun for observer}$$
$$= \text{local time at place of observation.}$$

Also, angle QPM — angle DPM — angle QPD = longitude of place.

Hence *the longitude of the place of observation equals the difference* of local times between the standard meridian and the place in question.* Or, in general, we have the following

Rule for finding longitude : *The observer's longitude is the amount by which noon at Greenwich is earlier or later than noon at the place of observation. If Greenwich has the earlier time, the longitude of the observer is east ; if it has the later time, then the longitude is west.*

We have already shown (p. 248) how the observer may find his own local time. It then remains *to determine the Greenwich time without going there.* The two methods which follow are those in general use.

First method. *Find Greenwich time by telegraph (wire or wireless).* By far the best method, whenever it is available, is to make a direct telegraphic comparison between the clock of the observer and that of some station the longitude of which is known. The difference between the two clocks will be the difference in longitude of the two places.

Ex. 1. The navigator on a battleship has determined his local time to be 2 hr. 25 min. P.M. By wireless he finds the mean solar time at Greenwich to be 4 hr. 30 min. P.M. What is the longitude of the ship?

Solution. Greenwich having the later time,

$$\begin{array}{ll} & 4 \text{ hr. } 30 \text{ min.} \\ & 2 \text{ hr. } 25 \text{ min.} \\ \hline \text{Subtracting,} & 2 \text{ hr. } 5 \text{ min.} = \text{west longitude of the ship.} \end{array}$$

Reducing this to degrees and minutes of arc,

$$\begin{array}{ll} & 2 \text{ hr. } 5 \text{ min.} \\ & 15 \\ \hline \text{Multiplying,} & 31° \ 15' \quad = \text{west longitude of ship.} \end{array}$$

Second method. *Find Greenwich time from a Greenwich chronometer.* The chronometer is merely a very accurate watch. It has been set to Greenwich time at some place whose longitude is known, and thereafter keeps that time wherever carried.

---

* This difference in time is not taken greater than 12 hours. If a difference in time between the two places is calculated to be more than 12 hours, we subtract it from 24 hours and use the remainder instead as the difference.

Ex. 2   An exploring party have calculated their local time to be 10 hr. A.M.
The Greenwich chronometer which they carry gives the time as 8 hr. 30 min. A.M.
What is their longitude ?

Solution. Greenwich has here the earlier time.

	10 hr.
	8 hr. 30 min.
Subtracting,	1 hr. 30 min. $= 22°\ 30' =$ east longitude.

EXAMPLES

1. In the following examples we have given the local time of the observer
and the Greenwich time at the same instant. Find the longitude of the observer
in each case.

Observer's local time	Corresponding Greenwich time	Longitude of observer
(a) Noon.	3 hr. 30 min. P.M.	*Ans.* 52° 30′ W.
(b) Noon.	7 hr. 20 min. A.M.	70° E.
(c) Midnight.	10 hr. 15 min. P.M.	26° 15′ E.
(d) 4 hr. 10 min. P.M.	Noon.	62° 30′ E.
(e) 8 hr. 25 min. A.M.	Noon.	53° 45′ W.
(f) 9 hr. 40 min. P.M.	Midnight.	35° W.
(g) 2 hr. 15 min. P.M.	11 hr. 20 min. A.M.	43° 45′ E.
(h) 10 hr. 26 min. A.M.	5 hr. 16 min. A.M.	77° 30′ E.
(i) 1 hr. 30 min. P.M.	7 hr. 45 min. P.M.	93° 45′ W.
(j) Noon.	Midnight.	180° W. or E.
(k) 6 hr. P.M.	6 hr. A.M.	180° E. or W.
(l) 5 hr. 45 min. A.M.	7 hr. 30 min. P.M.	153° 45′ E.
(m) 10 hr. 55 min. P.M.	8 hr. 35 min. A.M.	145° W.

2. If the Greenwich time is 9 hr. 20 min. P.M., January 24, at the same instant
that the time is 3 hr. 40 min. A.M., January 25, at the place of observation, what
is the observer's longitude ?   *Ans.* 95° E.

3. The local time is 4 hr. 40 min. A.M., March 4, and the corresponding Greenwich time is 8 hr. P.M., March 3. What is the longitude of the place ?

Ans. 130° E.

4. In the following examples we have given the local time of the observer
and the local time at the same instant of some other place whose longitude is
known. Find the longitude of the observer in each case.

Observer's local time	Corresponding time and longitude of the other place	Longitude of observer
(a) 2 hr. P.M.	5 hr. P.M. at Havana (long. 82° 23′ W.)	*Ans.* 127° 23′ W.
(b) 10 hr. A.M.	3 hr. P.M. at Yokohama (long. 139° 41′ E.)	64° 41′ E.
(c) 5 hr. 20 min. P.M.	11 hr. 30 min. P.M. at Glasgow (long. 4° 16′ W.)	96° 46′ W.
(d) 8 hr. 25 min. A.M.	6 hr. 35 min. A.M. at Vera Cruz (long. 96° 9′ W.)	68° 39′ W.
(e) 9 hr. 45 min. P.M.	Midnight at Batavia (long. 106° 52′ E.)	73° 7′ E.
(f) 7 hr. 40 min. P.M.	Noon at Gibraltar (long. 5° 21′ W.)	109° 39′ E.
(g) 4 hr. 50 min. P.M.	Noon at Auckland (long. 174° 50′ E.)	112° 40′ W.

5. What is the longitude of each place mentioned in the examples on p. 249, the Greenwich time for the same instant being given below ?

Example, p. 249	*Greenwich time*	*Longitude of place*
(a) Ex. 3	2 hr. 12 min. P.M.	*Ans.* 57° W. long. (vessel)
(b) Ex. 4	4 hr. 52 min. P.M.	31° 45′ W. long. (vessel)
(c) Ex. 5	5 hr. 9 min. A.M.	50° E. long. (observer)
(d) Ex. 7	10 hr. 33 min. P.M.	84° W. long. (surveyor)
(e) Ex. 8	6 hr. 25 min. P.M.	26° W. long. (ship)

37. The ecliptic and the equinoxes. The earth makes a complete circuit around the sun in one year. To us, however, it appears as if the sun moved and the earth stood still, the (apparent) yearly path of the sun among the stars being a great circle of the celestial sphere which we call the **ecliptic.** Evidently the plane of the earth's orbit

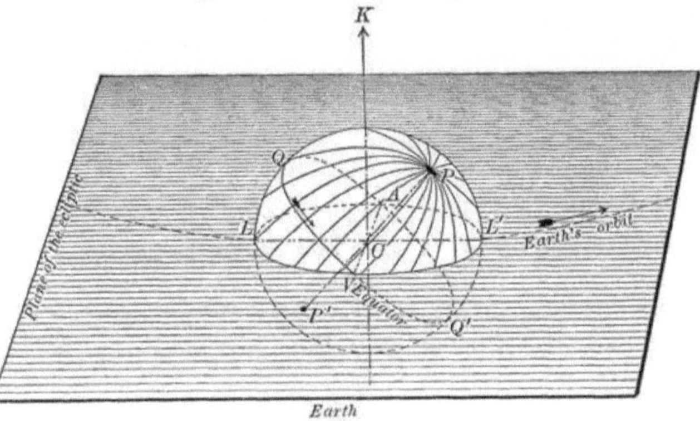

Earth

cuts the celestial sphere in the ecliptic. The plane of the equator and the plane of the ecliptic are inclined to each other at an angle of about $23\tfrac{1}{2}°$ ($= e$), called the *obliquity of the ecliptic* (angle LVQ in figure).

The points where the ecliptic intersects the celestial equator are called the **equinoxes.** The point where the sun crosses the celestial equator when moving northward (in the spring, about March 21) is called the **vernal equinox,** and the point where it crosses the celestial equator when moving southward (in the fall, about September 21) is called the **autumnal equinox.**

If we project the points V and A in our figure on the celestial sphere, the point V will be projected in the vernal equinox and the point A in the autumnal equinox.

38. The equator and hour circle of vernal equinox system.* The two fixed and mutually perpendicular great circles of reference are in

* Sometimes called the *equator system.*

this case the celestial equator (QVQ') and the hour circle of the vernal equinox (PVP'), also called the *equinoctial colure* ; and the spher-

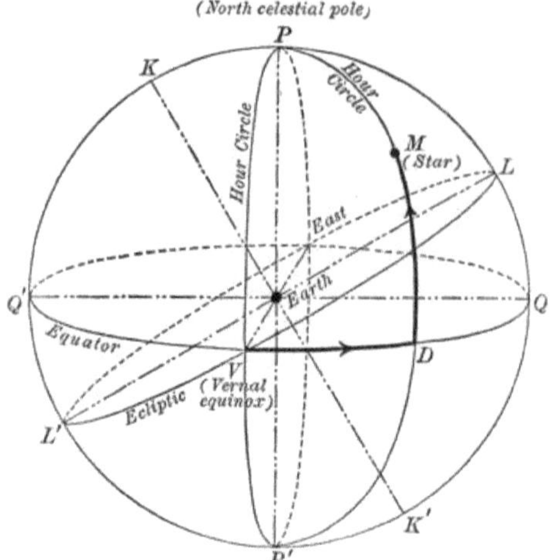

ical coordinates of a heavenly body are its *declination* and *right ascension.*

The **declination** of a heavenly body has already been defined on p. 240 as its angular distance north or south of the celestial equator measured on the hour circle of the body from 0° to 90°, positive if north and negative if south. In the figure DM is the north declination of the star M.

The **right ascension** of a heavenly body is the angle between the hour circle of the body and the hour circle of the vernal equinox measured *eastward* from the latter circle from 0° to 360°, or in hours from 0 to 24. In the figure, the angle VPD (or the arc VD) is the right ascension of the star M. The right ascensions of the sun, moon, and planets are continually changing.* The angle LVQ ($= e$) is the obliquity of the ecliptic ($= 23\frac{1}{2}°$).

Ex. 1. In each of the following examples draw a figure of the celestial sphere and locate the body from the given spherical coordinates.

Right ascension	Declination	Right ascension	Declination
(a) 0°	0°	(j) 90°	0°
(b) 180°	0°	(k) 270°	0°
(c) 90°	N. 90°	(l) 90°	S. 90°
(d) 45°	N. 45°	(m) 45°	S. 45°
(e) 60°	N. 60°	(n) 90°	S. 30°
(f) 120°	+ 30°	(o) 240°	+ 60°
(g) 300°	− 60°	(p) 330°	− 45°
(h) 12 hr.	+ 45°	(q) 6 hr.	+ 15°
(i) 20 hr.	0°	(r) 9 hr.	− 75°

---

* The right ascensions of the sun, moon, and planets may be found in the Nautical Almanac for any time of the year.

Ex. 2. The right ascension of a planet is 10 hr. 40 min. and its declination S. 6°. Find the angular distance from this planet to a fixed star whose right ascension is 3 hr. 20 min. and declination N. 48°.

Solution. Locate the planet and the star on the celestial sphere. Draw the spherical triangle whose vertices are at the north pole, the planet, and the fixed star. Then

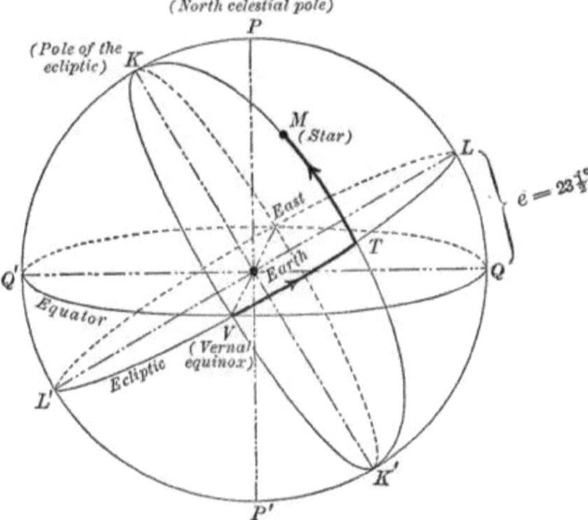

Angle A = difference of right ascensions
    = 10 hr. 40 min. − 3 hr. 20 min.
    = 7 hr. 20 min. = 110°.
Side b = co-declination of star
    = 90° − 48° = 42°.
Side c = co-declination of planet
    = 90° − (− 6°) = 96°. To find side a.

As we have two sides and the included angle given, the solution of this triangle comes under Case II, (a), p. 219. Since a only is required, the shortest method is that illustrated on p. 220, the solution depending on the solution of right spherical triangles. On solving, we get $a = 107°\,48'$. *Ans.*

39. The system having for reference circles the ecliptic and the great circle KVK' passing through the pole of the ecliptic and the vernal

equinox.* The spherical coördinates of a heavenly body in this case are its *latitude and longitude.*†

The **latitude** of a heavenly body is its angular distance north or south of the ecliptic, measured on the great circle passing through

---

* Sometimes called the *ecliptic system.*

† Sometimes called *celestial latitude and longitude* in contradistinction to the latitude and longitude of places on the earth's surface (*terrestrial latitude and longitude*), which were defined on p. 231, and which have different meanings.

the body and the pole of the ecliptic. Thus, in the figure, the arc TM measures the north latitude of the star M.

The **longitude** of a heavenly body is the angle between the great circle passing through the body and the pole of the ecliptic, and the great circle passing through the vernal equinox and the pole of the ecliptic, measured *eastward* from the latter circle from 0° to 360°. In the figure, the angle VKT (or the arc VT) is the longitude of the star M. The latitudes and longitudes of the sun, moon, and planets are continually changing. The angle LVQ ($= e$) is the obliquity of the ecliptic ($= 23\tfrac{1}{2}°$ = arc KP).

Since the ecliptic is the apparent yearly path of the sun, the celestial latitude of the sun is always zero. The declination of the sun, however, varies from N. $23\tfrac{1}{2}°$ ($=$ arc QL) on the longest day of the year in the northern hemisphere (June 21), the sun being then the highest in the sky (at L), to S. $23\tfrac{1}{2}°$ (arc $Q'L'$) on the shortest day of the year (December 22), the sun being then the lowest in the sky (at L'). The declination of the sun is zero at the equinoxes (March 21 and September 21).

Ex. 1.　In each of the following examples draw a figure of the celestial sphere and locate the body from the given spherical coordinates.

Celestial longitude	Celestial latitude	Celestial longitude	Celestial latitude
(a) 0°	0°	(j) 90°	0°
(b) 90°	N. 90°	(k) 180°	0°
(c) 180°	N. 45°	(l) 0°	S. 60°
(d) 270°	0°	(m) 60°	N. 30°
(e) 45°	S. 30°	(n) 120°	N. 45°
(f) 135°	+ 15°	(o) 270°	− 75°
(g) 315°	+ 60°	(p) 30°	− 60°
(h) 6 hr.	− 45°	(q) 9 hr.	0°
(i) 15 hr.	+ 45°	(r) 18 hr.	+ 30°

Ex. 2.　Given the right ascension of a star 2 hr. 40 min. and its declination 24° 20′ N., find its celestial latitude and longitude.

Solution. Locate the star on the celestial sphere. Consider the spherical triangle KPM on the next page.

$$\text{Angle } KPM = \angle Q'PV + \angle VPD$$
$$= 90° + \text{right ascension}$$
$$= 90° + 2 \text{ hr. } 40 \text{ min.}$$
$$= 90° + 40° = 130°.$$
$$\text{Side } PM = \text{co-declension}$$
$$= 90° - 24° 20'$$
$$= 65° 40'.$$

Side $KP = LQ = e = 23° 30'$.

To find side $KM =$ co-latitude of the star,

and   angle $PKM =$ co-longitude of the star.

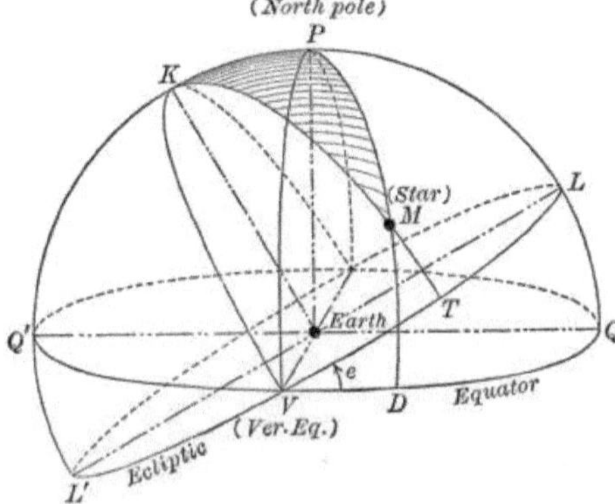

As we have two sides and the included angle given, the solution of this triangle comes under Case II, (a), p. 219. Solving, we get

Side $KM = 81° 52'$ and $\angle PKM = 44° 52'$.

$\therefore\ 90° - KM = 90° - 81° 52' = 8° 8' = TM =$ latitude of star,

and $90° - \angle PKM = 90° - 44° 52' = 45° 8' = VT =$ longitude of star.

EXAMPLES

1. Find the distance in degrees between the sun and the moon when their right ascensions are respectively 12 hr. 39 min., 6 hr. 56 min., and their declinations are 9° 23′ S., 22° 50′ N.   *Ans.* 90°.

2. Find the distance between Regulus and Antares, the right ascensions being 10 hr. and 16 hr. 20 min., and the polar distances 77° 19′ and 116° 6′.

Ans. 99° 56′.

3. Find the distance in degrees between the sun and the moon when their right ascensions are respectively 15 hr. 12 min., 4 hr. 45 min., and their declinations are 21° 30′ S., 5° 30′ N.   *Ans.* 154° 19′.

4. The right ascension of Sirius is 6 hr. 39 min., and his declination is 16° 31′ S.; the right ascension of Aldebaran is 4 hr. 27 min., and his declination is 16° 12′ N. Find the angular distance between the stars.   *Ans.* 46° 2′.

5. Given the right ascension of a star 10 hr. 50 min., and its declination 12° 30′ N., find its latitude and longitude. Take $e = 23° 30'$.

Ans. Latitude = 18° 24′ N., longitude = 281° 7′.

6. If the moon's right ascension is 4 hr. 15 min. and its declination 6° 20′ N., what is its latitude and longitude ?

Ans. Latitude = 14° 43′ N., longitude = 62° 58′.

7. The sun's longitude was 59° 40′. What was its right ascension and declination ? Take $e = 23° 27′$.

Ans. Right ascension = 3 hr. 50 min., declination = 20° 38′ N.

Hint. The latitude of the sun is always zero, since it moves in the ecliptic. Hence in the triangle *KPM* (figure, p. 257), *KM* = 90°, and it is a quadrantal triangle. This triangle may then be solved by the method explained on p. 204.

8. Given the sun's declination 16° 1′ N., find the sun's right ascension and longitude. Take $e = 23° 27′$.

Ans. Right ascension = 9 hr. 14 min., longitude = 136° 6′.

9. The sun's right ascension is 14 hr. 8 min.; find its longitude and declination. Take $e = 23° 27′$.   *Ans.* Longitude = 214° 16′, declination = 12° 56′ S.

10. Find the length of the longest day of the year in latitude 42° 17′ N.

Ans. 15 hr. 6 min.

Hint. This will be the time from sunrise to sunset when the sun is the highest in the sky, that is, when its declination is 23° 27′ N.

11. Find the length of the shortest day in lat. 42° 17′ N.   *Ans.* 8 hr. 54 min.

Hint. The sun will then be the lowest in the sky, that is, its declination will be 23° 27′ S.

12. Find the length of the longest day in New Haven (lat. 41° 19′ N.). Take $e = 23° 27′$.   *Ans.* 15 hr.

13. Find the length of the shortest day in New Haven.   *Ans.* 9 hr.

14. Find the length of the longest day in Stockholm (lat. 59° 21′ N.). Take $e = 23° 27′$.   *Ans.* 18 hr. 16 min.

15. Find the length of the shortest day in Stockholm.   *Ans.* 5 hr. 48 min.

40. The astronomical triangle. We have seen that many of our most important astronomical problems depend on the solution of

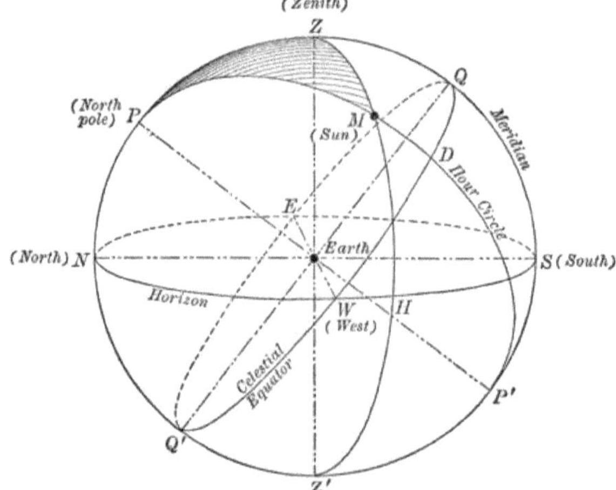

the astronomical triangle *PZM*. In any such problem the first thing to do is to ascertain which parts of the astronomical triangle are given or can be obtained directly from the given data, and which are required. The different magnitudes which may enter into such problems are

$$HM = \text{altitude of the heavenly body,}$$
$$DM = \text{declination of the heavenly body,}$$
$$\text{angle } ZPM = \text{hour angle of the heavenly body,}$$
$$\text{angle } SZM = \text{azimuth of the heavenly body,}$$
$$NP = \text{altitude of the celestial pole}$$
$$= \text{latitude of the observer.}$$

As parts of the astronomical triangle PZM we then have

$$\text{side } MZ = 90° - HM = \text{co-altitude,}$$
$$\text{side } PM = 90° - DM = \text{co-declination,}$$
$$\text{side } PZ = 90° - NP = \text{co-latitude,}$$
$$\text{angle } ZPM = \text{hour angle,}$$
$$\text{angle } PZM = 180° - \text{azimuth (angle } SZM\text{).}*$$

The student should be given practice in picking out the known and unknown parts in examples involving the astronomical triangle, and in indicating the case under which the solution of the triangle comes.

For instance, let us take Ex. 15, p. 261.

$$
\textit{Given parts} \quad
\begin{cases}
\text{Latitude} \quad = 51°\ 32'\ \text{N.} \\
\therefore \text{ side } PZ = 90° - 51°\ 32' = 38°\ 28'. \\
\text{Altitude} \quad = 35°\ 15'. \\
\therefore \text{ side } MZ = 90° - 35°\ 15' = 54°\ 45'. \\
\text{Declination} = 21°\ 27'\ \text{N.} \\
\therefore \text{ side } MP = 90° - 21°\ 27' = 68°\ 33'.
\end{cases}
$$

Required :   Local time = hour angle = angle ZPM.

Since we have three sides given to find an angle, the solution of the triangle comes under Case I, (a), p. 217. This gives angle ZPM $= 59°\ 45' = 3$ hr. 59 min. P.M.

41. Errors arising in the measurement of physical quantities.† Errors of some sort will enter into all data obtained by measurement. For instance, if the length of a line is measured by a steel tape, account must be taken of the expansion due to heat as well as the sagging of the tape under various tensions. Or, suppose the navigator of a ship

---

* When the heavenly body is situated as in the figure. If the body is east of the observer's meridian, we would have angle $PZM = $ azimuth $- 180°$.

† In this connection the student is advised to read § 93 in Granville's *Plane Trigonometry*.

at sea is measuring the altitude of the sun by means of a sextant. The observed altitude should be corrected for errors due to the following causes :

1. Dip. Owing to the observer's elevation above the sea level (on the deck or bridge of the ship), the observed altitude will be too great on account of the dip (or lowering) of the horizon.

2. Index error of sextant. As no instrument is perfect in construction, each one is subject to a certain constant error which is determined by experiment.

3. Refraction of light. Celestial bodies appear higher than they really are because of the refraction of light by the earth's atmosphere. This refraction will depend on the height of the celestial body above the horizon, and also on the state of the barometer and thermometer, since changes in the pressure and temperature of the air affect its density.

4. Semidiameter of the sun. As the observer cannot be sure where the center of the sun is, the altitude of (say) the lower edge of the sun is observed and to that is added the known semidiameter of the sun for that day found from the Nautical Almanac.

5. Parallax. The parallax of a celestial body is the angle subtended by the radius of the earth passing through the observer, as seen from the body. As viewed from the earth's surface, a celestial body appears lower than it would be if viewed from the center, and this may be shown to depend on the parallax of the body.

We shall not enter into the detail connected with these corrections, as that had better be left to works on Field Astronomy ; our purpose here is merely to call the attention of the student to the necessity of eliminating as far as possible the errors that arise when measuring physical quantities.

For the sake of simplicity we have assumed that the necessary corrections have been applied to the data given in the examples found in this book.

MISCELLANEOUS EXAMPLES

1. The continent of Asia has nearly the shape of an equilateral triangle. Assuming each side to be 4800 geographical miles and the radius of the earth to be 3440 geographical miles, find the area of Asia.

Ans. About 13,333,000 sq. mi.

2. The distance between Paris (lat. 48° 50′ N.) and Berlin (lat. 52° 30′ N.) is 472 geographical miles, measured on the arc of a great circle. What time is it at Berlin when it is noon at Paris ?   *Ans.* 44 min. past noon.

3. The altitude of the north pole is 45°, and the azimuth of a star on the horizon is 135°. Find the polar distance of the star.   *Ans.* 60°.

4. What will be the altitude of the sun at 9 A.M. in Mexico City (lat. 19° 25′ N.), if its declination at that time is 8° 23′ N.?   *Ans.* 37° 41′.

5. Find the altitude of the sun at 6 hr. A.M. at Munich (lat. 48° 9′ N.) on the longest day of the year.   *Ans.* Altitude = 17° 15′.

6. Find the time of day when the sun bears due east and due west on the longest day of the year at St. Petersburg (lat. 59° 56′ N.).

Ans. 6 hr. 58 min. A.M., 5 hr. 2 min. P.M.

7. What is the direction of a wall in lat. 52° 30′ N. which casts no shadow at 6 A.M. on the longest day of the year?

Ans. 75° 11′, reckoned from the north point of the horizon.

8. Find the latitude of the place at which the sun rises exactly in the north-east on the longest day of the year.   *Ans.* 55° 45′ N.

9. Find the latitude of the place at which the sun sets at 10 hr. P.M. on the longest day.   *Ans.* 63° 23′ N. or S.

10. Given the latitude of the place of observation 52° 30′ N., the declination of a star 38°, its hour angle 28° 17′. Find the altitude of the star.

Ans. Altitude = 65° 33′.

11. Given the latitude of the place of observation 51° 19′ N., the polar distance of a star 67° 59′, its hour angle 15° 8′. Find the altitude and azimuth of the star.   *Ans.* Altitude = 58° 22′, azimuth = 27° 48′.

12. Given the declination of a star 7° 54′ N., its altitude 22° 45′, its azimuth 50° 14′. Find the hour angle of the star and the latitude of the observer.

Ans. Hour angle = 45° 41′, latitude = 67° 59′ N.

13. The latitude of a star is 51° N., and its longitude 315°. Find its declination. Take $e = 23° 27′$.   *Ans.* Declination = 32° 23′ N.

14. Given the latitude of the observer 44° 50′ N., the azimuth of a star 41° 2′, its hour angle 20°. Find its declination.   *Ans.* Declination = 20° 49′ N.

15. Given the latitude of the place of observation 51° 32′ N., the altitude of the sun west of the meridian 35° 15′, its declination 21° 27′ N. Find the local time.   *Ans.* 3 hr. 59 min. P.M.

CHAPTER IV

RECAPITULATION OF FORMULAS

Spherical Trigonometry

42. Right spherical triangles, pp. 196–197.

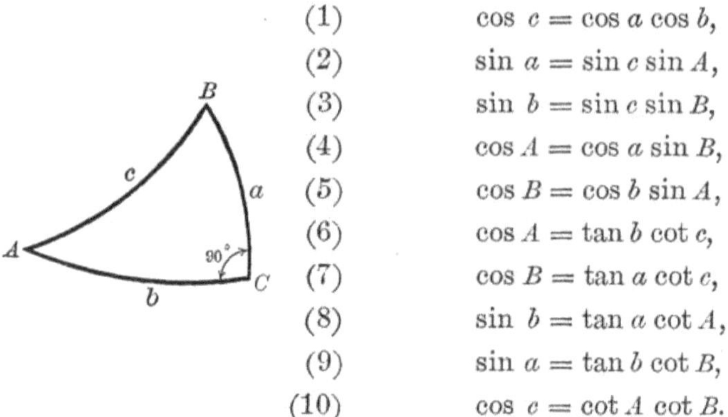

$$(1) \qquad \cos c = \cos a \cos b,$$
$$(2) \qquad \sin a = \sin c \sin A,$$
$$(3) \qquad \sin b = \sin c \sin B,$$
$$(4) \qquad \cos A = \cos a \sin B,$$
$$(5) \qquad \cos B = \cos b \sin A,$$
$$(6) \qquad \cos A = \tan b \cot c,$$
$$(7) \qquad \cos B = \tan a \cot c,$$
$$(8) \qquad \sin b = \tan a \cot A,$$
$$(9) \qquad \sin a = \tan b \cot B,$$
$$(10) \qquad \cos c = \cot A \cot B.$$

General directions for solving right spherical triangles by Napier's rules of circular parts are given on p. 200.

Spherical isosceles and quadrantal triangles are discussed on p. 204.

43. Relations between the sides and angles of oblique spherical triangles, pp. 206–216.

$$\alpha = 180° - A, \qquad \beta = 180° - B, \qquad \gamma = 180° - C.$$

$$s = \tfrac{1}{2}(a + b + c), \qquad \sigma = \tfrac{1}{2}(\alpha + \beta + \gamma).$$

$d =$ diameter of inscribed circle.

$\delta = 180° -$ diameter of circumscribed circle.

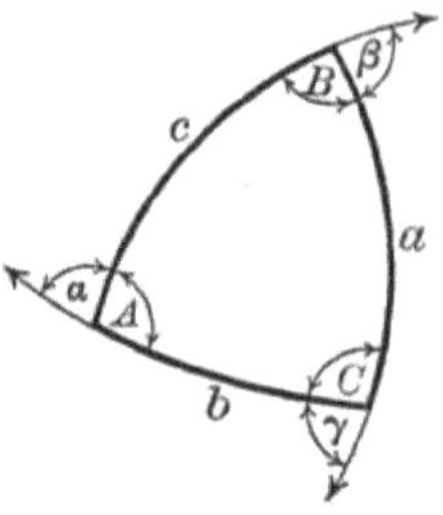

Law of sines, p. 207.

$$(11) \qquad \frac{\sin a}{\sin A} = \frac{\sin b}{\sin B} = \frac{\sin c}{\sin C},$$

or,
$$\frac{\sin a}{\sin \alpha} = \frac{\sin b}{\sin \beta} = \frac{\sin c}{\sin \gamma}.$$

Law of cosines for the sides, p. 209.

$$(12) \qquad \cos a = \cos b \cos c - \sin b \sin c \cos \alpha.$$

Law of cosines for the angles, p. 209.

$$(15) \qquad \cos \alpha = \cos \beta \cos \gamma - \sin \beta \sin \gamma \cos a.$$

Functions of $\tfrac{1}{2}\alpha$, $\tfrac{1}{2}\beta$, $\tfrac{1}{2}\gamma$ in terms of the sides, pp. 211–213.

$$(18) \qquad \sin \tfrac{1}{2}\alpha = \sqrt{\frac{\sin s \sin (s - a)}{\sin b \sin c}}.$$

$$(19) \qquad \cos \tfrac{1}{2}\alpha = \sqrt{\frac{\sin (s - b) \sin (s - c)}{\sin b \sin c}}.$$

$$(20) \qquad \tan \tfrac{1}{2}\alpha = \sqrt{\frac{\sin s \sin (s - a)}{\sin (s - b) \sin (s - c)}}.$$

$$(27) \qquad \tan \tfrac{1}{2}d = \sqrt{\frac{\sin (s - a) \sin (s - b) \sin (s - c)}{\sin s}}.$$

$$(28) \qquad \tan \tfrac{1}{2}\alpha = \frac{\sin (s - a)}{\tan \tfrac{1}{2}d}.$$

$$(29) \qquad \tan \tfrac{1}{2}\beta = \frac{\sin (s - b)}{\tan \tfrac{1}{2}d}.$$

$$(30) \qquad \tan \tfrac{1}{2}\gamma = \frac{\sin (s - c)}{\tan \tfrac{1}{2}d}.$$

Functions of the half sides in terms of α, β, γ, p. 214.

$$(31) \qquad \sin \tfrac{1}{2}a = \sqrt{\frac{\sin \sigma \sin (\sigma - \alpha)}{\sin \beta \sin \gamma}}.$$

$$(32) \qquad \cos \tfrac{1}{2}a = \sqrt{\frac{\sin (\sigma - \beta) \sin (\sigma - \gamma)}{\sin \beta \sin \gamma}}.$$

$$(33) \qquad \tan \tfrac{1}{2}a = \sqrt{\frac{\sin \sigma \sin (\sigma - \alpha)}{\sin (\sigma - \beta) \sin (\sigma - \gamma)}}.$$

$$(40) \qquad \tan \tfrac{1}{2}\delta = \sqrt{\frac{\sin (\sigma - \alpha) \sin (\sigma - \beta) \sin (\sigma - \gamma)}{\sin \sigma}}.$$

$$(41) \qquad \tan \tfrac{1}{2}a = \frac{\sin (\sigma - \alpha)}{\tan \tfrac{1}{2}\delta}.$$

$$(42) \qquad \tan \tfrac{1}{2}b = \frac{\sin (\sigma - \beta)}{\tan \tfrac{1}{2}\delta}.$$

$$(43) \qquad \tan \tfrac{1}{2}c = \frac{\sin (\sigma - \gamma)}{\tan \tfrac{1}{2}\delta}.$$

Napier's Analogies, p. 215.

$$(44) \qquad \tan \tfrac{1}{2}(a - b) = - \frac{\sin \tfrac{1}{2}(\alpha - \beta)}{\sin \tfrac{1}{2}(\alpha + \beta)} \tan \tfrac{1}{2} c.$$

$$(45) \qquad \tan \tfrac{1}{2}(a + b) = - \frac{\cos \tfrac{1}{2}(\alpha - \beta)}{\cos \tfrac{1}{2}(\alpha + \beta)} \tan \tfrac{1}{2} c.$$

$$(46) \qquad \tan \tfrac{1}{2}(\alpha - \beta) = - \frac{\sin \tfrac{1}{2}(a - b)}{\sin \tfrac{1}{2}(a + b)} \tan \tfrac{1}{2} \gamma.$$

$$(47) \qquad \tan \tfrac{1}{2}(\alpha + \beta) = - \frac{\cos \tfrac{1}{2}(a - b)}{\cos \tfrac{1}{2}(a + b)} \tan \tfrac{1}{2} \gamma.$$

44. General directions for the solution of oblique spherical triangles, pp. 216–227.

CASE   I. (*a*) *Given the three sides, p. 217.*

        (*b*) *Given the three angles, p. 218.*

CASE II. (*a*) *Given two sides and their included angle, p. 219.*

        (*b*) *Given two angles and their included side, p. 222.*

CASE III. (*a*) *Given two sides and the angle opposite one of them, p. 224.*

        (*b*) *Given two angles and the side opposite one of them, p. 226.*

45. Length of an arc of a circle in linear units, p. 228.

$$(52) \qquad L = \frac{\pi R N}{180}.$$

$N = $ number of degrees in angle.

46. Area of a spherical triangle, p. 229.

$$(54) \qquad \text{Area} = \frac{\pi R^2 E}{180}.$$

$$E = A + B + C - 180°.$$

$$(55) \qquad \tan \tfrac{1}{4} E = \sqrt{\tan \tfrac{1}{2} s \, \tan \tfrac{1}{2}(s - a) \, \tan \tfrac{1}{2}(s - b) \, \tan \tfrac{1}{2}(s - c)}$$

FOUR-PLACE TABLES OF LOGARITHMS

COMPILED BY

WILLIAM ANTHONY GRANVILLE, Ph.D.

SHEFFIELD SCIENTIFIC SCHOOL, YALE UNIVERSITY

GINN AND COMPANY

BOSTON · NEW YORK · CHICAGO · LONDON

The Athenæum Press
GINN AND COMPANY · PRO-
PRIETORS · BOSTON · U.S.A.

CONTENTS

Table I

FOUR–PLACE LOGARITHMS OF NUMBERS

This table gives the mantissas of the common logarithms (base 10) of the natural numbers (integers) from 1 to 2000, calculated to four places of decimals.

A logarithm found from this table by interpolation may be in error by one unit in the last decimal place.

No.	0	1	2	3	4	5	6	7	8	9	Prop. Parts	
100	0000	0004	0009	0013	0017	0022	0026	0030	0035	0039		
101	0043	0048	0052	0056	0060	0065	0069	0073	0077	0082		
102	0086	0090	0095	0099	0103	0107	0111	0116	0120	0124	Extra digit	Difference
103	0128	0133	0137	0141	0145	0149	0154	0158	0162	0166		
104	0170	0175	0179	0183	0187	0191	0195	0199	0204	0208		
105	0212	0216	0220	0224	0228	0233	0237	0241	0245	0249		**5**
106	0253	0257	0261	0265	0269	0273	0278	0282	0286	0290	1	0.5
107	0294	0298	0302	0306	0310	0314	0318	0322	0326	0330	2	1.0
108	0334	0338	0342	0346	0350	0354	0358	0362	0366	0370	3	1.5
109	0374	0378	0382	0386	0390	0394	0398	0402	0406	0410	4	2.0
110	0414	0418	0422	0426	0430	0434	0438	0441	0445	0449	5	2.5
111	0453	0457	0461	0465	0469	0473	0477	0481	0484	0488	6	3.0
112	0492	0496	0500	0504	0508	0512	0515	0519	0523	0527	7	3.5
113	0531	0535	0538	0542	0546	0550	0554	0558	0561	0565	8	4.0
114	0569	0573	0577	0580	0584	0588	0592	0596	0599	0603	9	4.5
115	0607	0611	0615	0618	0622	0626	0630	0633	0637	0641		
116	0645	0648	0652	0656	0660	0663	0667	0671	0674	0678		
117	0682	0686	0689	0693	0697	0700	0704	0708	0711	0715		**4**
118	0719	0722	0726	0730	0734	0737	0741	0745	0748	0752	1	0.4
119	0755	0759	0763	0766	0770	0774	0777	0781	0785	0788	2	0.8
120	0792	0795	0799	0803	0806	0810	0813	0817	0821	0824	3	1.2
121	0828	0831	0835	0839	0842	0846	0849	0853	0856	0860	4	1.6
122	0864	0867	0871	0874	0878	0881	0885	0888	0892	0896	5	2.0
123	0899	0903	0906	0910	0913	0917	0920	0924	0927	0931	6	2.4
124	0934	0938	0941	0945	0948	0952	0955	0959	0962	0966	7	2.8
125	0969	0973	0976	0980	0983	0986	0990	0993	0997	1000	8	3.2
126	1004	1007	1011	1014	1017	1021	1024	1028	1031	1035	9	3.6
127	1038	1041	1045	1048	1052	1055	1059	1062	1065	1069		
128	1072	1075	1079	1082	1086	1089	1093	1096	1099	1103		
129	1106	1109	1113	1116	1119	1123	1126	1129	1133	1136		**3**
130	1139	1143	1146	1149	1153	1156	1159	1163	1166	1169	1	0.3
131	1173	1176	1179	1183	1186	1189	1193	1196	1199	1202	2	0.6
132	1206	1209	1212	1216	1219	1222	1225	1229	1232	1235	3	0.9
133	1239	1242	1245	1248	1252	1255	1258	1261	1265	1268	4	1.2
134	1271	1274	1278	1281	1284	1287	1290	1294	1297	1300	5	1.5
135	1303	1307	1310	1313	1316	1319	1323	1326	1329	1332	6	1.8
136	1335	1339	1342	1345	1348	1351	1355	1358	1361	1364	7	2.1
137	1367	1370	1374	1377	1380	1383	1386	1389	1392	1396	8	2.4
138	1399	1402	1405	1408	1411	1414	1418	1421	1424	1427	9	2.7
139	1430	1433	1436	1440	1443	1446	1449	1452	1455	1458		
140	1461	1464	1467	1471	1474	1477	1480	1483	1486	1489		**2**
141	1492	1495	1498	1501	1504	1508	1511	1514	1517	1520	1	0.2
142	1523	1526	1529	1532	1535	1538	1541	1544	1547	1550	2	0.4
143	1553	1556	1559	1562	1565	1569	1572	1575	1578	1581	3	0.6
144	1584	1587	1590	1593	1596	1599	1602	1605	1608	1611	4	0.8
145	1614	1617	1620	1623	1626	1629	1632	1635	1638	1641	5	1.0
146	1644	1647	1649	1652	1655	1658	1661	1664	1667	1670	6	1.2
147	1673	1676	1679	1682	1685	1688	1691	1694	1697	1700	7	1.4
148	1703	1706	1708	1711	1714	1717	1720	1723	1726	1729	8	1.6
149	1732	1735	1738	1741	1744	1746	1749	1752	1755	1758	9	1.8
150	1761	1764	1767	1770	1772	1775	1778	1781	1784	1787		
No.	0	1	2	3	4	5	6	7	8	9		

No.	0	1	2	3	4	5	6	7	8	9	Prop. Parts	
150	1761	1764	1767	1770	1772	1775	1778	1781	1784	1787	**Extra digit**	**Difference**
151	1790	1793	1796	1798	1801	1804	1807	1810	1813	1816		
152	1818	1821	1824	1827	1830	1833	1836	1838	1841	1844		
153	1847	1850	1853	1855	1858	1861	1864	1867	1870	1872		
154	1875	1878	1881	1884	1886	1889	1892	1895	1898	1901		**3**
155	1903	1906	1909	1912	1915	1917	1920	1923	1926	1928	1	0.3
156	1931	1934	1937	1940	1942	1945	1948	1951	1953	1956	2	0.6
157	1959	1962	1965	1967	1970	1973	1976	1978	1981	1984	3	0.9
158	1987	1989	1992	1995	1998	2000	2003	2006	2009	2011	4	1.2
159	2014	2017	2019	2022	2025	2028	2030	2033	2036	2038	5	1.5
160	2041	2044	2047	2049	2052	2055	2057	2060	2063	2066	6	1.8
161	2068	2071	2074	2076	2079	2082	2084	2087	2090	2092	7	2.1
162	2095	2098	2101	2103	2106	2109	2111	2114	2117	2119	8	2.4
163	2122	2125	2127	2130	2133	2135	2138	2140	2143	2146	9	2.7
164	2148	2151	2154	2156	2159	2162	2164	2167	2170	2172		
165	2175	2177	2180	2183	2185	2188	2191	2193	2196	2198		
166	2201	2204	2206	2209	2212	2214	2217	2219	2222	2225		
167	2227	2230	2232	2235	2238	2240	2243	2245	2248	2251		
168	2253	2256	2258	2261	2263	2266	2269	2271	2274	2276		
169	2279	2281	2284	2287	2289	2292	2294	2297	2299	2302		
170	2304	2307	2310	2312	2315	2317	2320	2322	2325	2327		
171	2330	2333	2335	2338	2340	2343	2345	2348	2350	2353		
172	2355	2358	2360	2363	2365	2368	2370	2373	2375	2378		
173	2380	2383	2385	2388	2390	2393	2395	2398	2400	2403		
174	2405	2408	2410	2413	2415	2418	2420	2423	2425	2428		
175	2430	2433	2435	2438	2440	2443	2445	2448	2450	2453		
176	2455	2458	2460	2463	2465	2467	2470	2472	2475	2477		
177	2480	2482	2485	2487	2490	2492	2494	2497	2499	2502		**2**
178	2504	2507	2509	2512	2514	2516	2519	2521	2524	2526	1	0.2
179	2529	2531	2533	2536	2538	2541	2543	2545	2548	2550	2	0.4
180	2553	2555	2558	2560	2562	2565	2567	2570	2572	2574	3	0.6
181	2577	2579	2582	2584	2586	2589	2591	2594	2596	2598	4	0.8
182	2601	2603	2605	2608	2610	2613	2615	2617	2620	2622	5	1.0
183	2625	2627	2629	2632	2634	2636	2639	2641	2643	2646	6	1.2
184	2648	2651	2653	2655	2658	2660	2662	2665	2667	2669	7	1.4
185	2672	2674	2676	2679	2681	2683	2686	2688	2690	2693	8	1.6
186	2695	2697	2700	2702	2704	2707	2709	2711	2714	2716	9	1.8
187	2718	2721	2723	2725	2728	2730	2732	2735	2737	2739		
188	2742	2744	2746	2749	2751	2753	2755	2758	2760	2762		
189	2765	2767	2769	2772	2774	2776	2778	2781	2783	2785		
190	2788	2790	2792	2794	2797	2799	2801	2804	2806	2808		
191	2810	2813	2815	2817	2819	2822	2824	2826	2828	2831		
192	2833	2835	2838	2840	2842	2844	2847	2849	2851	2853		
193	2856	2858	2860	2862	2865	2867	2869	2871	2874	2876		
194	2878	2880	2883	2885	2887	2889	2891	2894	2896	2898		
195	2900	2903	2905	2907	2909	2911	2914	2916	2918	2920		
196	2923	2925	2927	2929	2931	2934	2936	2938	2940	2942		
197	2945	2947	2949	2951	2953	2956	2958	2960	2962	2964		
198	2967	2969	2971	2973	2975	2978	2980	2982	2984	2986		
199	2989	2991	2993	2995	2997	2999	3002	3004	3006	3008		
200	3010	3012	3015	3017	3019	3021	3023	3025	3028	3030		
No.	0	1	2	3	4	5	6	7	8	9		

No.	0	1	2	3	4	5	6	7	8	9
20	3010	3032	3054	3075	3096	3118	3139	3160	3181	3201
21	3222	3243	3263	3284	3304	3324	3345	3365	3385	3404
22	3424	3444	3464	3483	3502	3522	3541	3560	3579	3598
23	3617	3636	3655	3674	3692	3711	3729	3747	3766	3784
24	3802	3820	3838	3856	3874	3892	3909	3927	3945	3962
25	3979	3997	4014	4031	4048	4065	4082	4099	4116	4133
26	4150	4166	4183	4200	4216	4232	4249	4265	4281	4298
27	4314	4330	4346	4362	4378	4393	4409	4425	4440	4456
28	4472	4487	4502	4518	4533	4548	4564	4579	4594	4609
29	4624	4639	4654	4669	4683	4698	4713	4728	4742	4757
30	4771	4786	4800	4814	4829	4843	4857	4871	4886	4900
31	4914	4928	4942	4955	4969	4983	4997	5011	5024	5038
32	5051	5065	5079	5092	5105	5119	5132	5145	5159	5172
33	5185	5198	5211	5224	5237	5250	5263	5276	5289	5302
34	5315	5328	5340	5353	5366	5378	5391	5403	5416	5428
35	5441	5453	5465	5478	5490	5502	5514	5527	5539	5551
36	5563	5575	5587	5599	5611	5623	5635	5647	5658	5670
37	5682	5694	5705	5717	5729	5740	5752	5763	5775	5786
38	5798	5809	5821	5832	5843	5855	5866	5877	5888	5900
39	5911	5922	5933	5944	5955	5966	5977	5988	5999	6010
40	6021	6031	6042	6053	6064	6075	6085	6096	6107	6117
41	6128	6138	6149	6160	6170	6180	6191	6201	6212	6222
42	6232	6243	6253	6263	6274	6284	6294	6304	6314	6325
43	6335	6345	6355	6365	6375	6385	6395	6405	6415	6425
44	6435	6444	6454	6464	6474	6484	6493	6503	6513	6522
45	6532	6542	6551	6561	6571	6580	6590	6599	6609	6618
46	6628	6637	6646	6656	6665	6675	6684	6693	6702	6712
47	6721	6730	6739	6749	6758	6767	6776	6785	6794	6803
48	6812	6821	6830	6839	6848	6857	6866	6875	6884	6893
49	6902	6911	6920	6928	6937	6946	6955	6964	6972	6981
50	6990	6998	7007	7016	7024	7033	7042	7050	7059	7067
51	7076	7084	7093	7101	7110	7118	7126	7135	7143	7152
52	7160	7168	7177	7185	7193	7202	7210	7218	7226	7235
53	7243	7251	7259	7267	7275	7284	7292	7300	7308	7316
54	7324	7332	7340	7348	7356	7364	7372	7380	7388	7396
55	7404	7412	7419	7427	7435	7443	7451	7459	7466	7474
56	7482	7490	7497	7505	7513	7520	7528	7536	7543	7551
57	7559	7566	7574	7582	7589	7597	7604	7612	7619	7627
58	7634	7642	7649	7657	7664	7672	7679	7686	7694	7701
59	7709	7716	7723	7731	7738	7745	7752	7760	7767	7774
60	7782	7789	7796	7803	7810	7818	7825	7832	7839	7846
61	7853	7860	7868	7875	7882	7889	7896	7903	7910	7917
62	7924	7931	7938	7945	7952	7959	7966	7973	7980	7987
63	7993	8000	8007	8014	8021	8028	8035	8041	8048	8055
64	8062	8069	8075	8082	8089	8096	8102	8109	8116	8122
65	8129	8136	8142	8149	8156	8162	8169	8176	8182	8189
66	8195	8202	8209	8215	8222	8228	8235	8241	8248	8254
67	8261	8267	8274	8280	8287	8293	8299	8306	8312	8319
68	8325	8331	8338	8344	8351	8357	8363	8370	8376	8382
69	8388	8395	8401	8407	8414	8420	8426	8432	8439	8445
70	8451	8457	8463	8470	8476	8482	8488	8494	8500	8506
No.	0	1	2	3	4	5	6	7	8	9

Prop. Parts

Extra digit — Difference

Extra digit	22	21		20	19		18	17		16	15		14	13		12	11
1	2.2	2.1		2.0	1.9		1.8	1.7		1.6	1.5		1.4	1.3		1.2	1.1
2	4.4	4.2		4.0	3.8		3.6	3.4		3.2	3.0		2.8	2.6		2.4	2.2
3	6.6	6.3		6.0	5.7		5.4	5.1		4.8	4.5		4.2	3.9		3.6	3.3
4	8.8	8.4		8.0	7.6		7.2	6.8		6.4	6.0		5.6	5.2		4.8	4.4
5	11.0	10.5		10.0	9.5		9.0	8.5		8.0	7.5		7.0	6.5		6.0	5.5
6	13.2	12.6		12.0	11.4		10.8	10.2		9.6	9.0		8.4	7.8		7.2	6.6
7	15.4	14.7		14.0	13.3		12.6	11.9		11.2	10.5		9.8	9.1		8.4	7.7
8	17.6	16.8		16.0	15.2		14.4	13.6		12.8	12.0		11.2	10.4		9.6	8.8
9	19.8	18.9		18.0	17.1		16.2	15.3		14.4	13.5		12.6	11.7		10.8	9.9

No.	0	1	2	3	4	5	6	7	8	9	Prop. Parts		
70	8451	8457	8463	8470	8476	8482	8488	8494	8500	8506	Ex. dig.	Difference	
71	8513	8519	8525	8531	8537	8543	8549	8555	8561	8567			
72	8573	8579	8585	8591	8597	8603	8609	8615	8621	8627		10	9
73	8633	8639	8645	8651	8657	8663	8669	8675	8681	8686	1	1.0	0.9
											2	2.0	1.8
74	8692	8698	8704	8710	8716	8722	8727	8733	8739	8745	3	3.0	2.7
75	8751	8756	8762	8768	8774	8779	8785	8791	8797	8802	4	4.0	3.6
76	8808	8814	8820	8825	8831	8837	8842	8848	8854	8859	5	5.0	4.5
											6	6.0	5.4
											7	7.0	6.3
77	8865	8871	8876	8882	8887	8893	8899	8904	8910	8915	8	8.0	7.2
78	8921	8927	8932	8938	8943	8949	8954	8960	8965	8971	9	9.0	8.1
79	8976	8982	8987	8993	8998	9004	9009	9015	9020	9025		8	7
80	9031	9036	9042	9047	9053	9058	9063	9069	9074	9079	1	0.8	0.7
81	9085	9090	9096	9101	9106	9112	9117	9122	9128	9133	2	1.6	1.4
82	9138	9143	9149	9154	9159	9165	9170	9175	9180	9186	3	2.4	2.1
83	9191	9196	9201	9206	9212	9217	9222	9227	9232	9238	4	3.2	2.8
											5	4.0	3.5
											6	4.8	4.2
84	9243	9248	9253	9258	9263	9269	9274	9279	9284	9289	7	5.6	4.9
85	9294	9299	9304	9309	9315	9320	9325	9330	9335	9340	8	6.4	5.6
86	9345	9350	9355	9360	9365	9370	9375	9380	9385	9390	9	7.2	6.3
												6	5
87	9395	9400	9405	9410	9415	9420	9425	9430	9435	9440	1	0.6	0.5
88	9445	9450	9455	9460	9465	9469	9474	9479	9484	9489	2	1.2	1.0
89	9494	9499	9504	9509	9513	9518	9523	9528	9533	9538	3	1.8	1.5
											4	2.4	2.0
90	9542	9547	9552	9557	9562	9566	9571	9576	9581	9586	5	3.0	2.5
											6	3.6	3.0
91	9590	9595	9600	9605	9609	9614	9619	9624	9628	9633	7	4.2	3.5
92	9638	9643	9647	9652	9657	9661	9666	9671	9675	9680	8	4.8	4.0
93	9685	9689	9694	9699	9703	9708	9713	9717	9722	9727	9	5.4	4.5
94	9731	9736	9741	9745	9750	9754	9759	9763	9768	9773		4	
95	9777	9782	9786	9791	9795	9800	9805	9809	9814	9818	1	0.4	
96	9823	9827	9832	9836	9841	9845	9850	9854	9859	9863	2	0.8	
											3	1.2	
											4	1.6	
97	9868	9872	9877	9881	9886	9890	9894	9899	9903	9908	5	2.0	
98	9912	9917	9921	9926	9930	9934	9939	9943	9948	9952	6	2.4	
99	9956	9961	9965	9969	9974	9978	9983	9987	9991	9996	7	2.8	
100	0000	0004	0009	0013	0017	0022	0026	0030	0035	0039	8	3.2	
											9	3.6	
No.	0	1	2	3	4	5	6	7	8	9			

RULES FOR FINDING THE LOGARITHMS OF THE TRIGONOMETRIC FUNCTIONS OF ANGLES NEAR 0° AND 90°

The derivation of the following rules will be found on page 182, Granville's *Plane Trigonometry*.

If the angle is given in degrees, minutes, and seconds, it should first be reduced to degrees and the decimal part of a degree. For this purpose use the conversion table on page 17.

Rule I. *To find the Logarithms of the Functions of an Angle near 0°.**

$$\log \sin x° = \bar{2}.2419 + \log x.$$
$$\log \tan x° = 2.2419 + \log x.$$
$$\log \cot x° = 1.7581 - \log x.$$

log cos x° is found from the tables in the usual way.

Rule II. *To find the Logarithms of the Functions of an Angle near 90°.*†

$$\log \cos x° = \bar{2}.2419 + \log (90 - x).$$
$$\log \cot x° = 2.2419 + \log (90 - x).$$
$$\log \tan x° = 1.7581 - \log (90 - x).$$

log sin x° is found from the tables in the usual way.

These rules will give results accurate to four decimal places for all angles between 0° and 1.1° and between 88.9° and 90°.

---

* Example 1, page 182, Granville's *Plane Trigonometry*, illustrates the application of this rule.

† Example 2, page 183, Granville's *Plane Trigonometry*, illustrates the application of this rule.

TABLE II

FOUR–PLACE LOGARITHMS OF TRIGONOMETRIC FUNCTIONS, THE ANGLE BEING EXPRESSED IN DEGREES AND MINUTES

This table gives the common logarithms (base 10) of the sines, cosines, tangents, and cotangents of all angles from 0° to 5° and from 85° to 90° for each minute; and from 5° to 85° at intervals of 10 minutes, all calculated to four places of decimals. In order to avoid the printing of negative characteristics, the number 10 has been added to every logarithm in the first, second, and fourth columns (those having **log sin, log tan,** and **log cos** at the top). Hence in writing down any logarithm taken from these three columns — 10 should be written after it. Logarithms taken from the third column (having **log cot** at the top) should be used as printed.

A logarithm found from this table by interpolation may be in error by one unit in the last decimal place, except for angles between 0° and 18′ or between 89° 42′ and 90°, when the error may be larger. In the latter cases the table refers the student to the formulas on page 6 for more accurate results.

7

0°

Angle	log sin	diff. 1'	log tan	com. diff. 1'	log cot	log cos	
0° 0'	———		———		———	10.0000	90° 00'
0° 1'	6.4637		6.4637		3.5363	10.0000	89° 59'
0° 2'	6.7648		6.7648		3.2352	10.0000	89° 58'
0° 3'	6.9408		6.9408		3.0592	10.0000	89° 57'
0° 4'	7.0658		7.0658		2.9342	10.0000	89° 56'
0° 5'	7.1627		7.1627		2.8373	10.0000	89° 55'
0° 6'	7.2419		7.2419		2.7581	10.0000	89° 54'
0° 7'	7.3088		7.3088		2.6912	10.0000	89° 53'
0° 8'	7.3668		7.3668		2.6332	10.0000	89° 52'
0° 9'	7.4180		7.4180		2.5820	10.0000	89° 51'
0° 10'	7.4637		7.4637		2.5363	10.0000	89° 50'
0° 11'	7.5051		7.5051		2.4949	10.0000	89° 49'
0° 12'	7.5429		7.5429		2.4571	10.0000	89° 48'
0° 13'	7.5777		7.5777		2.4223	10.0000	89° 47'
0° 14'	7.6099		7.6099		2.3901	10.0000	89° 46'
0° 15'	7.6398		7.6398		2.3602	10.0000	89° 45'
0° 16'	7.6678		7.6678		2.3322	10.0000	89° 44'
0° 17'	7.6942		7.6942		2.3058	10.0000	89° 43'
0° 18'	7.7190		7.7190		2.2810	10.0000	89° 42'
0° 19'	7.7425	235	7.7425	235	2.2575	10.0000	89° 41'
0° 20'	7.7648	223	7.7648	223	2.2352	10.0000	89° 40'
0° 21'	7.7859	211	7.7860	212	2.2140	10.0000	89° 39'
0° 22'	7.8061	202	7.8062	202	2.1938	10.0000	89° 38'
0° 23'	7.8255	194	7.8255	193	2.1745	10.0000	89° 37'
0° 24'	7.8439	184	7.8439	184	2.1561	10.0000	89° 36'
0° 25'	7.8617	178	7.8617	178	2.1383	10.0000	89° 35'
0° 26'	7.8787	170	7.8787	170	2.1213	10.0000	89° 34'
0° 27'	7.8951	164	7.8951	164	2.1049	10.0000	89° 33'
0° 28'	7.9109	158	7.9109	158	2.0891	10.0000	89° 32'
0° 29'	7.9261	152	7.9261	152	2.0739	10.0000	89° 31'
0° 30'	7.9408	147	7.9409	148	2.0591	10.0000	89° 30'
0° 31'	7.9551	143	7.9551	142	2.0449	10.0000	89° 29'
0° 32'	7.9689	138	7.9689	138	2.0311	10.0000	89° 28'
0° 33'	7.9822	133	7.9823	134	2.0177	10.0000	89° 27'
0° 34'	7.9952	130	7.9952	129	2.0048	10.0000	89° 26'
0° 35'	8.0078	126	8.0078	126	1.9922	10.0000	89° 25'
0° 36'	8.0200	122	8.0200	122	1.9800	10.0000	89° 24'
0° 37'	8.0319	119	8.0319	119	1.9681	10.0000	89° 23'
0° 38'	8.0435	116	8.0435	116	1.9565	10.0000	89° 22'
0° 39'	8.0548	113	8.0548	113	1.9452	10.0000	89° 21'
0° 40'	8.0658	110	8.0658	110	1.9342	10.0000	89° 20'
0° 41'	8.0765	107	8.0765	107	1.9235	10.0000	89° 19'
0° 42'	8.0870	105	8.0870	105	1.9130	10.0000	89° 18'
0° 43'	8.0972	102	8.0972	102	1.9028	10.0000	89° 17'
0° 44'	8.1072	100	8.1072	100	1.8928	10.0000	89° 16'
0° 45'	8.1169	97	8.1170	98	1.8830	10.0000	89° 15'
0° 46'	8.1265	96	8.1265	95	1.8735	10.0000	89° 14'
0° 47'	8.1358	93	8.1359	94	1.8641	10.0000	89° 13'
0° 48'	8.1450	92	8.1450	91	1.8550	10.0000	89° 12'
0° 49'	8.1539	89	8.1540	90	1.8460	10.0000	89° 11'
0° 50'	8.1627	88	8.1627	87	1.8373	10.0000	89° 10'
0° 51'	8.1713	86	8.1713	86	1.8287	10.0000	89° 9'
0° 52'	8.1797	84	8.1798	85	1.8202	10.0000	89° 8'
0° 53'	8.1880	83	8.1880	82	1.8120	9.9999	89° 7'
0° 54'	8.1961	81	8.1962	82	1.8038	9.9999	89° 6'
0° 55'	8.2041	80	8.2041	79	1.7959	9.9999	89° 5'
0° 56'	8.2119	78	8.2120	79	1.7880	9.9999	89° 4'
0° 57'	8.2196	77	8.2196	76	1.7804	9.9999	89° 3'
0° 58'	8.2271	75	8.2272	76	1.7728	9.9999	89° 2'
0° 59'	8.2346	75	8.2346	74	1.7654	9.9999	89° 1'
0° 60'	8.2419	73	8.2419	73	1.7581	9.9999	89° 0'
	log cos	diff. 1'	log cot	com. diff. 1'	log tan	log sin	Angle

Note in diff. 1' and com. diff. 1' columns (0° 0' to 0° 20'): *Ordinary interpolation here will in general give inaccurate results. Instead use formulas on p. 6.*

89°

1°

Angle	log sin	diff.1'	log tan	com. diff.1'	log cot	log cos	
1° 0'	8.2419		8.2419		1.7581	9.9999	88° 60'
1° 1'	8.2490	71	8.2491	72	1.7509	9.9999	88° 59'
1° 2'	8.2561	71	8.2562	71	1.7438	9.9999	88° 58'
1° 3'	8.2630	69	8.2631	69	1.7369	9.9999	88° 57'
1° 4'	8.2699	69	8.2700	69	1.7300	9.9999	88° 56'
1° 5'	8.2766	67	8.2767	67	1.7233	9.9999	88° 55'
1° 6'	8.2832	66	8.2833	66	1.7167	9.9999	88° 54'
1° 7'	8.2898	66	8.2899	66	1.7101	9.9999	88° 53'
1° 8'	8.2962	64	8.2963	64	1.7037	9.9999	88° 52'
1° 9'	8.3025	63	8.3026	63	1.6974	9.9999	88° 51'
1° 10'	8.3088	63	8.3089	63	1.6911	9.9999	88° 50'
1° 11'	8.3150	62	8.3150	61	1.6850	9.9999	88° 49'
1° 12'	8.3210	60	8.3211	61	1.6789	9.9999	88° 48'
1° 13'	8.3270	60	8.3271	60	1.6729	9.9999	88° 47'
1° 14'	8.3329	59	8.3330	59	1.6670	9.9999	88° 46'
1° 15'	8.3388	59	8.3389	59	1.6611	9.9999	88° 45'
1° 16'	8.3445	57	8.3446	57	1.6554	9.9999	88° 44'
1° 17'	8.3502	57	8.3503	56	1.6497	9.9999	88° 43'
1° 18'	8.3558	56	8.3559	56	1.6441	9.9999	88° 42'
1° 19'	8.3613	55	8.3614	55	1.6386	9.9999	88° 41'
1° 20'	8.3668	55	8.3669	55	1.6331	9.9999	88° 40'
1° 21'	8.3722	54	8.3723	54	1.6277	9.9999	88° 39'
1° 22'	8.3775	53	8.3776	53	1.6224	9.9999	88° 38'
1° 23'	8.3828	53	8.3829	53	1.6171	9.9999	88° 37'
1° 24'	8.3880	52	8.3881	52	1.6119	9.9999	88° 36'
1° 25'	8.3931	51	8.3932	51	1.6068	9.9999	88° 35'
1° 26'	8.3982	51	8.3983	51	1.6017	9.9999	88° 34'
1° 27'	8.4032	50	8.4033	50	1.5967	9.9999	88° 33'
1° 28'	8.4082	50	8.4083	50	1.5917	9.9999	88° 32'
1° 29'	8.4131	49	8.4132	49	1.5868	9.9999	88° 31'
1° 30'	8.4179	49	8.4181	49	1.5819	9.9999	88° 30'
1° 31'	8.4227	48	8.4229	48	1.5771	9.9998	88° 29'
1° 32'	8.4275	48	8.4276	47	1.5724	9.9998	88° 28'
1° 33'	8.4322	47	8.4323	47	1.5677	9.9998	88° 27'
1° 34'	8.4368	46	8.4370	47	1.5630	9.9998	88° 26'
1° 35'	8.4414	46	8.4416	46	1.5584	9.9998	88° 25'
1° 36'	8.4459	45	8.4461	45	1.5539	9.9998	88° 24'
1° 37'	8.4504	45	8.4506	45	1.5494	9.9998	88° 23'
1° 38'	8.4549	45	8.4551	45	1.5449	9.9998	88° 22'
1° 39'	8.4593	44	8.4595	44	1.5405	9.9998	88° 21'
1° 40'	8.4637	44	8.4638	43	1.5362	9.9998	88° 20'
1° 41'	8.4680	43	8.4682	44	1.5318	9.9998	88° 19'
1° 42'	8.4723	43	8.4725	43	1.5275	9.9998	88° 18'
1° 43'	8.4765	42	8.4767	42	1.5233	9.9998	88° 17'
1° 44'	8.4807	42	8.4809	42	1.5191	9.9998	88° 16'
1° 45'	8.4848	41	8.4851	42	1.5149	9.9998	88° 15'
1° 46'	8.4890	42	8.4892	41	1.5108	9.9998	88° 14'
1° 47'	8.4930	40	8.4933	41	1.5067	9.9998	88° 13'
1° 48'	8.4971	41	8.4973	40	1.5027	9.9998	88° 12'
1° 49'	8.5011	40	8.5013	40	1.4987	9.9998	88° 11'
1° 50'	8.5050	39	8.5053	40	1.4947	9.9998	88° 10'
1° 51'	8.5090	40	8.5092	39	1.4908	9.9998	88° 9'
1° 52'	8.5129	39	8.5131	39	1.4869	9.9998	88° 8'
1° 53'	8.5167	38	8.5170	39	1.4830	9.9998	88° 7'
1° 54'	8.5206	39	8.5208	38	1.4792	9.9998	88° 6'
1° 55'	8.5243	37	8.5246	38	1.4754	9.9998	88° 5'
1° 56'	8.5281	38	8.5283	37	1.4717	9.9998	88° 4'
1° 57'	8.5318	37	8.5321	38	1.4679	9.9997	88° 3'
1° 58'	8.5355	37	8.5358	37	1.4642	9.9997	88° 2'
1° 59'	8.5392	37	8.5394	36	1.4606	9.9997	88° 1'
1° 60'	8.5428	36	8.5431	37	1.4569	9.9997	88° 0'
	log cos	diff.1'	log cot	com. diff.1'	log tan	log sin	Angle

88°

2°

Angle	log sin	diff. 1′	log tan	com. diff. 1′	log cot	log cos	
2° 0′	8.5428		8.5431		1.4569	9.9997	**87° 60′**
		36		36			
2° 1′	8.5464		8.5467		1.4533	9.9997	87° 59′
		36		36			
2° 2′	8.5500		8.5503		1.4497	9.9997	87° 58′
		35		35			
2° 3′	8.5535		8.5538		1.4462	9.9997	87° 57′
		36		35			
2° 4′	8.5571		8.5573		1.4427	9.9997	87° 56′
		34		35			
2° 5′	8.5605		8.5608		1.4392	9.9997	87° 55′
		35		35			
2° 6′	8.5640		8.5643		1.4357	9.9997	87° 54′
		34		34			
2° 7′	8.5674		8.5677		1.4323	9.9997	87° 53′
		34		34			
2° 8′	8.5708		8.5711		1.4289	9.9997	87° 52′
		34		34			
2° 9′	8.5742		8.5745		1.4255	9.9997	87° 51′
		34		34			
2° 10′	8.5776		8.5779		1.4221	9.9997	**87° 50′**
		33		33			
2° 11′	8.5809		8.5812		1.4188	9.9997	87° 49′
		33		33			
2° 12′	8.5842		8.5845		1.4155	9.9997	87° 48′
		33		33			
2° 13′	8.5875		8.5878		1.4122	9.9997	87° 47′
		32		33			
2° 14′	8.5907		8.5911		1.4089	9.9997	87° 46′
		32		32			
2° 15′	8.5939		8.5943		1.4057	9.9997	87° 45′
		33		32			
2° 16′	8.5972		8.5975		1.4025	9.9997	87° 44′
		31		32			
2° 17′	8.6003		8.6007		1.3993	9.9997	87° 43′
		32		31			
2° 18′	8.6035		8.6038		1.3962	9.9997	87° 42′
		31		32			
2° 19′	8.6066		8.6070		1.3930	9.9996	87° 41′
		31		31			
2° 20′	8.6097		8.6101		1.3899	9.9996	**87° 40′**
		31		31			
2° 21′	8.6128		8.6132		1.3868	9.9996	87° 39′
		31		31			
2° 22′	8.6159		8.6163		1.3837	9.9996	87° 38′
		30		30			
2° 23′	8.6189		8.6193		1.3807	9.9996	87° 37′
		31		30			
2° 24′	8.6220		8.6223		1.3777	9.9996	87° 36′
		30		31			
2° 25′	8.6250		8.6254		1.3746	9.9996	87° 35′
		29		29			
2° 26′	8.6279		8.6283		1.3717	9.9996	87° 34′
		30		30			
2° 27′	8.6309		8.6313		1.3687	9.9996	87° 33′
		30		30			
2° 28′	8.6339		8.6343		1.3657	9.9996	87° 32′
		29		29			
2° 29′	8.6368		8.6372		1.3628	9.9996	87° 31′
		29		29			
2° 30′	8.6397		8.6401		1.3599	9.9996	**87° 30′**
		29		29			
2° 31′	8.6426		8.6430		1.3570	9.9996	87° 29′
		28		29			
2° 32′	8.6454		8.6459		1.3541	9.9996	87° 28′
		29		28			
2° 33′	8.6483		8.6487		1.3513	9.9996	87° 27′
		28		28			
2° 34′	8.6511		8.6515		1.3485	9.9996	87° 26′
		28		29			
2° 35′	8.6539		8.6544		1.3456	9.9996	87° 25′
		28		27			
2° 36′	8.6567		8.6571		1.3429	9.9996	87° 24′
		28		28			
2° 37′	8.6595		8.6599		1.3401	9.9995	87° 23′
		27		28			
2° 38′	8.6622		8.6627		1.3373	9.9995	87° 22′
		28		27			
2° 39′	8.6650		8.6654		1.3346	9.9995	87° 21′
		27		28			
2° 40′	8.6677		8.6682		1.3318	9.9995	**87° 20′**
		27		27			
2° 41′	8.6704		8.6709		1.3291	9.9995	87° 19′
		27		27			
2° 42′	8.6731		8.6736		1.3264	9.9995	87° 18′
		27		26			
2° 43′	8.6758		8.6762		1.3238	9.9995	87° 17′
		26		27			
2° 44′	8.6784		8.6789		1.3211	9.9995	87° 16′
		26		26			
2° 45′	8.6810		8.6815		1.3185	9.9995	87° 15′
		27		27			
2° 46′	8.6837		8.6842		1.3158	9.9995	87° 14′
		26		26			
2° 47′	8.6863		8.6868		1.3132	9.9995	87° 13′
		26		26			
2° 48′	8.6889		8.6894		1.3106	9.9995	87° 12′
		25		26			
2° 49′	8.6914		8.6920		1.3080	9.9995	87° 11′
		26		25			
2° 50′	8.6940		8.6945		1.3055	9.9995	**87° 10′**
		25		26			
2° 51′	8.6965		8.6971		1.3029	9.9995	87° 9′
		26		25			
2° 52′	8.6991		8.6996		1.3004	9.9995	87° 8′
		25		25			
2° 53′	8.7016		8.7021		1.2979	9.9995	87° 7′
		25		25			
2° 54′	8.7041		8.7046		1.2954	9.9994	87° 6′
		25		25			
2° 55′	8.7066		8.7071		1.2929	9.9994	87° 5′
		24		25			
2° 56′	8.7090		8.7096		1.2904	9.9994	87° 4′
		25		25			
2° 57′	8.7115		8.7121		1.2879	9.9994	87° 3′
		25		24			
2° 58′	8.7140		8.7145		1.2855	9.9994	87° 2′
		24		25			
2° 59′	8.7164		8.7170		1.2830	9.9994	87° 1′
		24		24			
2° 60′	8.7188		8.7194		1.2806	9.9994	**87° 0′**
	log cos	diff. 1′	log cot	com. diff. 1′	log tan	log sin	Angle

87°

3°

Angle	log sin	diff.1′	log tan	com. diff.1′	log cot	log cos	
3° 0′	8.7188	24	8.7194	24	1.2806	9.9994	86° 60′
3° 1′	8.7212	24	8.7218	24	1.2782	9.9994	86° 59′
3° 2′	8.7236	24	8.7242	24	1.2758	9.9994	86° 58′
3° 3′	8.7260	23	8.7266	24	1.2734	9.9994	86° 57′
3° 4′	8.7283	24	8.7290	24	1.2710	9.9994	86° 56′
3° 5′	8.7307	23	8.7313	23	1.2687	9.9994	86° 55′
3° 6′	8.7330	24	8.7337	24	1.2663	9.9994	86° 54′
3° 7′	8.7354	23	8.7360	23	1.2640	9.9994	86° 53′
3° 8′	8.7377	23	8.7383	23	1.2617	9.9994	86° 52′
3° 9′	8.7400	23	8.7406	23	1.2594	9.9993	86° 51′
3° 10′	8.7423	22	8.7429	23	1.2571	9.9993	86° 50′
3° 11′	8.7445	23	8.7452	23	1.2548	9.9993	86° 49′
3° 12′	8.7468	23	8.7475	22	1.2525	9.9993	86° 48′
3° 13′	8.7491	22	8.7497	23	1.2503	9.9993	86° 47′
3° 14′	8.7513	22	8.7520	22	1.2480	9.9993	86° 46′
3° 15′	8.7535	22	8.7542	23	1.2458	9.9993	86° 45′
3° 16′	8.7557	23	8.7565	22	1.2435	9.9993	86° 44′
3° 17′	8.7580	22	8.7587	22	1.2413	9.9993	86° 43′
3° 18′	8.7602	21	8.7609	22	1.2391	9.9993	86° 42′
3° 19′	8.7623	22	8.7631	21	1.2369	9.9993	86° 41′
3° 20′	8.7645	22	8.7652	22	1.2348	9.9993	86° 40′
3° 21′	8.7667	21	8.7674	22	1.2326	9.9993	86° 39′
3° 22′	8.7688	22	8.7696	21	1.2304	9.9993	86° 38′
3° 23′	8.7710	21	8.7717	22	1.2283	9.9992	86° 37′
3° 24′	8.7731	21	8.7739	21	1.2261	9.9992	86° 36′
3° 25′	8.7752	21	8.7760	21	1.2240	9.9992	86° 35′
3° 26′	8.7773	21	8.7781	21	1.2219	9.9992	86° 34′
3° 27′	8.7794	21	8.7802	21	1.2198	9.9992	86° 33′
3° 28′	8.7815	21	8.7823	21	1.2177	9.9992	86° 32′
3° 29′	8.7836	21	8.7844	21	1.2156	9.9992	86° 31′
3° 30′	8.7857	20	8.7865	21	1.2135	9.9992	86° 30′
3° 31′	8.7877	21	8.7886	20	1.2114	9.9992	86° 29′
3° 32′	8.7898	20	8.7906	21	1.2094	9.9992	86° 28′
3° 33′	8.7918	21	8.7927	20	1.2073	9.9992	86° 27′
3° 34′	8.7939	20	8.7947	20	1.2053	9.9992	86° 26′
3° 35′	8.7959	20	8.7967	20	1.2033	9.9992	86° 25′
3° 36′	8.7979	20	8.7988	21	1.2012	9.9991	86° 24′
3° 37′	8.7999	20	8.8008	20	1.1992	9.9991	86° 23′
3° 38′	8.8019	20	8.8028	20	1.1972	9.9991	86° 22′
3° 39′	8.8039	20	8.8048	20	1.1952	9.9991	86° 21′
3° 40′	8.8059	19	8.8067	20	1.1933	9.9991	86° 20′
3° 41′	8.8078	20	8.8087	20	1.1913	9.9991	86° 19′
3° 42′	8.8098	19	8.8107	19	1.1893	9.9991	86° 18′
3° 43′	8.8117	20	8.8126	20	1.1874	9.9991	86° 17′
3° 44′	8.8137	19	8.8146	19	1.1854	9.9991	86° 16′
3° 45′	8.8156	19	8.8165	20	1.1835	9.9991	86° 15′
3° 46′	8.8175	19	8.8185	19	1.1815	9.9991	86° 14′
3° 47′	8.8194	19	8.8204	19	1.1796	9.9991	86° 13′
3° 48′	8.8213	19	8.8223	19	1.1777	9.9990	86° 12′
3° 49′	8.8232	19	8.8242	19	1.1758	9.9990	86° 11′
3° 50′	8.8251	19	8.8261	19	1.1739	9.9990	86° 10′
3° 51′	8.8270	19	8.8280	19	1.1720	9.9990	86° 9′
3° 52′	8.8289	18	8.8299	18	1.1701	9.9990	86° 8′
3° 53′	8.8307	19	8.8317	19	1.1683	9.9990	86° 7′
3° 54′	8.8326	19	8.8336	19	1.1664	9.9990	86° 6′
3° 55′	8.8345	18	8.8355	18	1.1645	9.9990	86° 5′
3° 56′	8.8363	18	8.8373	19	1.1627	9.9990	86° 4′
3° 57′	8.8381	19	8.8392	18	1.1608	9.9990	86° 3′
3° 58′	8.8400	18	8.8410	18	1.1590	9.9990	86° 2′
3° 59′	8.8418	18	8.8428	18	1.1572	9.9990	86° 1′
3° 60′	8.8436		8.8446		1.1554	9.9989	86° 0′
	log cos	diff.1′	log cot	com. diff.1′	log tan	log sin	Angle

86°

4°							
Angle	log sin	diff. 1'	log tan	com. diff. 1'	log cot	log cos	
4° 0'	8.8436		8.8446		1.1554	9.9989	85° 60'
		18		19			
4° 1'	8.8454		8.8465		1.1535	9.9989	85° 59'
		18		18			
4° 2'	8.8472		8.8483		1.1517	9.9989	85° 58'
		18		18			
4° 3'	8.8490		8.8501		1.1499	9.9989	85° 57'
		18		17			
4° 4'	8.8508		8.8518		1.1482	9.9989	85° 56'
		17		18			
4° 5'	8.8525		8.8536		1.1464	9.9989	85° 55'
		18		18			
4° 6'	8.8543		8.8554		1.1446	9.9989	85° 54'
		17		18			
4° 7'	8.8560		8.8572		1.1428	9.9989	85° 53'
		18		17			
4° 8'	8.8578		8.8589		1.1411	9.9989	85° 52'
		17		18			
4° 9'	8.8595		8.8607		1.1393	9.9989	85° 51'
		18		17			
4° 10'	8.8613		8.8624		1.1376	9.9989	85° 50'
		17		18			
4° 11'	8.8630		8.8642		1.1358	9.9988	85° 49'
		17		17			
4° 12'	8.8647		8.8659		1.1341	9.9988	85° 48'
		18		17			
4° 13'	8.8665		8.8676		1.1324	9.9988	85° 47'
		17		18			
4° 14'	8.8682		8.8694		1.1306	9.9988	85° 46'
		17		17			
4° 15'	8.8699		8.8711		1.1289	9.9988	85° 45'
		17		17			
4° 16'	8.8716		8.8728		1.1272	9.9988	85° 44'
		17		17			
4° 17'	8.8733		8.8745		1.1255	9.9988	85° 43'
		16		17			
4° 18'	8.8749		8.8762		1.1238	9.9988	85° 42'
		17		16			
4° 19'	8.8766		8.8778		1.1222	9.9988	85° 41'
		17		17			
4° 20'	8.8783		8.8795		1.1205	9.9988	85° 40'
		16		17			
4° 21'	8.8799		8.8812		1.1188	9.9987	85° 39'
		17		17			
4° 22'	8.8816		8.8829		1.1171	9.9987	85° 38'
		17		16			
4° 23'	8.8833		8.8845		1.1155	9.9987	85° 37'
		16		17			
4° 24'	8.8849		8.8862		1.1138	9.9987	85° 36'
		16		16			
4° 25'	8.8865		8.8878		1.1122	9.9987	85° 35'
		17		17			
4° 26'	8.8882		8.8895		1.1105	9.9987	85° 34'
		16		16			
4° 27'	8.8898		8.8911		1.1089	9.9987	85° 33'
		16		16			
4° 28'	8.8914		8.8927		1.1073	9.9987	85° 32'
		16		17			
4° 29'	8.8930		8.8944		1.1056	9.9987	85° 31'
		16		16			
4° 30'	8.8946		8.8960		1.1040	9.9987	85° 30'
		16		16			
4° 31'	8.8962		8.8976		1.1024	9.9986	85° 29'
		16		16			
4° 32'	8.8978		8.8992		1.1008	9.9986	85° 28'
		16		16			
4° 33'	8.8994		8.9008		1.0992	9.9986	85° 27'
		16		16			
4° 34'	8.9010		8.9024		1.0976	9.9986	85° 26'
		16		16			
4° 35'	8.9026		8.9040		1.0960	9.9986	85° 25'
		16		16			
4° 36'	8.9042		8.9056		1.0944	9.9986	85° 24'
		15		15			
4° 37'	8.9057		8.9071		1.0929	9.9986	85° 23'
		16		16			
4° 38'	8.9073		8.9087		1.0913	9.9986	85° 22'
		16		16			
4° 39'	8.9089		8.9103		1.0897	9.9986	85° 21'
		15		15			
4° 40'	8.9104		8.9118		1.0882	9.9986	85° 20'
		15		16			
4° 41'	8.9119		8.9134		1.0866	9.9985	85° 19'
		16		16			
4° 42'	8.9135		8.9150		1.0850	9.9985	85° 18'
		15		15			
4° 43'	8.9150		8.9165		1.0835	9.9985	85° 17'
		16		15			
4° 44'	8.9166		8.9180		1.0820	9.9985	85° 16'
		15		16			
4° 45'	8.9181		8.9196		1.0804	9.9985	85° 15'
		15		15			
4° 46'	8.9196		8.9211		1.0789	9.9985	85° 14'
		15		15			
4° 47'	8.9211		8.9226		1.0774	9.9985	85° 13'
		15		15			
4° 48'	8.9226		8.9241		1.0759	9.9985	85° 12'
		15		15			
4° 49'	8.9241		8.9256		1.0744	9.9985	85° 11'
		15		16			
4° 50'	8.9256		8.9272		1.0728	9.9985	85° 10'
		15		15			
4° 51'	8.9271		8.9287		1.0713	9.9984	85° 9'
		15		15			
4° 52'	8.9286		8.9302		1.0698	9.9984	85° 8'
		15		14			
4° 53'	8.9301		8.9316		1.0684	9.9984	85° 7'
		14		15			
4° 54'	8.9315		8.9331		1.0669	9.9984	85° 6'
		15		15			
4° 55'	8.9330		8.9346		1.0654	9.9984	85° 5'
		15		15			
4° 56'	8.9345		8.9361		1.0639	9.9984	85° 4'
		14		15			
4° 57'	8.9359		8.9376		1.0624	9.9984	85° 3'
		15		14			
4° 58'	8.9374		8.9390		1.0610	9.9984	85° 2'
		15		15			
4° 59'	8.9388		8.9405		1.0595	9.9984	85° 1'
		15		15			
4° 60'	8.9403		8.9420		1.0580	9.9983	85° 0'
	log cos	diff. 1'	log cot	com. diff. 1'	log tan	log sin	Angle

85°

5°–15°

Angle	log sin	diff.1′	log tan	com. diff.1′	log cot	log cos	diff.1′	Angle
5° 0′	8.9403	14.2	8.9420	14.3	1.0580	9.9983	.1	85° 0′
5° 10′	8.9545	13.7	8.9563	13.8	1.0437	9.9982	.1	84° 50′
5° 20′	8.9682	13.4	8.9701	13.5	1.0299	9.9981	.1	84° 40′
5° 30′	8.9816	12.9	8.9836	13.0	1.0164	9.9980	.1	84° 30′
5° 40′	8.9945	12.5	8.9966	12.7	1.0034	9.9979	.2	84° 20′
5° 50′	9.0070	12.2	9.0093	12.3	0.9907	9.9977	.1	84° 10′
6° 0′	9.0192	11.9	9.0216	12.0	0.9784	9.9976	.1	84° 0′
6° 10′	9.0311	11.5	9.0336	11.7	0.9664	9.9975	.2	83° 50′
6° 20′	9.0426	11.3	9.0453	11.4	0.9547	9.9973	.1	83° 40′
6° 30′	9.0539	10.9	9.0567	11.1	0.9433	9.9972	.1	83° 30′
6° 40′	9.0648	10.7	9.0678	10.8	0.9322	9.9971	.1	83° 20′
6° 50′	9.0755	10.4	9.0786	10.5	0.9214	9.9969	.2	83° 10′
7° 0′	9.0859	10.2	9.0891	10.4	0.9109	9.9968	.1	83° 0′
7° 10′	9.0961	9.9	9.0995	10.1	0.9005	9.9966	.2	82° 50′
7° 20′	9.1060	9.7	9.1096	9.8	0.8904	9.9964	.2	82° 40′
7° 30′	9.1157	9.5	9.1194	9.7	0.8806	9.9963	.1	82° 30′
7° 40′	9.1252	9.3	9.1291	9.4	0.8709	9.9961	.2	82° 20′
7° 50′	9.1345	9.1	9.1385	9.3	0.8615	9.9959	.2	82° 10′
8° 0′	9.1436	8.9	9.1478	9.1	0.8522	9.9958	.1	82° 0′
8° 10′	9.1525	8.7	9.1569	8.9	0.8431	9.9956	.2	81° 50′
8° 20′	9.1612	8.5	9.1658	8.7	0.8342	9.9954	.2	81° 40′
8° 30′	9.1697	8.4	9.1745	8.6	0.8255	9.9952	.2	81° 30′
8° 40′	9.1781	8.2	9.1831	8.4	0.8169	9.9950	.2	81° 20′
8° 50′	9.1863	8.0	9.1915	8.2	0.8085	9.9948	.2	81° 10′
9° 0′	9.1943	7.9	9.1997	8.1	0.8003	9.9946	.2	81° 0′
9° 10′	9.2022	7.8	9.2078	8.0	0.7922	9.9944	.2	80° 50′
9° 20′	9.2100	7.6	9.2158	7.8	0.7842	9.9942	.2	80° 40′
9° 30′	9.2176	7.5	9.2236	7.7	0.7764	9.9940	.2	80° 30′
9° 40′	9.2251	7.3	9.2313	7.6	0.7687	9.9938	.2	80° 20′
9° 50′	9.2324	7.3	9.2389	7.4	0.7611	9.9936	.2	80° 10′
10° 0′	9.2397	7.1	9.2463	7.3	0.7537	9.9934	.3	80° 0′
10° 10′	9.2468	7.0	9.2536	7.3	0.7464	9.9931	.2	79° 50′
10° 20′	9.2538	6.8	9.2609	7.1	0.7391	9.9929	.2	79° 40′
10° 30′	9.2606	6.8	9.2680	7.0	0.7320	9.9927	.3	79° 30′
10° 40′	9.2674	6.6	9.2750	6.9	0.7250	9.9924	.2	79° 20′
10° 50′	9.2740	6.6	9.2819	6.8	0.7181	9.9922	.3	79° 10′
11° 0′	9.2806	6.4	9.2887	6.6	0.7113	9.9919	.2	79° 0′
11° 10′	9.2870	6.4	9.2953	6.7	0.7047	9.9917	.3	78° 50′
11° 20′	9.2934	6.3	9.3020	6.5	0.6980	9.9914	.2	78° 40′
11° 30′	9.2997	6.1	9.3085	6.4	0.6915	9.9912	.3	78° 30′
11° 40′	9.3058	6.1	9.3149	6.3	0.6851	9.9909	.2	78° 20′
11° 50′	9.3119	6.0	9.3212	6.3	0.6788	9.9907	.3	78° 10′
12° 0′	9.3179	5.9	9.3275	6.1	0.6725	9.9904	.3	78° 0′
12° 10′	9.3238	5.8	9.3336	6.1	0.6664	9.9901	.2	77° 50′
12° 20′	•9.3296	5.7	9.3397	6.1	0.6603	9.9899	.3	77° 40′
12° 30′	9.3353	5.7	9.3458	5.9	0.6542	9.9896	.3	77° 30′
12° 40′	9.3410	5.6	9.3517	5.9	0.6483	9.9893	.3	77° 20′
12° 50′	9.3466	5.5	9.3576	5.8	0.6424	9.9890	.3	77° 10′
13° 0′	9.3521	5.4	9.3634	5.7	0.6366	9.9887	.3	77° 0′
13° 10′	9.3575	5.4	9.3691	5.7	0.6309	9.9884	.3	76° 50′
13° 20′	9.3629	5.3	9.3748	5.6	0.6252	9.9881	.3	76° 40′
13° 30′	9.3682	5.2	9.3804	5.5	0.6196	9.9878	.3	76° 30′
13° 40′	9.3734	5.2	9.3859	5.5	0.6141	9.9875	.3	76° 20′
13° 50′	9.3786	5.1	9.3914	5.4	0.6086	9.9872	.3	76° 10′
14° 0′	9.3837	5.0	9.3968	5.3	0.6032	9.9869	.3	76° 0′
14° 10′	9.3887	5.0	9.4021	5.3	0.5979	9.9866	.3	75° 50′
14° 20′	9.3937	4.9	9.4074	5.3	0.5926	9.9863	.4	75° 40′
14° 30′	9.3986	4.9	9.4127	5.1	0.5873	9.9859	.3	75° 30′
14° 40′	9.4035	4.8	9.4178	5.2	0.5822	9.9856	.3	75° 20′
14° 50′	9.4083	4.7	9.4230	5.1	0.5770	9.9853	.4	75° 10′
15° 0′	9.4130		9.4281		0.5719	9.9849		75° 0′
	log cos	diff.1′	log cot	com. diff.1′	log tan	log sin	diff.1′	Angle

75°–85°

15°–25°

Angle	log sin	diff.1	log tan	com. diff.1′	log cot	log cos	diff.1′	Angle
15° 0′	9.4130		9.4281		0.5719	9.9849		75° 0′
15° 10′	9.4177	4.7	9.4331	5.0	0.5669	9.9846	.3	74° 50′
15° 20′	9.4223	4.6	9.4381	5.0	0.5619	9.9843	.3	74° 40′
15° 30′	9.4269	4.6	9.4430	4.9	0.5570	9.9839	.4	74° 30′
15° 40′	9.4314	4.5	9.4479	4.9	0.5521	9.9836	.3	74° 20′
15° 50′	9.4359	4.5	9.4527	4.8	0.5473	9.9832	.4	74° 10′
16° 0′	9.4403	4.4	9.4575	4.8	0.5425	9.9828	.4	74° 0′
16° 10′	9.4447	4.4	9.4622	4.7	0.5378	9.9825	.3	73° 50′
16° 20′	9.4491	4.4	9.4669	4.7	0.5331	9.9821	.4	73° 40′
16° 30′	9.4533	4.2	9.4716	4.7	0.5284	9.9817	.4	73° 30′
16° 40′	9.4576	4.3	9.4762	4.6	0.5238	9.9814	.3	73° 20′
16° 50′	9.4618	4.2	9.4808	4.6	0.5192	9.9810	.4	73° 10′
17° 0′	9.4659	4.1	9.4853	4.5	0.5147	9.9806	.4	73° 0′
17° 10′	9.4700	4.1	9.4898	4.5	0.5102	9.9802	.4	72° 50′
17° 20′	9.4741	4.1	9.4943	4.5	0.5057	9.9798	.4	72° 40′
17° 30′	9.4781	4.0	9.4987	4.4	0.5013	9.9794	.4	72° 30′
17° 40′	9.4821	4.0	9.5031	4.4	0.4969	9.9790	.4	72° 20′
17° 50′	9.4861	4.0	9.5075	4.4	0.4925	9.9786	.4	72° 10′
18° 0′	9.4900	3.9	9.5118	4.3	0.4882	9.9782	.4	72° 0′
18° 10′	9.4939	3.9	9.5161	4.3	0.4839	9.9778	.4	71° 50′
18° 20′	9.4977	3.8	9.5203	4.2	0.4797	9.9774	.4	71° 40′
18° 30′	9.5015	3.8	9.5245	4.2	0.4755	9.9770	.4	71° 30′
18° 40′	9.5052	3.7	9.5287	4.2	0.4713	9.9765	.5	71° 20′
18° 50′	9.5090	3.8	9.5329	4.2	0.4671	9.9761	.4	71° 10′
19° 0′	9.5126	3.6	9.5370	4.1	0.4630	9.9757	.4	71° 0′
19° 10′	9.5163	3.7	9.5411	4.1	0.4589	9.9752	.5	70° 50′
19° 20′	9.5199	3.6	9.5451	4.0	0.4549	9.9748	.4	70° 40′
19° 30′	9.5235	3.6	9.5491	4.0	0.4509	9.9743	.5	70° 30′
19° 40′	9.5270	3.5	9.5531	4.0	0.4469	9.9739	.4	70° 20′
19° 50′	9.5306	3.6	9.5571	4.0	0.4429	9.9734	.5	70° 10′
20° 0′	9.5341	3.5	9.5611	4.0	0.4389	9.9730	.4	70° 0′
20° 10′	9.5375	3.4	9.5650	3.9	0.4350	9.9725	.5	69° 50′
20° 20′	9.5409	3.4	9.5689	3.9	0.4311	9.9721	.4	69° 40′
20° 30′	9.5443	3.4	9.5727	3.8	0.4273	9.9716	.5	69° 30′
20° 40′	9.5477	3.4	9.5766	3.9	0.4234	9.9711	.5	69° 20′
20° 50′	9.5510	3.3	9.5804	3.8	0.4196	9.9706	.5	69° 10′
21° 0′	9.5543	3.3	9.5842	3.8	0.4158	9.9702	.4	69° 0′
21° 10′	9.5576	3.3	9.5879	3.7	0.4121	9.9697	.5	68° 50′
21° 20′	9.5609	3.3	9.5917	3.8	0.4083	9.9692	.5	68° 40′
21° 30′	9.5641	3.2	9.5954	3.7	0.4046	9.9687	.5	68° 30′
21° 40′	9.5673	3.2	9.5991	3.7	0.4009	9.9682	.5	68° 20′
21° 50′	9.5704	3.1	9.6028	3.7	0.3972	9.9677	.5	68° 10′
22° 0′	9.5736	3.2	9.6064	3.6	0.3936	9.9672	.5	68° 0′
22° 10′	9.5767	3.1	9.6100	3.6	0.3900	9.9667	.5	67° 50′
22° 20′	9.5798	3.1	9.6136	3.6	0.3864	9.9661	.6	67° 40′
22° 30′	9.5828	3.0	9.6172	3.6	0.3828	9.9656	.5	67° 30′
22° 40′	9.5859	3.1	9.6208	3.6	0.3792	9.9651	.5	67° 20′
22° 50′	9.5889	3.0	9.6243	3.5	0.3757	9.9646	.5	67° 10′
23° 0′	9.5919	3.0	9.6279	3.6	0.3721	9.9640	.6	67° 0′
23° 10′	9.5948	2.9	9.6314	3.5	0.3686	9.9635	.5	66° 50′
23° 20′	9.5978	3.0	9.6348	3.4	0.3652	9.9629	.6	66° 40′
23° 30′	9.6007	2.9	9.6383	3.5	0.3617	9.9624	.5	66° 30′
23° 40′	9.6036	2.9	9.6417	3.4	0.3583	9.9618	.6	66° 20′
23° 50′	9.6065	2.9	9.6452	3.5	0.3548	9.9613	.5	66° 10′
24° 0′	9.6093	2.8	9.6486	3.4	0.3514	9.9607	.6	66° 0′
24° 10′	9.6121	2.8	9.6520	3.4	0.3480	9.9602	.5	65° 50′
24° 20′	9.6149	2.8	9.6553	3.3	0.3447	9.9596	.6	65° 40′
24° 30′	9.6177	2.8	9.6587	3.4	0.3413	9.9590	.6	65° 30′
24° 40′	9.6205	2.8	9.6620	3.3	0.3380	9.9584	.6	65° 20′
24° 50′	9.6232	2.7	9.6654	3.4	0.3346	9.9579	.5	65° 10′
25° 0′	9.6259	2.7	9.6687	3.3	0.3313	9.9573	.6	65° 0′
	log cos	diff.1′	log cot	com. diff.1′	log tan	log sin	diff.1′	Angle

65°–75°

25°–35°

Angle	log sin	diff.1′	log tan	com. diff.1′	log cot	log cos	diff.1′	Angle
25° 0′	9.6259	2.7	9.6687	3.3	0.3313	9.9573	.6	65° 0′
25° 10′	9.6286	2.7	9.6720	3.2	0.3280	9.9567	.6	64° 50′
25° 20′	9.6313	2.7	9.6752	3.3	0.3248	9.9561	.6	64° 40′
25° 30′	9.6340	2.6	9.6785	3.2	0.3215	9.9555	.6	64° 30′
25° 40′	9.6366	2.6	9.6817	3.3	0.3183	9.9549	.6	64° 20′
25° 50′	9.6392	2.6	9.6850	3.2	0.3150	9.9543	.6	64° 10′
26° 0′	9.6418	2.6	9.6882	3.2	0.3118	9.9537	.7	64° 0′
26° 10′	9.6444	2.6	9.6914	3.2	0.3086	9.9530	.6	63° 50′
26° 20′	9.6470	2.6	9.6946	3.1	0.3054	9.9524	.6	63° 40′
26° 30′	9.6495	2.5	9.6977	3.2	0.3023	9.9518	.6	63° 30′
26° 40′	9.6521	2.6	9.7009	3.1	0.2991	9.9512	.7	63° 20′
26° 50′	9.6546	2.5	9.7040	3.2	0.2960	9.9505	.6	63° 10′
27° 0′	9.6570	2.4	9.7072	3.1	0.2928	9.9499	.7	63° 0′
27° 10′	9.6595	2.5	9.7103	3.1	0.2897	9.9492	.6	62° 50′
27° 20′	9.6620	2.5	9.7134	3.1	0.2866	9.9486	.7	62° 40′
27° 30′	9.6644	2.4	9.7165	3.1	0.2835	9.9479	.6	62° 30′
27° 40′	9.6668	2.4	9.7196	3.1	0.2804	9.9473	.7	62° 20′
27° 50′	9.6692	2.4	9.7226	3.0	0.2774	9.9466	.7	62° 10′
28° 0′	9.6716	2.4	9.7257	3.1	0.2743	9.9459	.6	62° 0′
28° 10′	9.6740	2.4	9.7287	3.0	0.2713	9.9453	.7	61° 50′
28° 20′	9.6763	2.3	9.7317	3.0	0.2683	9.9446	.7	61° 40′
28° 30′	9.6787	2.4	9.7348	3.1	0.2652	9.9439	.7	61° 30′
28° 40′	9.6810	2.3	9.7378	3.0	0.2622	9.9432	.7	61° 20′
28° 50′	9.6833	2.3	9.7408	3.0	0.2592	9.9425	.7	61° 10′
29° 0′	9.6856	2.3	9.7438	3.0	0.2562	9.9418	.7	61° 0′
29° 10′	9.6878	2.2	9.7467	2.9	0.2533	9.9411	.7	60° 50′
29° 20′	9.6901	2.3	9.7497	3.0	0.2503	9.9404	.7	60° 40′
29° 30′	9.6923	2.2	9.7526	2.9	0.2474	9.9397	.7	60° 30′
29° 40′	9.6946	2.3	9.7556	3.0	0.2444	9.9390	.7	60° 20′
29° 50′	9.6968	2.2	9.7585	2.9	0.2415	9.9383	.7	60° 10′
30° 0′	9.6990	2.2	9.7614	2.9	0.2386	9.9375	.8	60° 0′
30° 10′	9.7012	2.2	9.7644	2.9	0.2356	9.9368	.7	59° 50′
30° 20′	9.7033	2.1	9.7673	2.9	0.2327	9.9361	.7	59° 40′
30° 30′	9.7055	2.2	9.7701	2.8	0.2299	9.9353	.8	59° 30′
30° 40′	9.7076	2.1	9.7730	2.9	0.2270	9.9346	.7	59° 20′
30° 50′	9.7097	2.1	9.7759	2.9	0.2241	9.9338	.8	59° 10′
31° 0′	9.7118	2.1	9.7788	2.8	0.2212	9.9331	.7	59° 0′
31° 10′	9.7139	2.1	9.7816	2.9	0.2184	9.9323	.8	58° 50′
31° 20′	9.7160	2.1	9.7845	2.8	0.2155	9.9315	.8	58° 40′
31° 30′	9.7181	2.1	9.7873	2.9	0.2127	9.9308	.7	58° 30′
31° 40′	9.7201	2.0	9.7902	2.8	0.2098	9.9300	.8	58° 20′
31° 50′	9.7222	2.1	9.7930	2.8	0.2070	9.9292	.8	58° 10′
32° 0′	9.7242	2.0	9.7958	2.8	0.2042	9.9284	.8	58° 0′
32° 10′	9.7262	2.0	9.7986	2.8	0.2014	9.9276	.8	57° 50′
32° 20′	9.7282	2.0	9.8014	2.8	0.1986	9.9268	.8	57° 40′
32° 30′	9.7302	2.0	9.8042	2.8	0.1958	9.9260	.8	57° 30′
32° 40′	9.7322	2.0	9.8070	2.8	0.1930	9.9252	.8	57° 20′
32° 50′	9.7342	2.0	9.8097	2.7	0.1903	9.9244	.8	57° 10′
33° 0′	9.7361	1.9	9.8125	2.8	0.1875	9.9236	.8	57° 0′
33° 10′	9.7380	1.9	9.8153	2.8	0.1847	9.9228	.9	56° 50′
33° 20′	9.7400	2.0	9.8180	2.7	0.1820	9.9219	.8	56° 40′
33° 30′	9.7419	1.9	9.8208	2.8	0.1792	9.9211	.8	56° 30′
33° 40′	9.7438	1.9	9.8235	2.7	0.1765	9.9203	.9	56° 20′
33° 50′	9.7457	1.9	9.8263	2.8	0.1737	9.9194	.8	56° 10′
34° 0′	9.7476	1.9	9.8290	2.7	0.1710	9.9186	.9	56° 0′
34° 10′	9.7494	1.8	9.8317	2.7	0.1683	9.9177	.9	55° 50′
34° 20′	9.7513	1.9	9.8344	2.7	0.1656	9.9169	.8	55° 40′
34° 30′	9.7531	1.8	9.8371	2.7	0.1629	9.9160	.9	55° 30′
34° 40′	9.7550	1.9	9.8398	2.7	0.1602	9.9151	.9	55° 20′
34° 50′	9.7568	1.8	9.8425	2.7	0.1575	9.9142	.9	55° 10′
35° 0′	9.7586		9.8452		0.1548	9.9134	.8	55° 0′
	log cos	diff.1′	log cot	com. diff.1′	log tan	log sin	diff.1′	Angle

55°–65°

35°–45°

Angle	log sin	diff. 1′	log tan	com. diff. 1′	log cot	log cos	diff. 1′	
35° 0′	9.7586	1.8	9.8452	2.7	0.1548	9.9134	.9	55° 0′
35° 10′	9.7604	1.8	9.8479	2.7	0.1521	9.9125	.9	54° 50′
35° 20′	9.7622	1.8	9.8506	2.7	0.1494	9.9116	.9	54° 40′
35° 30′	9.7640	1.7	9.8533	2.6	0.1467	9.9107	.9	54° 30′
35° 40′	9.7657	1.8	9.8559	2.7	0.1441	9.9098	.9	54° 20′
35° 50′	9.7675	1.7	9.8586	2.7	0.1414	9.9089	.9	54° 10′
36° 0′	9.7692	1.8	9.8613	2.6	0.1387	9.9080	1.0	54° 0′
36° 10′	9.7710	1.7	9.8639	2.7	0.1361	9.9070	.9	53° 50′
36° 20′	9.7727	1.7	9.8666	2.6	0.1334	9.9061	.9	53° 40′
36° 30′	9.7744	1.7	9.8692	2.6	0.1308	9.9052	1.0	53° 30′
36° 40′	9.7761	1.7	9.8718	2.7	0.1282	9.9042	.9	53° 20′
36° 50′	9.7778	1.7	9.8745	2.6	0.1255	9.9033	1.0	53° 10′
37° 0′	9.7795	1.6	9.8771	2.6	0.1229	9.9023	.9	53° 0′
37° 10′	9.7811	1.7	9.8797	2.7	0.1203	9.9014	1.0	52° 50′
37° 20′	9.7828	1.6	9.8824	2.6	0.1176	9.9004	.9	52° 40′
37° 30′	9.7844	1.7	9.8850	2.6	0.1150	9.8995	1.0	52° 30′
37° 40′	9.7861	1.6	9.8876	2.6	0.1124	9.8985	1.0	52° 20′
37° 50′	9.7877	1.6	9.8902	2.6	0.1098	9.8975	1.0	52° 10′
38° 0′	9.7893	1.7	9.8928	2.6	0.1072	9.8965	1.0	52° 0′
38° 10′	9.7910	1.6	9.8954	2.6	0.1046	9.8955	1.0	51° 50′
38° 20′	9.7926	1.5	9.8980	2.6	0.1020	9.8945	1.0	51° 40′
38° 30′	9.7941	1.6	9.9006	2.6	0.0994	9.8935	1.0	51° 30′
38° 40′	9.7957	1.6	9.9032	2.6	0.0968	9.8925	1.0	51° 20′
38° 50′	9.7973	1.6	9.9058	2.6	0.0942	9.8915	1.0	51° 10′
39° 0′	9.7989	1.5	9.9084	2.6	0.0916	9.8905	1.0	51° 0′
39° 10′	9.8004	1.6	9.9110	2.5	0.0890	9.8895	1.1	50° 50′
39° 20′	9.8020	1.5	9.9135	2.6	0.0865	9.8884	1.0	50° 40′
39° 30′	9.8035	1.5	9.9161	2.6	0.0839	9.8874	1.0	50° 30′
39° 40′	9.8050	1.6	9.9187	2.5	0.0813	9.8864	1.1	50° 20′
39° 50′	9.8066	1.5	9.9212	2.6	0.0788	9.8853	1.0	50° 10′
40° 0′	9.8081	1.5	9.9238	2.6	0.0762	9.8843	1.1	50° 0′
40° 10′	9.8096	1.5	9.9264	2.5	0.0736	9.8832	1.1	49° 50′
40° 20′	9.8111	1.4	9.9289	2.6	0.0711	9.8821	1.1	49° 40′
40° 30′	9.8125	1.5	9.9315	2.6	0.0685	9.8810	1.0	49° 30′
40° 40′	9.8140	1.5	9.9341	2.5	0.0659	9.8800	1.1	49° 20′
40° 50′	9.8155	1.4	9.9366	2.6	0.0634	9.8789	1.1	49° 10′
41° 0′	9.8169	1.5	9.9392	2.5	0.0608	9.8778	1.1	49° 0′
41° 10′	9.8184	1.4	9.9417	2.6	0.0583	9.8767	1.1	48° 50′
41° 20′	9.8198	1.5	9.9443	2.5	0.0557	9.8756	1.1	48° 40′
41° 30′	9.8213	1.4	9.9468	2.6	0.0532	9.8745	1.2	48° 30′
41° 40′	9.8227	1.4	9.9494	2.5	0.0506	9.8733	1.1	48° 20′
41° 50′	9.8241	1.4	9.9519	2.5	0.0481	9.8722	1.1	48° 10′
42° 0′	9.8255	1.4	9.9544	2.6	0.0456	9.8711	1.2	48° 0′
42° 10′	9.8269	1.4	9.9570	2.5	0.0430	9.8699	1.1	47° 50′
42° 20′	9.8283	1.4	9.9595	2.6	0.0405	9.8688	1.2	47° 40′
42° 30′	9.8297	1.4	9.9621	2.5	0.0379	9.8676	1.1	47° 30′
42° 40′	9.8311	1.3	9.9646	2.5	0.0354	9.8665	1.2	47° 20′
42° 50′	9.8324	1.4	9.9671	2.6	0.0329	9.8653	1.2	47° 10′
43° 0′	9.8338	1.3	9.9697	2.5	0.0303	9.8641	1.2	47° 0′
43° 10′	9.8351	1.4	9.9722	2.5	0.0278	9.8629	1.1	46° 50′
43° 20′	9.8365	1.3	9.9747	2.5	0.0253	9.8618	1.2	46° 40′
43° 30′	9.8378	1.3	9.9772	2.6	0.0228	9.8606	1.2	46° 30′
43° 40′	9.8391	1.4	9.9798	2.5	0.0202	9.8594	1.2	46° 20′
43° 50′	9.8405	1.3	9.9823	2.5	0.0177	9.8582	1.3	46° 10′
44° 0′	9.8418	1.3	9.9848	2.6	0.0152	9.8569	1.2	46° 0′
44° 10′	9.8431	1.3	9.9874	2.5	0.0126	9.8557	1.2	45° 50′
44° 20′	9.8444	1.3	9.9899	2.5	0.0101	9.8545	1.3	45° 40′
44° 30′	9.8457	1.2	9.9924	2.5	0.0076	9.8532	1.2	45° 30′
44° 40′	9.8469	1.3	9.9949	2.6	0.0051	9.8520	1.3	45° 20′
44° 50′	9.8482	1.3	9.9975	2.5	0.0025	9.8507	1.2	45° 10′
45° 0′	9.8495		0.0000		0.0000	9.8495		45° 0′
	log cos	diff. 1′	log cot	com. diff. 1′	log tan	log sin	diff. 1′	Angle

45°–55°

To change from Minutes and Seconds into the Decimal
Parts of a Degree or into Radians

From seconds	From minutes	From degrees into radians
1″ = 0.00028° = 0.0000048 Rad.	1′ = 0.017° = 0.00029 Rad.	1° = 0.01745 Rad.
2″ = 0.00056° = 0.0000097 "	2′ = 0.033° = 0.00058 "	2° = 0.03491 "
3″ = 0.00083° = 0.0000145 "	3′ = 0.050° = 0.00087 "	3° = 0.05236 "
4″ = 0.00111° = 0.0000194 "	4′ = 0.067° = 0.00116 "	4° = 0.06981 "
5″ = 0.00139° = 0.0000242 "	5′ = 0.083° = 0.00145 "	5° = 0.08727 "
6″ = 0.00167° = 0.0000291 "	6′ = 0.100° = 0.00175 "	6° = 0.10472 "
7″ = 0.00194° = 0.0000339 "	7′ = 0.117° = 0.00204 "	7° = 0.12217 "
8″ = 0.00222° = 0.0000388 "	8′ = 0.133° = 0.00233 "	8° = 0.13963 "
9″ = 0.00250° = 0.0000436 "	9′ = 0.150° = 0.00262 "	9° = 0.15708 "
10″ = 0.00278° = 0.0000485 "	10′ = 0.167° = 0.00291 "	10° = 0.17453 "
20″ = 0.00556° = 0.0000970 "	20′ = 0.333° = 0.00582 "	20° = 0.34907 "
30″ = 0.00833° = 0.0001454 "	30′ = 0.500° = 0.00873 "	30° = 0.52360 "
40″ = 0.01111° = 0.0001939 "	40′ = 0.667° = 0.01164 "	40° = 0.69813 "
50″ = 0.01389° = 0.0002424 "	50′ = 0.833° = 0.01454 "	50° = 0.87266 "

To change from Decimal Parts of a Degree into Minutes
and Seconds

0.0000° =	0.000′ =	0″	0.20° = 12.0′ = 12′	0.60° = 36.0′ = 36′
0.0001° =	0.006′ =	0.36″	0.21° = 12.6′ = 12′ 36″	0.61° = 36.6′ = 36′ 36″
0.0002° =	0.012′ =	0.72″	0.22° = 13.2′ = 13′ 12″	0.62° = 37.2′ = 37′ 12″
0.0003° =	0.018′ =	1.08″	0.23° = 13.8′ = 13′ 48″	0.63° = 37.8′ = 37′ 48″
0.0004° =	0.024′ =	1.44″	0.24° = 14.4′ = 14′ 24″	0.64° = 38.4′ = 38′ 24″
0.0005° =	0.030′ =	1.80″	0.25° = 15.0′ = 15′	0.65° = 39.0′ = 39′
0.0006° =	0.036′ =	2.16″	0.26° = 15.6′ = 15′ 36″	0.66° = 39.6′ = 39′ 36″
0.0007° =	0.042′ =	2.52″	0.27° = 16.2′ = 16′ 12″	0.67° = 40.2′ = 40′ 12″
0.0008° =	0.048′ =	2.88″	0.28° = 16.8′ = 16′ 48″	0.68° = 40.8′ = 40′ 48″
0.0009° =	0.054′ =	3.24″	0.29° = 17.4′ = 17′ 24″	0.69° = 41.4′ = 41′ 24″
0.0010° =	0.060′ =	3.60″	0.30° = 18.0′ = 18′	0.70° = 42.0′ = 42′
0.001° =	0.06′ =	3.6″	0.31° = 18.6′ = 18′ 36″	0.71° = 42.6′ = 42′ 36″
0.002° =	0.12′ =	7.2″	0.32° = 19.2′ = 19′ 12″	0.72° = 43.2′ = 43′ 12″
0.003° =	0.18′ =	10.8″	0.33° = 19.8′ = 19′ 48″	0.73° = 43.8′ = 43′ 48″
0.004° =	0.24′ =	14.4″	0.34° = 20.4′ = 20′ 24″	0.74° = 44.4′ = 44′ 24″
0.005° =	0.30′ =	18.0″	0.35° = 21.0′ = 21′	0.75° = 45.0′ = 45′
0.006° =	0.36′ =	21.6″	0.36° = 21.6′ = 21′ 36″	0.76° = 45.6′ = 45′ 36″
0.007° =	0.42′ =	25.2″	0.37° = 22.2′ = 22′ 12″	0.77° = 46.2′ = 46′ 12″
0.008° =	0.48′ =	28.8″	0.38° = 22.8′ = 22′ 48″	0.78° = 46.8′ = 46′ 48″
0.009° =	0.54′ =	32.4″	0.39° = 23.4′ = 23′ 24″	0.79° = 47.4′ = 47′ 24″
0.010° =	0.60′ =	36.0″	0.40° = 24.0′ = 24′	0.80° = 48.0′ = 48′
0.01° =	0.6′ =	36″	0.41° = 24.6′ = 24′ 36″	0.81° = 48.6′ = 48′ 36″
0.02° =	1.2′ =	1′ 12″	0.42° = 25.2′ = 25′ 12″	0.82° = 49.2′ = 49′ 12″
0.03° =	1.8′ =	1′ 48″	0.43° = 25.8′ = 25′ 48″	0.83° = 49.8′ = 49′ 48″
0.04° =	2.4′ =	2′ 24″	0.44° = 26.4′ = 26′ 24″	0.84° = 50.4′ = 50′ 24″
0.05° =	3.0′ =	3′	0.45° = 27.0′ = 27′	0.85° = 51.0′ = 51′
0.06° =	3.6′ =	3′ 36″	0.46° = 27.6′ = 27′ 36″	0.86° = 51.6′ = 51′ 36″
0.07° =	4.2′ =	4′ 12″	0.47° = 28.2′ = 28′ 12″	0.87° = 52.2′ = 52′ 12″
0.08° =	4.8′ =	4′ 48″	0.48° = 28.8′ = 28′ 48″	0.88° = 52.8′ = 52′ 48″
0.09° =	5.4′ =	5′ 24″	0.49° = 29.4′ = 29′ 24″	0.89° = 53.4′ = 53′ 24″
0.10° =	6.0′ =	6′	0.50° = 30.0′ = 30′	0.90° = 54.0′ = 54′
0.11° =	6.6′ =	6′ 36″	0.51° = 30.6′ = 30′ 36″	0.91° = 54.6′ = 54′ 36″
0.12° =	7.2′ =	7′ 12″	0.52° = 31.2′ = 31′ 12″	0.92° = 55.2′ = 55′ 12″
0.13° =	7.8′ =	7′ 48″	0.53° = 31.8′ = 31′ 48″	0.93° = 55.8′ = 55′ 48″
0.14° =	8.4′ =	8′ 24″	0.54° = 32.4′ = 32′ 24″	0.94° = 56.4′ = 56′ 24″
0.15° =	9.0′ =	9′	0.55° = 33.0′ = 33′	0.95° = 57.0′ = 57′
0.16° =	9.6′ =	9′ 36″	0.56° = 33.6′ = 33′ 36″	0.96° = 57.6′ = 57′ 36″
0.17° =	10.2′ =	10′ 12″	0.57° = 34.2′ = 34′ 12″	0.97° = 58.2′ = 58′ 12″
0.18° =	10.8′ =	10′ 48″	0.58° = 34.8′ = 34′ 48″	0.98° = 58.8′ = 58′ 48″
0.19° =	11.4′ =	11′ 24″	0.59° = 35.4′ = 35′ 24″	0.99° = 59.4′ = 59′ 24″
0.20° =	12.0′ =	12′	0.60° = 36.0′ = 36′	1.00° = 60.0′ = 60′

FOUR–PLACE LOGARITHMS OF TRIGONOMETRIC FUNCTIONS, THE ANGLE BEING EXPRESSED IN DEGREES AND THE DECIMAL PART OF A DEGREE

This table gives the common logarithms (base 10) of the sines, cosines, tangents, and cotangents of all angles from 0° to 5°, and from 85° to 90° for every hundredth part of a degree, and from 5° to 85° for every tenth of a degree, all calculated to four places of decimals. In order to avoid the printing of negative characteristics, the number 10 has been added to every logarithm in the first, second, and fourth columns (those having **log sin, log tan,** and **log cos** at the top). Hence in writing down any logarithm taken from these three columns − 10 should be written after it. Logarithms taken from the third column (having **log cot** at the top) should be used as printed.

A logarithm found from this table by interpolation may be in error by one unit in the last decimal place, except for angles between 0° and 0.3° or between 89.7° and 90°, when the error may be larger. In the latter cases the table refers the student to the formulas on page 6 for more accurate results.

0°

Angle	log sin	diff.	log tan	com. diff.	log cot	log cos	
0.00°	——		——		——	10.0000	**90.00°**
0.01°	6.2419		6.2419		3.7581	10.0000	89.99°
0.02°	6.5429		6.5429		3.4571	10.0000	89.98°
0.03°	6.7190		6.7190		3.2810	10.0000	89.97°
0.04°	6.8439		6.8439		3.1561	10.0000	89.96°
0.05°	6.9408		6.9408		3.0592	10.0000	89.95°
0.06°	7.0200		7.0200		2.9800	10.0000	89.94°
0.07°	7.0870		7.0870		2.9130	10.0000	89.93°
0.08°	7.1450		7.1450		2.8550	10.0000	89.92°
0.09°	7.1961		7.1961		2.8039	10.0000	89.91°
0.10°	7.2419		7.2419		2.7581	10.0000	**89.90°**
0.11°	7.2833		7.2833		2.7167	10.0000	89.89°
0.12°	7.3211		7.3211		2.6789	10.0000	89.88°
0.13°	7.3558		7.3558		2.6442	10.0000	89.87°
0.14°	7.3880		7.3880		2.6120	10.0000	89.86°
0.15°	7.4180		7.4180		2.5820	10.0000	89.85°
0.16°	7.4460		7.4460		2.5540	10.0000	89.84°
0.17°	7.4723		7.4723		2.5277	10.0000	89.83°
0.18°	7.4971		7.4972		2.5028	10.0000	89.82°
0.19°	7.5206		7.5206		2.4794	10.0000	89.81°
0.20°	7.5429		7.5429		2.4571	10.0000	**89.80°**
0.21°	7.5641		7.5641		2.4359	10.0000	89.79°
0.22°	7.5843		7.5843		2.4157	10.0000	89.78°
0.23°	7.6036		7.6036		2.3964	10.0000	89.77°
0.24°	7.6221		7.6221		2.3779	10.0000	89.76°
0.25°	7.6398		7.6398		2.3602	10.0000	89.75°
0.26°	7.6568		7.6569		2.3431	10.0000	89.74°
0.27°	7.6732		7.6732		2.3268	10.0000	89.73°
0.28°	7.6890		7.6890		2.3110	10.0000	89.72°
0.29°	7.7043		7.7043		2.2957	10.0000	89.71°
0.30°	7.7190		7.7190		2.2810	10.0000	**89.70°**
0.31°	7.7332	142	7.7332	142	2.2668	10.0000	89.69°
0.32°	7.7470	138	7.7470	138	2.2530	10.0000	89.68°
0.33°	7.7604	134	7.7604	134	2.2396	10.0000	89.67°
0.34°	7.7734	130	7.7734	130	2.2266	10.0000	89.66°
0.35°	7.7859	125	7.7860	126	2.2140	10.0000	89.65°
0.36°	7.7982	123	7.7982	122	2.2018	10.0000	89.64°
0.37°	7.8101	119	7.8101	119	2.1899	10.0000	89.63°
0.38°	7.8217	116	7.8217	116	2.1783	10.0000	89.62°
0.39°	7.8329	112	7.8329	112	2.1671	10.0000	89.61°
0.40°	7.8439	110	7.8439	110	2.1561	10.0000	**89.60°**
0.41°	7.8547	108	7.8547	108	2.1453	10.0000	89.59°
0.42°	7.8651	104	7.8651	104	2.1349	10.0000	89.58°
0.43°	7.8753	102	7.8754	103	2.1246	10.0000	89.57°
0.44°	7.8853	100	7.8853	99	2.1147	10.0000	89.56°
0.45°	7.8951	98	7.8951	98	2.1049	10.0000	89.55°
0.46°	7.9046	95	7.9046	95	2.0954	10.0000	89.54°
0.47°	7.9140	94	7.9140	94	2.0860	10.0000	89.53°
0.48°	7.9231	91	7.9231	91	2.0769	10.0000	89.52°
0.49°	7.9321	90	7.9321	90	2.0678	10.0000	89.51°
0.50°	7.9408	87	7.9409	88	2.0591	10.0000	**89.50°**
	log cos	diff.	log cot	com. diff.	log tan	log sin	Angle

(In the diff. column, printed sideways:) Ordinary interpolation here will in general give inaccurate results. Instead use formulas on p. 6.

(In the com. diff. column, printed sideways:) Ordinary interpolation here will in general give inaccurate results. Instead use formulas on p. 6.

89°

Prop. Parts

(Extra digit — Difference)

Extra digit	79	78	77
1	7.9	7.8	7.7
2	15.8	15.6	15.4
3	23.7	23.4	23.1
4	31.6	31.2	30.8
5	39.5	39.0	38.5
6	47.4	46.8	46.2
7	55.3	54.6	53.9
8	63.2	62.4	61.6
9	71.1	70.2	69.3

	76	75	74
1	7.6	7.5	7.4
2	15.2	15.0	14.8
3	22.8	22.5	22.2
4	30.4	30.0	29.6
5	38.0	37.5	37.0
6	45.6	45.0	44.4
7	53.2	52.5	51.8
8	60.8	60.0	59.2
9	68.4	67.5	66.6

	73	72	71
1	7.3	7.2	7.1
2	14.6	14.4	14.2
3	21.9	21.6	21.3
4	29.2	28.8	28.4
5	36.5	36.0	35.5
6	43.8	43.2	42.6
7	51.1	50.4	49.7
8	58.4	57.6	56.8
9	65.7	64.8	63.9

	69	68	67
1	6.9	6.8	6.7
2	13.8	13.6	13.4
3	20.7	20.4	20.1
4	27.6	27.2	26.8
5	34.5	34.0	33.5
6	41.4	40.8	40.2
7	48.3	47.6	46.9
8	55.2	54.4	53.6
9	62.1	61.2	60.3

	66	65	64
1	6.6	6.5	6.4
2	13.2	13.0	12.8
3	19.8	19.5	19.2
4	26.4	26.0	25.6
5	33.0	32.5	32.0
6	39.6	39.0	38.4
7	46.2	45.5	44.8
8	52.8	52.0	51.2
9	59.4	58.5	57.6

	63	62	61
1	6.3	6.2	6.1
2	12.6	12.4	12.2
3	18.9	18.6	18.3
4	25.2	24.8	24.4
5	31.5	31.0	30.5
6	37.8	37.2	36.6
7	44.1	43.4	42.7
8	50.4	49.6	48.8
9	56.7	55.8	54.9

0°

Angle	log sin	diff.	log tan	com. diff.	log cot	log cos	
0.50°	7.9408	86	7.9409	86	2.0591	10.0000	**89.50°**
0.51°	7.9494	85	7.9495	84	2.0505	10.0000	89.49°
0.52°	7.9579	82	7.9579	83	2.0421	10.0000	89.48°
0.53°	7.9661	82	7.9662	81	2.0338	10.0000	89.47°
0.54°	7.9743	79	7.9743	80	2.0257	10.0000	89.46°
0.55°	7.9822	79	7.9823	78	2.0177	10.0000	89.45°
0.56°	7.9901	76	7.9901	77	2.0099	10.0000	89.44°
0.57°	7.9977	76	7.9978	75	2.0022	10.0000	89.43°
0.58°	8.0053	74	8.0053	74	1.9947	10.0000	89.42°
0.59°	8.0127	73	8.0127	73	1.9873	10.0000	89.41°
0.60°	8.0200	72	8.0200	72	1.9800	10.0000	**89.40°**
0.61°	8.0272	71	8.0272	71	1.9728	10.0000	89.39°
0.62°	8.0343	69	8.0343	69	1.9657	10.0000	89.38°
0.63°	8.0412	68	8.0412	69	1.9588	10.0000	89.37°
0.64°	8.0480	68	8.0481	67	1.9519	10.0000	89.36°
0.65°	8.0548	66	8.0548	66	1.9452	10.0000	89.35°
0.66°	8.0614	65	8.0614	66	1.9386	10.0000	89.34°
0.67°	8.0679	65	8.0680	64	1.9320	10.0000	89.33°
0.68°	8.0744	63	8.0744	63	1.9256	10.0000	89.32°
0.69°	8.0807	63	8.0807	63	1.9193	10.0000	89.31°
0.70°	8.0870	61	8.0870	62	1.9130	10.0000	**89.30°**
0.71°	8.0931	61	8.0932	60	1.9068	10.0000	89.29°
0.72°	8.0992	60	8.0992	60	1.9008	10.0000	89.28°
0.73°	8.1052	59	8.1052	59	1.8948	10.0000	89.27°
0.74°	8.1111	58	8.1111	59	1.8889	10.0000	89.26°
0.75°	8.1169	58	8.1170	57	1.8830	10.0000	89.25°
0.76°	8.1227	57	8.1227	57	1.8773	10.0000	89.24°
0.77°	8.1284	56	8.1284	56	1.8716	10.0000	89.23°
0.78°	8.1340	55	8.1340	55	1.8660	10.0000	89.22°
0.79°	8.1395	55	8.1395	55	1.8605	10.0000	89.21°
0.80°	8.1450	53	8.1450	54	1.8550	10.0000	**89.20°**
0.81°	8.1503	54	8.1504	53	1.8496	10.0000	89.19°
0.82°	8.1557	52	8.1557	53	1.8443	10.0000	89.18°
0.83°	8.1609	52	8.1610	52	1.8390	10.0000	89.17°
0.84°	8.1661	52	8.1662	51	1.8338	10.0000	89.16°
0.85°	8.1713	51	8.1713	51	1.8287	10.0000	89.15°
0.86°	8.1764	50	8.1764	50	1.8236	10.0000	89.14°
0.87°	8.1814	49	8.1814	50	1.8186	9.9999	89.13°
0.88°	8.1863	49	8.1864	49	1.8136	9.9999	89.12°
0.89°	8.1912	49	8.1913	49	1.8087	9.9999	89.11°
0.90°	8.1961	48	8.1962	48	1.8038	9.9999	**89.10°**
0.91°	8.2009	47	8.2010	47	1.7990	9.9999	89.09°
0.92°	8.2056	47	8.2057	47	1.7943	9.9999	89.08°
0.93°	8.2103	47	8.2104	46	1.7896	9.9999	89.07°
0.94°	8.2150	46	8.2150	46	1.7850	9.9999	89.06°
0.95°	8.2196	45	8.2196	46	1.7804	9.9999	89.05°
0.96°	8.2241	45	8.2242	45	1.7758	9.9999	89.04°
0.97°	8.2286	45	8.2287	44	1.7713	9.9999	89.03°
0.98°	8.2331	44	8.2331	45	1.7669	9.9999	89.02°
0.99°	8.2375	44	8.2376	43	1.7624	9.9999	89.01°
1.00°	8.2419		8.2419		1.7581	9.9999	**89.00°**
	log cos	diff.	log cot	com. diff.	log tan	log sin	Angle

Prop. Parts

Extra digit — Difference

	60	59	58
1	6.0	5.9	5.8
2	12.0	11.8	11.6
3	18.0	17.7	17.4
4	24.0	23.6	23.2
5	30.0	29.5	29.0
6	36.0	35.4	34.8
7	42.0	41.3	40.6
8	48.0	47.2	46.4
9	54.0	53.1	52.2

	57	56	55
1	5.7	5.6	5.5
2	11.4	11.2	11.0
3	17.1	16.8	16.5
4	22.8	22.4	22.0
5	28.5	28.0	27.5
6	34.2	33.6	33.0
7	39.9	39.2	38.5
8	45.6	44.8	44.0
9	51.3	50.4	49.5

	54	53	52
1	5.4	5.3	5.2
2	10.8	10.6	10.4
3	16.2	15.9	15.6
4	21.6	21.2	20.8
5	27.0	26.5	26.0
6	32.4	31.8	31.2
7	37.8	37.1	36.4
8	43.2	42.4	41.6
9	48.6	47.7	46.8

	51	50	49
1	5.1	5.0	4.9
2	10.2	10.0	9.8
3	15.3	15.0	14.7
4	20.4	20.0	19.6
5	25.5	25.0	24.5
6	30.6	30.0	29.4
7	35.7	35.0	34.3
8	40.8	40.0	39.2
9	45.9	45.0	44.1

	48	47	46
1	4.8	4.7	4.6
2	9.6	9.4	9.2
3	14.4	14.1	13.8
4	19.2	18.8	18.4
5	24.0	23.5	23.0
6	28.8	28.2	27.6
7	33.6	32.9	32.2
8	38.4	37.6	36.8
9	43.2	42.3	41.4

	45	44	43
1	4.5	4.4	4.3
2	9.0	8.8	8.6
3	13.5	13.2	12.9
4	18.0	17.6	17.2
5	22.5	22.0	21.5
6	27.0	26.4	25.8
7	31.5	30.8	30.1
8	36.0	35.2	34.4
9	40.5	39.6	38.7

89°

1°

Angle	log sin	diff.	log tan	com. diff.	log cot	log cos	
1.00°	8.2419	43	8.2419	43	1.7581	9.9999	**89.00°**
1.01°	8.2462	43	8.2462	43	1.7538	9.9999	88.99°
1.02°	8.2505	42	8.2505	43	1.7495	9.9999	88.98°
1.03°	8.2547	42	8.2548	42	1.7452	9.9999	88.97°
1.04°	8.2589	41	8.2590	41	1.7410	9.9999	88.96°
1.05°	8.2630	42	8.2631	41	1.7369	9.9999	88.95°
1.06°	8.2672	40	8.2672	41	1.7328	9.9999	88.94°
1.07°	8.2712	41	8.2713	41	1.7287	9.9999	88.93°
1.08°	8.2753	40	8.2754	40	1.7246	9.9999	88.92°
1.09°	8.2793	39	8.2794	39	1.7206	9.9999	88.91°
1.10°	8.2832	40	8.2833	40	1.7167	9.9999	**88.90°**
1.11°	8.2872	39	8.2873	39	1.7127	9.9999	88.89°
1.12°	8.2911	38	8.2912	38	1.7088	9.9999	88.88°
1.13°	8.2949	39	8.2950	38	1.7050	9.9999	88.87°
1.14°	8.2988	37	8.2988	38	1.7012	9.9999	88.86°
1.15°	8.3025	38	8.3026	38	1.6974	9.9999	88.85°
1.16°	8.3063	37	8.3064	37	1.6936	9.9999	88.84°
1.17°	8.3100	37	8.3101	37	1.6899	9.9999	88.83°
1.18°	8.3137	37	8.3138	37	1.6862	9.9999	88.82°
1.19°	8.3174	36	8.3175	36	1.6825	9.9999	88.81°
1.20°	8.3210	36	8.3211	36	1.6789	9.9999	**88.80°**
1.21°	8.3246	36	8.3247	36	1.6753	9.9999	88.79°
1.22°	8.3282	35	8.3283	35	1.6717	9.9999	88.78°
1.23°	8.3317	36	8.3318	36	1.6682	9.9999	88.77°
1.24°	8.3353	35	8.3354	35	1.6646	9.9999	88.76°
1.25°	8.3388	34	8.3389	34	1.6611	9.9999	88.75°
1.26°	8.3422	34	8.3423	35	1.6577	9.9999	88.74°
1.27°	8.3456	35	8.3458	34	1.6542	9.9999	88.73°
1.28°	8.3491	33	8.3492	33	1.6508	9.9999	88.72°
1.29°	8.3524	34	8.3525	34	1.6475	9.9999	88.71°
1.30°	8.3558	33	8.3559	33	1.6441	9.9999	**88.70°**
1.31°	8.3591	33	8.3592	33	1.6408	9.9999	88.69°
1.32°	8.3624	33	8.3625	33	1.6375	9.9999	88.68°
1.33°	8.3657	32	8.3658	33	1.6342	9.9999	88.67°
1.34°	8.3689	33	8.3691	32	1.6309	9.9999	88.66°
1.35°	8.3722	32	8.3723	32	1.6277	9.9999	88.65°
1.36°	8.3754	32	8.3755	32	1.6245	9.9999	88.64°
1.37°	8.3786	31	8.3787	31	1.6213	9.9999	88.63°
1.38°	8.3817	31	8.3818	32	1.6182	9.9999	88.62°
1.39°	8.3848	32	8.3850	31	1.6150	9.9999	88.61°
1.40°	8.3880	31	8.3881	31	1.6119	9.9999	**88.60°**
1.41°	8.3911	30	8.3912	31	1.6088	9.9999	88.59°
1.42°	8.3941	31	8.3943	30	1.6057	9.9999	88.58°
1.43°	8.3972	30	8.3973	30	1.6027	9.9999	88.57°
1.44°	8.4002	30	8.4003	30	1.5997	9.9999	88.56°
1.45°	8.4032	30	8.4033	30	1.5967	9.9999	88.55°
1.46°	8.4062	29	8.4063	30	1.5937	9.9999	88.54°
1.47°	8.4091	30	8.4093	29	1.5907	9.9999	88.53°
1.48°	8.4121	29	8.4122	30	1.5878	9.9999	88.52°
1.49°	8.4150	29	8.4152	29	1.5848	9.9999	88.51°
1.50°	8.4179		8.4181		1.5819	9.9999	**88.50°**
	log cos	diff.	log cot	com. diff.	log tan	log sin	Angle

88°

Prop. Parts — Difference

Extra digit	43	42		41	40		39	38		37	36	35	34	33	32	31	30	29
1	4.3	4.2		4.1	4.0		3.9	3.8		3.7	3.6	3.5	3.4	3.3	3.2	3.1	3.0	2.9
2	8.6	8.4		8.2	8.0		7.8	7.6		7.4	7.2	7.0	6.8	6.6	6.4	6.2	6.0	5.8
3	12.9	12.6		12.3	12.0		11.7	11.4		11.1	10.8	10.5	10.2	9.9	9.6	9.3	9.0	8.7
4	17.2	16.8		16.4	16.0		15.6	15.2		14.8	14.4	14.0	13.6	13.2	12.8	12.4	12.0	11.6
5	21.5	21.0		20.5	20.0		19.5	19.0		18.5	18.0	17.5	17.0	16.5	16.0	15.5	15.0	14.5
6	25.8	25.2		24.6	24.0		23.4	22.8		22.2	21.6	21.0	20.4	19.8	19.2	18.6	18.0	17.4
7	30.1	29.4		28.7	28.0		27.3	26.6		25.9	25.2	24.5	23.8	23.1	22.4	21.7	21.0	20.3
8	34.4	33.6		32.8	32.0		31.2	30.4		29.6	28.8	28.0	27.2	26.4	25.6	24.8	24.0	23.2
9	38.7	37.8		36.9	36.0		35.1	34.2		33.3	32.4	31.5	30.6	29.7	28.8	27.9	27.0	26.1

1°

Angle	log sin	diff.	log tan	com. diff.	log cot	log cos	
1.50°	8.4179		8.4181		1.5819	9.9999	**88.50°**
1.51°	8.4208	29	8.4210	29	1.5790	9.9998	88.49°
1.52°	8.4237	29	8.4238	28	1.5762	9.9998	88.48°
1.53°	8.4265	28	8.4267	29	1.5733	9.9998	88.47°
1.54°	8.4293	28	8.4295	28	1.5705	9.9998	88.46°
1.55°	8.4322	29	8.4323	28	1.5677	9.9998	88.45°
1.56°	8.4349	27	8.4351	28	1.5649	9.9998	88.44°
1.57°	8.4377	28	8.4379	28	1.5621	9.9998	88.43°
1.58°	8.4405	28	8.4406	27	1.5594	9.9998	88.42°
1.59°	8.4432	27	8.4434	28	1.5566	9.9998	88.41°
1.60°	8.4459	27	8.4461	27	1.5539	9.9998	**88.40°**
1.61°	8.4486	27	8.4488	27	1.5512	9.9998	88.39°
1.62°	8.4513	27	8.4515	27	1.5485	9.9998	88.38°
1.63°	8.4540	27	8.4542	27	1.5458	9.9998	88.37°
1.64°	8.4567	27	8.4568	26	1.5432	9.9998	88.36°
1.65°	8.4593	26	8.4595	27	1.5405	9.9998	88.35°
1.66°	8.4619	26	8.4621	26	1.5379	9.9998	88.34°
1.67°	8.4645	26	8.4647	26	1.5353	9.9998	88.33°
1.68°	8.4671	26	8.4673	26	1.5327	9.9998	88.32°
1.69°	8.4697	26	8.4699	26	1.5301	9.9998	88.31°
1.70°	8.4723	26	8.4725	26	1.5275	9.9998	**88.30°**
1.71°	8.4748	25	8.4750	25	1.5250	9.9998	88.29°
1.72°	8.4773	25	8.4775	25	1.5225	9.9998	88.28°
1.73°	8.4799	26	8.4801	26	1.5199	9.9998	88.27°
1.74°	8.4824	25	8.4826	25	1.5174	9.9998	88.26°
1.75°	8.4848	24	8.4851	25	1.5149	9.9998	88.25°
1.76°	8.4873	25	8.4875	24	1.5125	9.9998	88.24°
1.77°	8.4898	25	8.4900	25	1.5100	9.9998	88.23°
1.78°	8.4922	24	8.4924	24	1.5076	9.9998	88.22°
1.79°	8.4947	25	8.4949	25	1.5051	9.9998	88.21°
1.80°	8.4971	24	8.4973	24	1.5027	9.9998	**88.20°**
1.81°	8.4995	24	8.4997	24	1.5003	9.9998	88.19°
1.82°	8.5019	24	8.5021	24	1.4979	9.9998	88.18°
1.83°	8.5043	24	8.5045	24	1.4955	9.9998	88.17°
1.84°	8.5066	23	8.5068	23	1.4932	9.9998	88.16°
1.85°	8.5090	24	8.5092	24	1.4908	9.9998	88.15°
1.86°	8.5113	23	8.5115	23	1.4885	9.9998	88.14°
1.87°	8.5136	23	8.5139	24	1.4861	9.9998	88.13°
1.88°	8.5160	24	8.5162	23	1.4838	9.9998	88.12°
1.89°	8.5183	23	8.5185	23	1.4815	9.9998	88.11°
1.90°	8.5206	23	8.5208	23	1.4792	9.9998	**88.10°**
1.91°	8.5228	22	8.5231	23	1.4769	9.9998	88.09°
1.92°	8.5251	23	8.5253	22	1.4747	9.9998	88.08°
1.93°	8.5274	23	8.5276	23	1.4724	9.9998	88.07°
1.94°	8.5296	22	8.5298	22	1.4702	9.9998	88.06°
1.95°	8.5318	22	8.5321	23	1.4679	9.9997	88.05°
1.96°	8.5340	22	8.5343	22	1.4657	9.9997	88.04°
1.97°	8.5363	23	8.5365	22	1.4635	9.9997	88.03°
1.98°	8.5385	22	8.5387	22	1.4613	9.9997	88.02°
1.99°	8.5406	21	8.5409	22	1.4591	9.9997	88.01°
2.00°	8.5428	22	8.5431	22	1.4569	9.9997	**88.00°**
	log cos	diff.	log cot	com. diff.	log tan	log sin	Angle

88°

Prop. Parts

Difference

Extra digit	29	28
1	2.9	2.8
2	5.8	5.6
3	8.7	8.4
4	11.6	11.2
5	14.5	14.0
6	17.4	16.8
7	20.3	19.6
8	23.2	22.4
9	26.1	25.2

Extra digit	27	26
1	2.7	2.6
2	5.4	5.2
3	8.1	7.8
4	10.8	10.4
5	13.5	13.0
6	16.2	15.6
7	18.9	18.2
8	21.6	20.8
9	24.3	23.4

Extra digit	25	24
1	2.5	2.4
2	5.0	4.8
3	7.5	7.2
4	10.0	9.6
5	12.5	12.0
6	15.0	14.4
7	17.5	16.8
8	20.0	19.2
9	22.5	21.6

Extra digit	23	22
1	2.3	2.2
2	4.6	4.4
3	6.9	6.6
4	9.2	8.8
5	11.5	11.0
6	13.8	13.2
7	16.1	15.4
8	18.4	17.6
9	20.7	19.8

Extra digit	21
1	2.1
2	4.2
3	6.3
4	8.4
5	10.5
6	12.6
7	14.7
8	16.8
9	18.9

2°

Angle	log sin	diff.	log tan	com. diff.	log cot	log cos	
2.00°	8.5428	22	8.5431	22	1.4569	9.9997	**88.00°**
2.01°	8.5450	21	8.5453	21	1.4547	9.9997	87.99°
2.02°	8.5471	22	8.5474	22	1.4526	9.9997	87.98°
2.03°	8.5493	21	8.5496	21	1.4504	9.9997	87.97°
2.04°	8.5514	21	8.5517	21	1.4483	9.9997	87.96°
2.05°	8.5535	22	8.5538	21	1.4462	9.9997	87.95°
2.06°	8.5557	21	8.5559	21	1.4441	9.9997	87.94°
2.07°	8.5578	20	8.5580	21	1.4420	9.9997	87.93°
2.08°	8.5598	21	8.5601	21	1.4399	9.9997	87.92°
2.09°	8.5619	21	8.5622	21	1.4378	9.9997	87.91°
2.10°	8.5640	21	8.5643	21	1.4357	9.9997	**87.90°**
2.11°	8.5661	20	8.5664	21	1.4336	9.9997	87.89°
2.12°	8.5681	21	8.5684	20	1.4316	9.9997	87.88°
2.13°	8.5702	20	8.5705	21	1.4295	9.9997	87.87°
2.14°	8.5722	20	8.5725	20	1.4275	9.9997	87.86°
2.15°	8.5742	20	8.5745	20	1.4255	9.9997	87.85°
2.16°	8.5762	20	8.5765	20	1.4235	9.9997	87.84°
2.17°	8.5782	20	8.5785	20	1.4215	9.9997	87.83°
2.18°	8.5802	20	8.5805	20	1.4195	9.9997	87.82°
2.19°	8.5822	20	8.5825	20	1.4175	9.9997	87.81°
2.20°	8.5842	20	8.5845	20	1.4155	9.9997	**87.80°**
2.21°	8.5862	19	8.5865	19	1.4135	9.9997	87.79°
2.22°	8.5881	20	8.5884	20	1.4116	9.9997	87.78°
2.23°	8.5901	19	8.5904	19	1.4096	9.9997	87.77°
2.24°	8.5920	19	8.5923	20	1.4077	9.9997	87.76°
2.25°	8.5939	20	8.5943	19	1.4057	9.9997	87.75°
2.26°	8.5959	19	8.5962	19	1.4038	9.9997	87.74°
2.27°	8.5978	19	8.5981	19	1.4019	9.9997	87.73°
2.28°	8.5997	19	8.6000	19	1.4000	9.9997	87.72°
2.29°	8.6016	19	8.6019	19	1.3981	9.9997	87.71°
2.30°	8.6035	19	8.6038	19	1.3962	9.9996	**87.70°**
2.31°	8.6054	18	8.6057	19	1.3943	9.9996	87.69°
2.32°	8.6072	19	8.6076	19	1.3924	9.9996	87.68°
2.33°	8.6091	19	8.6095	18	1.3905	9.9996	87.67°
2.34°	8.6110	18	8.6113	19	1.3887	9.9996	87.66°
2.35°	8.6128	19	8.6132	18	1.3868	9.9996	87.65°
2.36°	8.6147	18	8.6150	19	1.3850	9.9996	87.64°
2.37°	8.6165	18	8.6169	18	1.3831	9.9996	87.63°
2.38°	8.6183	18	8.6187	18	1.3813	9.9996	87.62°
2.39°	8.6201	19	8.6205	18	1.3795	9.9996	87.61°
2.40°	8.6220	18	8.6223	19	1.3777	9.9996	**87.60°**
2.41°	8.6238	18	8.6242	18	1.3758	9.9996	87.59°
2.42°	8.6256	18	8.6260	17	1.3740	9.9996	87.58°
2.43°	8.6274	17	8.6277	18	1.3723	9.9996	87.57°
2.44°	8.6291	18	8.6295	18	1.3705	9.9996	87.56°
2.45°	8.6309	18	8.6313	18	1.3687	9.9996	87.55°
2.46°	8.6327	17	8.6331	17	1.3669	9.9996	87.54°
2.47°	8.6344	18	8.6348	18	1.3652	9.9996	87.53°
2.48°	8.6362	17	8.6366	18	1.3634	9.9996	87.52°
2.49°	8.6379	18	8.6384	17	1.3616	9.9996	87.51°
2.50°	8.6397		8.6401		1.3599	9.9996	**87.50°**

log cos	diff.	log cot	com. diff.	log tan	log sin	Angle

87°

Prop. Parts

Extra digit / Difference:

22
1	2.2
2	4.4
3	6.6
4	8.8
5	11.0
6	13.2
7	15.4
8	17.6
9	19.8

21
1	2.1
2	4.2
3	6.3
4	8.4
5	10.5
6	12.6
7	14.7
8	16.8
9	18.9

20
1	2.0
2	4.0
3	6.0
4	8.0
5	10.0
6	12.0
7	14.0
8	16.0
9	18.0

19
1	1.9
2	3.8
3	5.7
4	7.6
5	9.5
6	11.4
7	13.3
8	15.2
9	17.1

18
1	1.8
2	3.6
3	5.4
4	7.2
5	9.0
6	10.8
7	12.6
8	14.4
9	16.2

17
1	1.7
2	3.4
3	5.1
4	6.8
5	8.5
6	10.2
7	11.9
8	13.6
9	15.3

2°

Angle	log sin	diff.	log tan	com. diff.	log cot	log cos	
2.50°	8.6397	17	8.6401	17	1.3599	9.9996	**87.50°**
2.51°	8.6414	17	8.6418	18	1.3582	9.9996	87.49°
2.52°	8.6431	18	8.6436	17	1.3564	9.9996	87.48°
2.53°	8.6449	17	8.6453	17	1.3547	9.9996	87.47°
2.54°	8.6466	17	8.6470	17	1.3530	9.9996	87.46°
2.55°	8.6483	17	8.6487	17	1.3513	9.9996	87.45°
2.56°	8.6500	17	8.6504	17	1.3496	9.9996	87.44°
2.57°	8.6517	17	8.6521	17	1.3479	9.9996	87.43°
2.58°	8.6534	16	8.6538	17	1.3462	9.9996	87.42°
2.59°	8.6550	17	8.6555	16	1.3445	9.9996	87.41°
2.60°	8.6567	17	8.6571	17	1.3429	9.9996	**87.40°**
2.61°	8.6584	16	8.6588	17	1.3412	9.9995	87.39°
2.62°	8.6600	17	8.6605	16	1.3395	9.9995	87.38°
2.63°	8.6617	16	8.6621	17	1.3379	9.9995	87.37°
2.64°	8.6633	17	8.6638	16	1.3362	9.9995	87.36°
2.65°	8.6650	16	8.6654	17	1.3346	9.9995	87.35°
2.66°	8.6666	16	8.6671	16	1.3329	9.9995	87.34°
2.67°	8.6682	17	8.6687	16	1.3313	9.9995	87.33°
2.68°	8.6699	16	8.6703	16	1.3297	9.9995	87.32°
2.69°	8.6715	16	8.6719	17	1.3281	9.9995	87.31°
2.70°	8.6731	16	8.6736	16	1.3264	9.9995	**87.30°**
2.71°	8.6747	16	8.6752	16	1.3248	9.9995	87.29°
2.72°	8.6763	16	8.6768	16	1.3232	9.9995	87.28°
2.73°	8.6779	16	8.6784	16	1.3216	9.9995	87.27°
2.74°	8.6795	15	8.6800	15	1.3200	9.9995	87.26°
2.75°	8.6810	16	8.6815	16	1.3185	9.9995	87.25°
2.76°	8.6826	16	8.6831	16	1.3169	9.9995	87.24°
2.77°	8.6842	16	8.6847	16	1.3153	9.9995	87.23°
2.78°	8.6858	15	8.6863	15	1.3137	9.9995	87.22°
2.79°	8.6873	16	8.6878	16	1.3122	9.9995	87.21°
2.80°	8.6889	15	8.6894	15	1.3106	9.9995	**87.20°**
2.81°	8.6904	16	8.6909	16	1.3091	9.9995	87.19°
2.82°	8.6920	15	8.6925	15	1.3075	9.9995	87.18°
2.83°	8.6935	15	8.6940	16	1.3060	9.9995	87.17°
2.84°	8.6950	15	8.6956	15	1.3044	9.9995	87.16°
2.85°	8.6965	16	8.6971	15	1.3029	9.9995	87.15°
2.86°	8.6981	15	8.6986	15	1.3014	9.9995	87.14°
2.87°	8.6996	15	8.7001	15	1.2999	9.9995	87.13°
2.88°	8.7011	15	8.7016	15	1.2984	9.9995	87.12°
2.89°	8.7026	15	8.7031	15	1.2969	9.9994	87.11°
2.90°	8.7041	15	8.7046	15	1.2954	9.9994	**87.10°**
2.91°	8.7056	15	8.7061	15	1.2939	9.9994	87.09°
2.92°	8.7071	15	8.7076	15	1.2924	9.9994	87.08°
2.93°	8.7086	14	8.7091	15	1.2909	9.9994	87.07°
2.94°	8.7100	15	8.7106	15	1.2894	9.9994	87.06°
2.95°	8.7115	15	8.7121	15	1.2879	9.9994	87.05°
2.96°	8.7130	14	8.7136	14	1.2864	9.9994	87.04°
2.97°	8.7144	15	8.7150	15	1.2850	9.9994	87.03°
2.98°	8.7159	15	8.7165	14	1.2835	9.9994	87.02°
2.99°	8.7174	14	8.7179	15	1.2821	9.9994	87.01°
3.00°	8.7188		8.7194		1.2806	9.9994	**87.00°**
	log cos	diff.	log cot	com. diff.	log tan	log sin	Angle

Prop. Parts

Extra digit	Difference
	18
1	1.8
2	3.6
3	5.4
4	7.2
5	9.0
6	10.8
7	12.6
8	14.4
9	16.2
	17
1	1.7
2	3.4
3	5.1
4	6.8
5	8.5
6	10.2
7	11.9
8	13.6
9	15.3
	16
1	1.6
2	3.2
3	4.8
4	6.4
5	8.0
6	9.6
7	11.2
8	12.8
9	14.4
	15
1	1.5
2	3.0
3	4.5
4	6.0
5	7.5
6	9.0
7	10.5
8	12.0
9	13.5
	14
1	1.4
2	2.8
3	4.2
4	5.6
5	7.0
6	8.4
7	9.8
8	11.2
9	12.6

87°

3°

Angle	log sin	diff.	log tan	com. diff.	log cot	log cos	
3.00°	8.7188	14	8.7194	14	1.2806	9.9994	**87.00°**
3.01°	8.7202	15	8.7208	15	1.2792	9.9994	86.99°
3.02°	8.7217	14	8.7223	14	1.2777	9.9994	86.98°
3.03°	8.7231	14	8.7237	15	1.2763	9.9994	86.97°
3.04°	8.7245	15	8.7252	14	1.2748	9.9994	86.96°
3.05°	8.7260	14	8.7266	14	1.2734	9.9994	86.95°
3.06°	8.7274	14	8.7280	14	1.2720	9.9994	86.94°
3.07°	8.7288	14	8.7294	14	1.2706	9.9994	86.93°
3.08°	8.7302	14	8.7308	15	1.2692	9.9994	86.92°
3.09°	8.7316	14	8.7323	14	1.2677	9.9994	86.91°
3.10°	8.7330	14	8.7337	14	1.2663	9.9994	**86.90°**
3.11°	8.7344	14	8.7351	14	1.2649	9.9994	86.89°
3.12°	8.7358	14	8.7365	14	1.2635	9.9994	86.88°
3.13°	8.7372	14	8.7379	13	1.2621	9.9994	86.87°
3.14°	8.7386	14	8.7392	14	1.2608	9.9993	86.86°
3.15°	8.7400	13	8.7406	14	1.2594	9.9993	86.85°
3.16°	8.7413	14	8.7420	14	1.2580	9.9993	86.84°
3.17°	8.7427	14	8.7434	14	1.2566	9.9993	86.83°
3.18°	8.7441	13	8.7448	13	1.2552	9.9993	86.82°
3.19°	8.7454	14	8.7461	14	1.2539	9.9993	86.81°
3.20°	8.7468	14	8.7475	13	1.2525	9.9993	**86.80°**
3.21°	8.7482	13	8.7488	14	1.2512	9.9993	86.79°
3.22°	8.7495	13	8.7502	13	1.2498	9.9993	86.78°
3.23°	8.7508	14	8.7515	14	1.2485	9.9993	86.77°
3.24°	8.7522	13	8.7529	13	1.2471	9.9993	86.76°
3.25°	8.7535	14	8.7542	14	1.2458	9.9993	86.75°
3.26°	8.7549	13	8.7556	13	1.2444	9.9993	86.74°
3.27°	8.7562	13	8.7569	13	1.2431	9.9993	86.73°
3.28°	8.7575	13	8.7582	14	1.2418	9.9993	86.72°
3.29°	8.7588	14	8.7596	13	1.2404	9.9993	86.71°
3.30°	8.7602	13	8.7609	13	1.2391	9.9993	**86.70°**
3.31°	8.7615	13	8.7622	13	1.2378	9.9993	86.69°
3.32°	8.7628	13	8.7635	13	1.2365	9.9993	86.68°
3.33°	8.7641	13	8.7648	13	1.2352	9.9993	86.67°
3.34°	8.7654	13	8.7661	13	1.2339	9.9993	86.66°
3.35°	8.7667	13	8.7674	13	1.2326	9.9993	86.65°
3.36°	8.7680	13	8.7687	13	1.2313	9.9993	86.64°
3.37°	8.7693	12	8.7700	13	1.2300	9.9992	86.63°
3.38°	8.7705	13	8.7713	13	1.2287	9.9992	86.62°
3.39°	8.7718	13	8.7726	13	1.2274	9.9992	86.61°
3.40°	8.7731	13	8.7739	12	1.2261	9.9992	**86.60°**
3.41°	8.7744	12	8.7751	13	1.2249	9.9992	86.59°
3.42°	8.7756	13	8.7764	13	1.2236	9.9992	86.58°
3.43°	8.7769	13	8.7777	13	1.2223	9.9992	86.57°
3.44°	8.7782	12	8.7790	12	1.2210	9.9992	86.56°
3.45°	8.7794	13	8.7802	13	1.2198	9.9992	86.55°
3.46°	8.7807	12	8.7815	12	1.2185	9.9992	86.54°
3.47°	8.7819	13	8.7827	13	1.2173	9.9992	86.53°
3.48°	8.7832	12	8.7840	12	1.2160	9.9992	86.52°
3.49°	8.7844	13	8.7852	13	1.2148	9.9992	86.51°
3.50°	8.7857		8.7865		1.2135	9.9992	**86.50°**
	log cos	diff.	log cot	com. diff.	log tan	log sin	Angle

Prop. Parts

Extra digit	Difference
	15
1	1.5
2	3.0
3	4.5
4	6.0
5	7.5
6	9.0
7	10.5
8	12.0
9	13.5
	14
1	1.4
2	2.8
3	4.2
4	5.6
5	7.0
6	8.4
7	9.8
8	11.2
9	12.6
	13
1	1.3
2	2.6
3	3.9
4	5.2
5	6.5
6	7.8
7	9.1
8	10.4
9	11.7
	12
1	1.2
2	2.4
3	3.6
4	4.8
5	6.0
6	7.2
7	8.4
8	9.6
9	10.8

86°

3°

Angle	log sin	diff.	log tan	com. diff.	log cot	log cos	
3.50°	8.7857	12	8.7865	12	1.2135	9.9992	**86.50°**
3.51°	8.7869	12	8.7877	13	1.2123	9.9992	86.49°
3.52°	8.7881	13	8.7890	12	1.2110	9.9992	86.48°
3.53°	8.7894	12	8.7902	12	1.2098	9.9992	86.47°
3.54°	8.7906	12	8.7914	13	1.2086	9.9992	86.46°
3.55°	8.7918	12	8.7927	12	1.2073	9.9992	86.45°
3.56°	8.7930	13	8.7939	12	1.2061	9.9992	86.44°
3.57°	8.7943	12	8.7951	12	1.2049	9.9992	86.43°
3.58°	8.7955	12	8.7963	12	1.2037	9.9992	86.42°
3.59°	8.7967	12	8.7975	13	1.2025	9.9991	86.41°
3.60°	8.7979	12	8.7988	12	1.2012	9.9991	**86.40°**
3.61°	8.7991	12	8.8000	12	1.2000	9.9991	86.39°
3.62°	8.8003	12	8.8012	12	1.1988	9.9991	86.38°
3.63°	8.8015	12	8.8024	12	1.1976	9.9991	86.37°
3.64°	8.8027	12	8.8036	12	1.1964	9.9991	86.36°
3.65°	8.8039	12	8.8048	11	1.1952	9.9991	86.35°
3.66°	8.8051	11	8.8059	12	1.1941	9.9991	86.34°
3.67°	8.8062	12	8.8071	12	1.1929	9.9991	86.33°
3.68°	8.8074	12	8.8083	12	1.1917	9.9991	86.32°
3.69°	8.8086	12	8.8095	12	1.1905	9.9991	86.31°
3.70°	8.8098	11	8.8107	12	1.1893	9.9991	**86.30°**
3.71°	8.8109	12	8.8119	11	1.1881	9.9991	86.29°
3.72°	8.8121	12	8.8130	12	1.1870	9.9991	86.28°
3.73°	8.8133	11	8.8142	12	1.1858	9.9991	86.27°
3.74°	8.8144	12	8.8154	11	1.1846	9.9991	86.26°
3.75°	8.8156	12	8.8165	12	1.1835	9.9991	86.25°
3.76°	8.8168	11	8.8177	11	1.1823	9.9991	86.24°
3.77°	8.8179	12	8.8188	12	1.1812	9.9991	86.23°
3.78°	8.8191	11	8.8200	12	1.1800	9.9991	86.22°
3.79°	8.8202	11	8.8212	11	1.1788	9.9990	86.21°
3.80°	8.8213	12	8.8223	11	1.1777	9.9990	**86.20°**
3.81°	8.8225	11	8.8234	12	1.1766	9.9990	86.19°
3.82°	8.8236	12	8.8246	11	1.1754	9.9990	86.18°
3.83°	8.8248	11	8.8257	12	1.1743	9.9990	86.17°
3.84°	8.8259	11	8.8269	11	1.1731	9.9990	86.16°
3.85°	8.8270	11	8.8280	11	1.1720	9.9990	86.15°
3.86°	8.8281	12	8.8291	11	1.1709	9.9990	86.14°
3.87°	8.8293	11	8.8302	12	1.1698	9.9990	86.13°
3.88°	8.8304	11	8.8314	11	1.1686	9.9990	86.12°
3.89°	8.8315	11	8.8325	11	1.1675	9.9990	86.11°
3.90°	8.8326	11	8.8336	11	1.1664	9.9990	**86.10°**
3.91°	8.8337	11	8.8347	11	1.1653	9.9990	86.09°
3.92°	8.8348	11	8.8358	12	1.1642	9.9990	86.08°
3.93°	8.8359	11	8.8370	11	1.1630	9.9990	86.07°
3.94°	8.8370	11	8.8381	11	1.1619	9.9990	86.06°
3.95°	8.8381	11	8.8392	11	1.1608	9.9990	86.05°
3.96°	8.8392	11	8.8403	11	1.1597	9.9990	86.04°
3.97°	8.8403	11	8.8414	11	1.1586	9.9990	86.03°
3.98°	8.8414	11	8.8425	11	1.1575	9.9990	86.02°
3.99°	8.8425	11	8.8436	11	1.1564	9.9989	86.01°
4.00°	8.8436		8.8446	10	1.1554	9.9989	**86.00°**
	log cos	diff.	log cot	com. diff.	log tan	log sin	Angle

Prop. Parts

Extra digit	Difference
	13
1	1.3
2	2.6
3	3.9
4	5.2
5	6.5
6	7.8
7	9.1
8	10.4
9	11.7
	12
1	1.2
2	2.4
3	3.6
4	4.8
5	6.0
6	7.2
7	8.4
8	9.6
9	10.8
	11
1	1.1
2	2.2
3	3.3
4	4.4
5	5.5
6	6.6
7	7.7
8	8.8
9	9.9
	10
1	1.0
2	2.0
3	3.0
4	4.0
5	5.0
6	6.0
7	7.0
8	8.0
9	9.0

86°

4°

Angle	log sin	diff.	log tan	com. diff.	log cot	log cos	
4.00°	8.8436		8.8446		1.1554	9.9989	**86.00°**
4.01°	8.8447	11	8.8457	11	1.1543	9.9989	85.99°
4.02°	8.8457	10	8.8468	11	1.1532	9.9989	85.98°
4.03°	8.8468	11	8.8479	11	1.1521	9.9989	85.97°
4.04°	8.8479	11	8.8490	11	1.1510	9.9989	85.96°
4.05°	8.8490	11	8.8501	11	1.1499	9.9989	85.95°
4.06°	8.8500	10	8.8511	10	1.1489	9.9989	85.94°
4.07°	8.8511	11	8.8522	11	1.1478	9.9989	85.93°
4.08°	8.8522	11	8.8533	11	1.1467	9.9989	85.92°
4.09°	8.8532	10	8.8543	10	1.1457	9.9989	85.91°
4.10°	8.8543	11	8.8554	11	1.1446	9.9989	**85.90°**
4.11°	8.8553	10	8.8565	11	1.1435	9.9989	85.89°
4.12°	8.8564	11	8.8575	10	1.1425	9.9989	85.88°
4.13°	8.8575	11	8.8586	11	1.1414	9.9989	85.87°
4.14°	8.8585	10	8.8596	10	1.1404	9.9989	85.86°
4.15°	8.8595	10	8.8607	11	1.1393	9.9989	85.85°
4.16°	8.8606	11	8.8617	10	1.1383	9.9989	85.84°
4.17°	8.8616	10	8.8628	11	1.1372	9.9988	85.83°
4.18°	8.8627	11	8.8638	10	1.1362	9.9988	85.82°
4.19°	8.8637	10	8.8649	11	1.1351	9.9988	85.81°
4.20°	8.8647	10	8.8659	10	1.1341	9.9988	**85.80°**
4.21°	8.8658	11	8.8669	10	1.1331	9.9988	85.79°
4.22°	8.8668	10	8.8680	11	1.1320	9.9988	85.78°
4.23°	8.8678	10	8.8690	10	1.1310	9.9988	85.77°
4.24°	8.8688	10	8.8700	10	1.1300	9.9988	85.76°
4.25°	8.8699	11	8.8711	11	1.1289	9.9988	85.75°
4.26°	8.8709	10	8.8721	10	1.1279	9.9988	85.74°
4.27°	8.8719	10	8.8731	10	1.1269	9.9988	85.73°
4.28°	8.8729	10	8.8741	10	1.1259	9.9988	85.72°
4.29°	8.8739	10	8.8751	10	1.1249	9.9988	85.71°
4.30°	8.8749	10	8.8762	11	1.1238	9.9988	**85.70°**
4.31°	8.8759	10	8.8772	10	1.1228	9.9988	85.69°
4.32°	8.8769	10	8.8782	10	1.1218	9.9988	85.68°
4.33°	8.8780	11	8.8792	10	1.1208	9.9988	85.67°
4.34°	8.8790	10	8.8802	10	1.1198	9.9988	85.66°
4.35°	8.8799	9	8.8812	10	1.1188	9.9987	85.65°
4.36°	8.8809	10	8.8822	10	1.1178	9.9987	85.64°
4.37°	8.8819	10	8.8832	10	1.1168	9.9987	85.63°
4.38°	8.8829	10	8.8842	10	1.1158	9.9987	85.62°
4.39°	8.8839	10	8.8852	10	1.1148	9.9987	85.61°
4.40°	8.8849	10	8.8862	10	1.1138	9.9987	**85.60°**
4.41°	8.8859	10	8.8872	10	1.1128	9.9987	85.59°
4.42°	8.8869	10	8.8882	10	1.1118	9.9987	85.58°
4.43°	8.8878	9	8.8891	9	1.1109	9.9987	85.57°
4.44°	8.8888	10	8.8901	10	1.1099	9.9987	85.56°
4.45°	8.8898	10	8.8911	10	1.1089	9.9987	85.55°
4.46°	8.8908	10	8.8921	10	1.1079	9.9987	85.54°
4.47°	8.8917	9	8.8931	10	1.1069	9.9987	85.53°
4.48°	8.8927	10	8.8940	9	1.1060	9.9987	85.52°
4.49°	8.8937	10	8.8950	10	1.1050	9.9987	85.51°
4.50°	8.8946	9	8.8960	10	1.1040	9.9987	**85.50°**

	log cos	diff.	log cot	com. diff.	log tan	log sin	Angle

Prop. Parts

Extra digit / Difference

Extra digit	11
1	1.1
2	2.2
3	3.3
4	4.4
5	5.5
6	6.6
7	7.7
8	8.8
9	9.9

Extra digit	10
1	1.0
2	2.0
3	3.0
4	4.0
5	5.0
6	6.0
7	7.0
8	8.0
9	9.0

Extra digit	9
1	0.9
2	1.8
3	2.7
4	3.6
5	4.5
6	5.4
7	6.3
8	7.2
9	8.1

85°

4°

Angle	log sin	diff.	log tan	com. diff.	log cot	log cos	
4.50°	8.8946	10	8.8960	10	1.1040	9.9987	**85.50°**
4.51°	8.8956	10	8.8970	9	1.1030	9.9987	85.49°
4.52°	8.8966	9	8.8979	10	1.1021	9.9986	85.48°
4.53°	8.8975	10	8.8989	9	1.1011	9.9986	85.47°
4.54°	8.8985	9	8.8998	10	1.1002	9.9986	85.46°
4.55°	8.8994	10	8.9008	10	1.0992	9.9986	85.45°
4.56°	8.9004	9	8.9018	9	1.0982	9.9986	85.44°
4.57°	8.9013	10	8.9027	10	1.0973	9.9986	85.43°
4.58°	8.9023	9	8.9037	9	1.0963	9.9986	85.42°
4.59°	8.9032	10	8.9046	10	1.0954	9.9986	85.41°
4.60°	8.9042	9	8.9056	9	1.0944	9.9986	**85.40°**
4.61°	8.9051	9	8.9065	10	1.0935	9.9986	85.39°
4.62°	8.9060	10	8.9075	9	1.0925	9.9986	85.38°
4.63°	8.9070	9	8.9084	9	1.0916	9.9986	85.37°
4.64°	8.9079	10	8.9093	10	1.0907	9.9986	85.36°
4.65°	8.9089	9	8.9103	9	1.0897	9.9986	85.35°
4.66°	8.9098	9	8.9112	10	1.0888	9.9986	85.34°
4.67°	8.9107	9	8.9122	9	1.0878	9.9986	85.33°
4.68°	8.9116	10	8.9131	9	1.0869	9.9985	85.32°
4.69°	8.9126	9	8.9140	10	1.0860	9.9985	85.31°
4.70°	8.9135	9	8.9150	9	1.0850	9.9985	**85.30°**
4.71°	8.9144	9	8.9159	9	1.0841	9.9985	85.29°
4.72°	8.9153	9	8.9168	9	1.0832	9.9985	85.28°
4.73°	8.9162	10	8.9177	9	1.0823	9.9985	85.27°
4.74°	8.9172	9	8.9186	10	1.0814	9.9985	85.26°
4.75°	8.9181	9	8.9196	9	1.0804	9.9985	85.25°
4.76°	8.9190	9	8.9205	9	1.0795	9.9985	85.24°
4.77°	8.9199	9	8.9214	9	1.0786	9.9985	85.23°
4.78°	8.9208	9	8.9223	9	1.0777	9.9985	85.22°
4.79°	8.9217	9	8.9232	9	1.0768	9.9985	85.21°
4.80°	8.9226	9	8.9241	9	1.0759	9.9985	**85.20°**
4.81°	8.9235	9	8.9250	10	1.0750	9.9985	85.19°
4.82°	8.9244	9	8.9260	9	1.0740	9.9985	85.18°
4.83°	8.9253	9	8.9269	9	1.0731	9.9985	85.17°
4.84°	8.9262	9	8.9278	9	1.0722	9.9984	85.16°
4.85°	8.9271	9	8.9287	9	1.0713	9.9984	85.15°
4.86°	8.9280	9	8.9296	9	1.0704	9.9984	85.14°
4.87°	8.9289	9	8.9305	8	1.0695	9.9984	85.13°
4.88°	8.9298	9	8.9313	9	1.0687	9.9984	85.12°
4.89°	8.9307	8	8.9322	9	1.0678	9.9984	85.11°
4.90°	8.9315	9	8.9331	9	1.0669	9.9984	**85.10°**
4.91°	8.9324	9	8.9340	9	1.0660	9.9984	85.09°
4.92°	8.9333	9	8.9349	9	1.0651	9.9984	85.08°
4.93°	8.9342	9	8.9358	9	1.0642	9.9984	85.07°
4.94°	8.9351	8	8.9367	9	1.0633	9.9984	85.06°
4.95°	8.9359	9	8.9376	8	1.0624	9.9984	85.05°
4.96°	8.9368	9	8.9384	9	1.0616	9.9984	85.04°
4.97°	8.9377	9	8.9393	9	1.0607	9.9984	85.03°
4.98°	8.9386	8	8.9402	9	1.0598	9.9984	85.02°
4.99°	8.9394	9	8.9411	9	1.0589	9.9984	85.01°
5.00°	8.9403		8.9420		1.0580	9.9983	**85.00°**
	log cos	diff.	log cot	com. diff.	log tan	log sin	Angle

Prop. Parts

Extra digit	Difference
	10
1	1.0
2	2.0
3	3.0
4	4.0
5	5.0
6	6.0
7	7.0
8	8.0
9	9.0
	9
1	0.9
2	1.8
3	2.7
4	3.6
5	4.5
6	5.4
7	6.3
8	7.2
9	8.1
	8
1	0.8
2	1.6
3	2.4
4	3.2
5	4.0
6	4.8
7	5.6
8	6.4
9	7.2

85°

5°–10°

Angle	log sin	diff.	log tan	com. diff.	log cot	log cos	diff.	
5.0°	8.9403	86	8.9420	86	1.0580	9.9983	0	**85.0°**
5.1°	8.9489	84	8.9506	85	1.0494	9.9983	1	84.9°
5.2°	8.9573	82	8.9591	83	1.0409	9.9982	1	84.8°
5.3°	8.9655	81	8.9674	82	1.0326	9.9981	0	84.7°
5.4°	8.9736	80	8.9756	80	1.0244	9.9981	1	84.6°
5.5°	8.9816	78	8.9836	79	1.0164	9.9980	1	84.5°
5.6°	8.9894	76	8.9915	77	1.0085	9.9979	1	84.4°
5.7°	8.9970	76	8.9992	76	1.0008	9.9978	0	84.3°
5.8°	9.0046	74	9.0068	75	0.9932	9.9978	1	84.2°
5.9°	9.0120	72	9.0143	73	0.9857	9.9977	1	84.1°
6.0°	9.0192	72	9.0216	73	0.9784	9.9976	1	**84.0°**
6.1°	9.0264	70	9.0289	71	0.9711	9.9975	0	83.9°
6.2°	9.0334	69	9.0360	70	0.9640	9.9975	1	83.8°
6.3°	9.0403	69	9.0430	69	0.9570	9.9974	1	83.7°
6.4°	9.0472	67	9.0499	68	0.9501	9.9973	1	83.6°
6.5°	9.0539	66	9.0567	66	0.9433	9.9972	1	83.5°
6.6°	9.0605	65	9.0633	66	0.9367	9.9971	1	83.4°
6.7°	9.0670	64	9.0699	65	0.9301	9.9970	1	83.3°
6.8°	9.0734	63	9.0764	64	0.9236	9.9969	1	83.2°
6.9°	9.0797	62	9.0828	63	0.9172	9.9968	0	83.1°
7.0°	9.0859	61	9.0891	63	0.9109	9.9968	1	**83.0°**
7.1°	9.0920	61	9.0954	61	0.9046	9.9967	1	82.9°
7.2°	9.0981	59	9.1015	61	0.8985	9.9966	1	82.8°
7.3°	9.1040	59	9.1076	59	0.8924	9.9965	1	82.7°
7.4°	9.1099	58	9.1135	59	0.8865	9.9964	1	82.6°
7.5°	9.1157	57	9.1194	58	0.8806	9.9963	1	82.5°
7.6°	9.1214	57	9.1252	58	0.8748	9.9962	1	82.4°
7.7°	9.1271	55	9.1310	57	0.8690	9.9961	1	82.3°
7.8°	9.1326	55	9.1367	56	0.8633	9.9960	1	82.2°
7.9°	9.1381	55	9.1423	55	0.8577	9.9959	1	82.1°
8.0°	9.1436	53	9.1478	55	0.8522	9.9958	2	**82.0°**
8.1°	9.1489	53	9.1533	54	0.8467	9.9956	1	81.9°
8.2°	9.1542	52	9.1587	53	0.8413	9.9955	1	81.8°
8.3°	9.1594	52	9.1640	53	0.8360	9.9954	1	81.7°
8.4°	9.1646	51	9.1693	52	0.8307	9.9953	1	81.6°
8.5°	9.1697	50	9.1745	52	0.8255	9.9952	1	81.5°
8.6°	9.1747	50	9.1797	51	0.8203	9.9951	1	81.4°
8.7°	9.1797	50	9.1848	50	0.8152	9.9950	1	81.3°
8.8°	9.1847	48	9.1898	50	0.8102	9.9949	2	81.2°
8.9°	9.1895	48	9.1948	49	0.8052	9.9947	1	81.1°
9.0°	9.1943	48	9.1997	49	0.8003	9.9946	1	**81.0°**
9.1°	9.1991	47	9.2046	48	0.7954	9.9945	1	80.9°
9.2°	9.2038	47	9.2094	48	0.7906	9.9944	1	80.8°
9.3°	9.2085	46	9.2142	47	0.7858	9.9943	2	80.7°
9.4°	9.2131	45	9.2189	47	0.7811	9.9941	1	80.6°
9.5°	9.2176	45	9.2236	46	0.7764	9.9940	1	80.5°
9.6°	9.2221	45	9.2282	46	0.7718	9.9939	2	80.4°
9.7°	9.2266	44	9.2328	46	0.7672	9.9937	1	80.3°
9.8°	9.2310	43	9.2374	45	0.7626	9.9936	1	80.2°
9.9°	9.2353	44	9.2419	44	0.7581	9.9935	1	80.1°
10.0°	9.2397		9.2463		0.7537	9.9934		**80.0°**
	log cos	diff.	log cot	com. diff.	log tan	log sin	diff.	Angle

80°–85°

Prop. Parts — Extra digit / Difference

Extra digit	62	61	60
1	6.2	6.1	6.0
2	12.4	12.2	12.0
3	18.6	18.3	18.0
4	24.8	24.4	24.0
5	31.0	30.5	30.0
6	37.2	36.6	36.0
7	43.4	42.7	42.0
8	49.6	48.8	48.0
9	55.8	54.9	54.0

Extra digit	59	58	57
1	5.9	5.8	5.7
2	11.8	11.6	11.4
3	17.7	17.4	17.1
4	23.6	23.2	22.8
5	29.5	29.0	28.5
6	35.4	34.8	34.2
7	41.3	40.6	39.9
8	47.2	46.4	45.6
9	53.1	52.2	51.3

Extra digit	56	55	54
1	5.6	5.5	5.4
2	11.2	11.0	10.8
3	16.8	16.5	16.2
4	22.4	22.0	21.6
5	28.0	27.5	27.0
6	33.6	33.0	32.4
7	39.2	38.5	37.8
8	44.8	44.0	43.2
9	50.4	49.5	48.6

Extra digit	53	52	51
1	5.3	5.2	5.1
2	10.6	10.4	10.2
3	15.9	15.6	15.3
4	21.2	20.8	20.4
5	26.5	26.0	25.5
6	31.8	31.2	30.6
7	37.1	36.4	35.7
8	42.4	41.6	40.8
9	47.7	46.8	45.9

Extra digit	50	49	48
1	5.0	4.9	4.8
2	10.0	9.8	9.6
3	15.0	14.7	14.4
4	20.0	19.6	19.2
5	25.0	24.5	24.0
6	30.0	29.4	28.8
7	35.0	34.3	33.6
8	40.0	39.2	38.4
9	45.0	44.1	43.2

Extra digit	47	46	45
1	4.7	4.6	4.5
2	9.4	9.2	9.0
3	14.1	13.8	13.5
4	18.8	18.4	18.0
5	23.5	23.0	22.5
6	28.2	27.6	27.0
7	32.9	32.2	31.5
8	37.6	36.8	36.0
9	42.3	41.4	40.5

10°–15°

Angle	log sin	diff.	log tan	com. diff.	log cot	log cos	diff.	
10.0°	9.2397	42	9.2463	44	0.7537	9.9934	2	**80.0°**
10.1°	9.2439	43	9.2507	44	0.7493	9.9932	1	79.9°
10.2°	9.2482	42	9.2551	43	0.7449	9.9931	2	79.8°
10.3°	9.2524	41	9.2594	43	0.7406	9.9929	1	79.7°
10.4°	9.2565	41	9.2637	43	0.7363	9.9928	1	79.6°
10.5°	9.2606	41	9.2680	42	0.7320	9.9927	2	79.5°
10.6°	9.2647	40	2.2722	42	0.7278	9.9925	1	79.4°
10.7°	9.2687	40	9.2764	41	0.7236	9.9924	2	79.3°
10.8°	9.2727	40	9.2805	41	0.7195	9.9922	1	79.2°
10.9°	9.2767	39	9.2846	41	0.7154	9.9921	2	79.1°
11.0°	9.2806	39	9.2887	40	0.7113	9.9919	1	**79.0°**
11.1°	9.2845	38	9.2927	40	0.7073	9.9918	2	78.9°
11.2°	9.2883	38	9.2967	39	0.7033	9.9916	1	78.8°
11.3°	9.2921	38	9.3006	40	0.6994	9.9915	2	78.7°
11.4°	9.2959	38	9.3046	39	0.6954	9.9913	1	78.6°
11.5°	9.2997	37	9.3085	38	0.6915	9.9912	2	78.5°
11.6°	9.3034	36	9.3123	39	0.6877	9.9910	1	78.4°
11.7°	9.3070	37	9.3162	38	0.6838	9.9909	2	78.3°
11.8°	9.3107	36	9.3200	37	0.6800	9.9907	1	78.2°
11.9°	9.3143	36	9.3237	38	0.6763	9.9906	2	78.1°
12.0°	9.3179	35	9.3275	37	0.6725	9.9904	2	**78.0°**
12.1°	9.3214	36	9.3312	37	0.6688	9.9902	1	77.9°
12.2°	9.3250	34	9.3349	36	0.6651	9.9901	2	77.8°
12.3°	9.3284	35	9.3385	37	0.6615	9.9899	2	77.7°
12.4°	9.3319	34	9.3422	36	0.6578	9.9897	1	77.6°
12.5°	9.3353	34	9.3458	35	0.6542	9.9896	2	77.5°
12.6°	9.3387	34	9.3493	36	0.6507	9.9894	2	77.4°
12.7°	9.3421	34	9.3529	35	0.6471	9.9892	1	77.3°
12.8°	9.3455	33	9.3564	35	0.6436	9.9891	2	77.2°
12.9°	9.3488	33	9.3599	35	0.6401	9.9889	2	77.1°
13.0°	9.3521	33	9.3634	34	0.6366	9.9887	2	**77.0°**
13.1°	9.3554	32	9.3668	34	0.6332	9.9885	1	76.9°
13.2°	9.3586	32	9.3702	34	0.6298	9.9884	2	76.8°
13.3°	9.3618	32	9.3736	34	0.6264	9.9882	2	76.7°
13.4°	9.3650	32	9.3770	34	0.6230	9.9880	2	76.6°
13.5°	9.3682	31	9.3804	33	0.6196	9.9878	2	76.5°
13.6°	9.3713	32	9.3837	33	0.6163	9.9876	1	76.4°
13.7°	9.3745	30	9.3870	33	0.6130	9.9875	2	76.3°
13.8°	9.3775	31	9.3903	32	0.6097	9.9873	2	76.2°
13.9°	9.3806	31	9.3935	33	0.6065	9.9871	2	76.1°
14.0°	9.3837	30	9.3968	32	0.6032	9.9869	2	**76.0°**
14.1°	9.3867	30	9.4000	32	0.6000	9.9867	2	75.9°
14.2°	9.3897	30	9.4032	32	0.5968	9.9865	2	75.8°
14.3°	9.3927	30	9.4064	31	0.5936	9.9863	2	75.7°
14.4°	9.3957	29	9.4095	32	0.5905	9.9861	2	75.6°
14.5°	9.3986	29	9.4127	31	0.5873	9.9859	2	75.5°
14.6°	9.4015	29	9.4158	31	0.5842	9.9857	2	75.4°
14.7°	9.4044	29	9.4189	31	0.5811	9.9855	2	75.3°
14.8°	9.4073	29	9.4220	30	0.5780	9.9853	2	75.2°
14.9°	9.4102	28	9.4250	31	0.5750	9.9851	2	75.1°
15.0°	9.4130		9.4281		0.5719	9.9849		**75.0°**
	log cos	diff.	log cot	com. diff.	log tan	log sin	diff.	Angle

Prop. Parts (Extra digit) — Difference

1	44	43	42
1	4.4	4.3	4.2
2	8.8	8.6	8.4
3	13.2	12.9	12.6
4	17.6	17.2	16.8
5	22.0	21.5	21.0
6	26.4	25.8	25.2
7	30.8	30.1	29.4
8	35.2	34.5	33.6
9	39.6	38.8	37.8

	41	40	39
1	4.1	4.0	3.9
2	8.2	8.0	7.8
3	12.3	12.0	11.7
4	16.4	16.0	15.6
5	20.5	20.0	19.5
6	24.6	24.0	23.4
7	28.7	28.0	27.3
8	32.8	32.0	31.2
9	36.9	36.0	35.1

	38	37	36
1	3.8	3.7	3.6
2	7.6	7.4	7.2
3	11.4	11.1	10.8
4	15.2	14.8	14.4
5	19.0	18.5	18.0
6	22.8	22.2	21.6
7	26.6	25.9	25.2
8	30.4	29.6	28.8
9	34.2	33.3	32.4

	35	34	33
1	3.5	3.4	3.3
2	7.0	6.8	6.6
3	10.5	10.2	9.9
4	14.0	13.6	13.2
5	17.5	17.0	16.5
6	21.0	20.4	19.8
7	24.5	23.8	23.1
8	28.0	27.2	26.4
9	31.5	30.6	29.7

	32	31	30
1	3.2	3.1	3.0
2	6.4	6.2	6.0
3	9.6	9.3	9.0
4	12.8	12.4	12.0
5	16.0	15.5	15.0
6	19.2	18.6	18.0
7	22.4	21.7	21.0
8	25.6	24.8	24.0
9	28.8	27.9	27.0

	29	28	2
1	2.9	2.8	0.2
2	5.8	5.6	0.4
3	8.7	8.4	0.6
4	11.6	11.2	0.8
5	14.5	14.0	1.0
6	17.4	16.8	1.2
7	20.3	19.6	1.4
8	23.2	22.4	1.6
9	26.1	25.2	1.8

75°–80°

15°–20°

Angle	log sin	diff.	log tan	com. diff.	log cot	log cos	diff.	
15.0°	9.4130	28	9.4281	30	0.5719	9.9849	2	**75.0°**
15.1°	9.4158	28	9.4311	30	0.5689	9.9847	2	74.9°
15.2°	9.4186	28	9.4341	30	0.5659	9.9845	2	74.8°
15.3°	9.4214	28	9.4371	29	0.5629	9.9843	2	74.7°
15.4°	9.4242	27	9.4400	30	0.5600	9.9841	2	74.6°
15.5°	9.4269	27	9.4430	29	0.5570	9.9839	2	74.5°
15.6°	9.4296	27	9.4459	29	0.5541	9.9837	2	74.4°
15.7°	9.4323	27	9.4488	29	0.5512	9.9835	2	74.3°
15.8°	9.4350	27	9.4517	29	0.5483	9.9833	2	74.2°
15.9°	9.4377	26	9.4546	29	0.5454	9.9831	3	74.1°
16.0°	9.4403	27	9.4575	28	0.5425	9.9828	2	**74.0°**
16.1°	9.4430	26	9.4603	29	0.5397	9.9826	2	73.9°
16.2°	9.4456	26	9.4632	28	0.5368	9.9824	2	73.8°
16.3°	9.4482	26	9.4660	28	0.5340	9.9822	2	73.7°
16.4°	9.4508	25	9.4688	28	0.5312	9.9820	3	73.6°
16.5°	9.4533	26	9.4716	28	0.5284	9.9817	2	73.5°
16.6°	9.4559	25	9.4744	27	0.5256	9.9815	2	73.4°
16.7°	9.4584	25	9.4771	28	0.5229	9.9813	2	73.3°
16.8°	9.4609	25	9.4799	27	0.5201	9.9811	3	73.2°
16.9°	9.4634	25	9.4826	27	0.5174	9.9808	2	73.1°
17.0°	9.4659	25	9.4853	27	0.5147	9.9806	2	**73.0°**
17.1°	9.4684	25	9.4880	27	0.5120	9.9804	3	72.9°
17.2°	9.4709	24	9.4907	27	0.5093	9.9801	2	72.8°
17.3°	9.4733	24	9.4934	27	0.5066	9.9799	2	72.7°
17.4°	9.4757	24	9.4961	26	0.5039	9.9797	3	72.6°
17.5°	9.4781	24	9.4987	27	0.5013	9.9794	2	72.5°
17.6°	9.4805	24	9.5014	26	0.4986	9.9792	3	72.4°
17.7°	9.4829	24	9.5040	26	0.4960	9.9789	2	72.3°
17.8°	9.4853	23	9.5066	26	0.4934	9.9787	2	72.2°
17.9°	9.4876	24	9.5092	26	0.4908	9.9785	3	72.1°
18.0°	9.4900	23	9.5118	25	0.4882	9.9782	2	**72.0°**
18.1°	9.4923	23	9.5143	26	0.4857	9.9780	3	71.9°
18.2°	9.4946	23	9.5169	26	0.4831	9.9777	2	71.8°
18.3°	9.4969	23	9.5195	25	0.4805	9.9775	3	71.7°
18.4°	9.4992	23	9.5220	25	0.4780	9.9772	2	71.6°
18.5°	9.5015	22	9.5245	25	0.4755	9.9770	3	71.5°
18.6°	9.5037	23	9.5270	25	0.4730	9.9767	3	71.4°
18.7°	9.5060	22	9.5295	25	0.4705	9.9764	2	71.3°
18.8°	9.5082	22	9.5320	25	0.4680	9.9762	3	71.2°
18.9°	9.5104	22	9.5345	25	0.4655	9.9759	2	71.1°
19.0°	9.5126	22	9.5370	24	0.4630	9.9757	3	**71.0°**
19.1°	9.5148	22	9.5394	25	0.4606	9.9754	3	70.9°
19.2°	9.5170	22	9.5419	24	0.4581	9.9751	2	70.8°
19.3°	9.5192	21	9.5443	24	0.4557	9.9749	3	70.7°
19.4°	9.5213	22	9.5467	24	0.4533	9.9746	3	70.6°
19.5°	9.5235	21	9.5491	25	0.4509	9.9743	2	70.5°
19.6°	9.5256	22	9.5516	23	0.4484	9.9741	3	70.4°
19.7°	9.5278	21	9.5539	24	0.4461	9.9738	3	70.3°
19.8°	9.5299	21	9.5563	24	0.4437	9.9735	2	70.2°
19.9°	9.5320	21	9.5587	24	0.4413	9.9733	3	70.1°
20.0°	9.5341		9.5611		0.4389	9.9730		**70.0°**
	log cos	diff.	log cot	com. diff.	log tan	log sin	diff.	Angle

Prop. Parts

Extra digit — Difference

Extra digit	30	29
1	3.0	2.9
2	6.0	5.8
3	9.0	8.7
4	12.0	11.6
5	15.0	14.5
6	18.0	17.4
7	21.0	20.3
8	24.0	23.2
9	27.0	26.1

Extra digit	28	27
1	2.8	2.7
2	5.6	5.4
3	8.4	8.1
4	11.2	10.8
5	14.0	13.5
6	16.8	16.2
7	19.6	18.9
8	22.4	21.6
9	25.2	24.3

Extra digit	26	25
1	2.6	2.5
2	5.2	5.0
3	7.8	7.5
4	10.4	10.0
5	13.0	12.5
6	15.6	15.0
7	18.2	17.5
8	20.8	20.0
9	23.4	22.5

Extra digit	24	23
1	2.4	2.3
2	4.8	4.6
3	7.2	6.9
4	9.6	9.2
5	12.0	11.5
6	14.4	13.8
7	16.8	16.1
8	19.2	18.4
9	21.6	20.7

Extra digit	22	21
1	2.2	2.1
2	4.4	4.2
3	6.6	6.3
4	8.8	8.4
5	11.0	10.5
6	13.2	12.6
7	15.4	14.7
8	17.6	16.8
9	19.8	18.9

70°–75°

20°–25°

Angle	log sin	diff.	log tan	com. diff.	log cot	log cos	diff.	
20.0°	9.5341	20	9.5611	23	0.4389	9.9730	3	**70.0°**
20.1°	9.5361	21	9.5634	24	0.4366	9.9727	3	69.9°
20.2°	9.5382	20	9.5658	23	0.4342	9.9724	2	69.8°
20.3°	9.5402	21	9.5681	23	0.4319	9.9722	3	69.7°
20.4°	9.5423	20	9.5704	23	0.4296	9.9719	3	69.6°
20.5°	9.5443	20	9.5727	23	0.4273	9.9716	3	69.5°
20.6°	9.5463	21	9.5750	23	0.4250	9.9713	3	69.4°
20.7°	9.5484	20	9.5773	23	0.4227	9.9710	3	69.3°
20.8°	9.5504	19	9.5796	23	0.4204	9.9707	3	69.2°
20.9°	9.5523	20	9.5819	23	0.4181	9.9704	2	69.1°
21.0°	9.5543	20	9.5842	22	0.4158	9.9702	3	**69.0°**
21.1°	9.5563	20	9.5864	23	0.4136	9.9699	3	68.9°
21.2°	9.5583	19	9.5887	22	0.4113	9.9696	3	68.8°
21.3°	9.5602	19	9.5909	23	0.4091	9.9693	3	68.7°
21.4°	9.5621	20	9.5932	22	0.4068	9.9690	3	68.6°
21.5°	9.5641	19	9.5954	22	0.4046	9.9687	3	68.5°
21.6°	9.5660	19	9.5976	22	0.4024	9.9684	3	68.4°
21.7°	9.5679	19	9.5998	22	0.4002	9.9681	3	68.3°
21.8°	9.5698	19	9.6020	22	0.3980	9.9678	3	68.2°
21.9°	9.5717	19	9.6042	22	0.3958	9.9675	3	68.1°
22.0°	9.5736	18	9.6064	22	0.3936	9.9672	3	**68.0°**
22.1°	9.5754	19	9.6086	22	0.3914	9.9669	3	67.9°
22.2°	9.5773	19	9.6108	21	0.3892	9.9666	4	67.8°
22.3°	9.5792	18	9.6129	22	0.3871	9.9662	3	67.7°
22.4°	9.5810	18	9.6151	21	0.3849	9.9659	3	67.6°
22.5°	9.5828	19	9.6172	22	0.3828	9.9656	3	67.5°
22.6°	9.5847	18	9.6194	21	0.3806	9.9653	3	67.4°
22.7°	9.5865	18	9.6215	21	0.3785	9.9650	3	67.3°
22.8°	9.5883	18	9.6236	21	0.3764	9.9647	4	67.2°
22.9°	9.5901	18	9.6257	22	0.3743	9.9643	3	67.1°
23.0°	9.5919	18	9.6279	21	0.3721	9.9640	3	**67.0°**
23.1°	9.5937	17	9.6300	21	0.3700	9.9637	3	66.9°
23.2°	9.5954	18	9.6321	20	0.3679	9.9634	3	66.8°
23.3°	9.5972	18	9.6341	21	0.3659	9.9631	4	66.7°
23.4°	9.5990	17	9.6362	21	0.3638	9.9627	3	66.6°
23.5°	9.6007	17	9.6383	21	0.3617	9.9624	3	66.5°
23.6°	9.6024	18	9.6404	20	0.3596	9.9621	4	66.4°
23.7°	9.6042	17	9.6424	21	0.3576	9.9617	3	66.3°
23.8°	9.6059	17	9.6445	20	0.3555	9.9614	3	66.2°
23.9°	9.6076	17	9.6465	21	0.3535	9.9611	4	66.1°
24.0°	9.6093	17	9.6486	20	0.3514	9.9607	3	**66.0°**
24.1°	9.6110	17	9.6506	21	0.3494	9.9604	3	65.9°
24.2°	9.6127	17	9.6527	20	0.3473	9.9601	4	65.8°
24.3°	9.6144	17	9.6547	20	0.3453	9.9597	3	65.7°
24.4°	9.6161	16	9.6567	20	0.3433	9.9594	4	65.6°
24.5°	9.6177	17	9.6587	20	0.3413	9.9590	3	65.5°
24.6°	9.6194	16	9.6607	20	0.3393	9.9587	4	65.4°
24.7°	9.6210	17	9.6627	20	0.3373	9.9583	3	65.3°
24.8°	9.6227	16	9.6647	20	0.3353	9.9580	4	65.2°
24.9°	9.6243	16	9.6667	20	0.3333	9.9576	3	65.1°
25.0°	9.6259	16	9.6687	20	0.3313	9.9573		**65.0°**
	log cos	diff.	log cot	com. diff.	log tan	log sin	diff.	Angle

Prop. Parts

Difference (Extra digit)

Extra digit	23	22
1	2.3	2.2
2	4.6	4.4
3	6.9	6.6
4	9.2	8.8
5	11.5	11.0
6	13.8	13.2
7	16.1	15.4
8	18.4	17.6
9	20.7	19.8

Extra digit	21	20
1	2.1	2.0
2	4.2	4.0
3	6.3	6.0
4	8.4	8.0
5	10.5	10.0
6	12.6	12.0
7	14.7	14.0
8	16.8	16.0
9	18.9	18.0

Extra digit	19	18
1	1.9	1.8
2	3.8	3.6
3	5.7	5.4
4	7.6	7.2
5	9.5	9.0
6	11.4	10.8
7	13.3	12.6
8	15.2	14.4
9	17.1	16.2

Extra digit	17	16
1	1.7	1.6
2	3.4	3.2
3	5.1	4.8
4	6.8	6.4
5	8.5	8.0
6	10.2	9.6
7	11.9	11.2
8	13.6	12.8
9	15.3	14.4

Extra digit	2
1	0.2
2	0.4
3	0.6
4	0.8
5	1.0
6	1.2
7	1.4
8	1.6
9	1.8

Extra digit	3	4
1	0.3	0.4
2	0.6	0.8
3	0.9	1.2
4	1.2	1.6
5	1.5	2.0
6	1.8	2.4
7	2.1	2.8
8	2.4	3.2
9	2.7	3.6

65°–70°

25°–30°

Angle	log sin	diff.	log tan	com. diff.	log cot	log cos	diff.	Angle
25.0°	9.6259	17	9.6687	19	0.3313	9.9573	4	**65.0°**
25.1°	9.6276	16	9.6706	20	0.3294	9.9569	3	64.9°
25.2°	9.6292	16	9.6726	20	0.3274	9.9566	4	64.8°
25.3°	9.6308	16	9.6746	19	0.3254	9.9562	4	64.7°
25.4°	9.6324	16	9.6765	20	0.3235	9.9558	3	64.6°
25.5°	9.6340	16	9.6785	19	0.3215	9.9555	4	64.5°
25.6°	9.6356	15	9.6804	20	0.3196	9.9551	3	64.4°
25.7°	9.6371	16	9.6824	19	0.3176	9.9548	4	64.3°
25.8°	9.6387	16	9.6843	20	0.3157	9.9544	4	64.2°
25.9°	9.6403	15	9.6863	19	0.3137	9.9540	3	64.1°
26.0°	9.6418	16	9.6882	19	0.3118	9.9537	4	**64.0°**
26.1°	9.6434	15	9.6901	19	0.3099	9.9533	4	63.9°
26.2°	9.6449	16	9.6920	19	0.3080	9.9529	4	63.8°
26.3°	9.6465	15	9.6939	19	0.3061	9.9525	3	63.7°
26.4°	9.6480	15	9.6958	19	0.3042	9.9522	4	63.6°
26.5°	9.6495	15	9.6977	19	0.3023	9.9518	4	63.5°
26.6°	9.6510	16	9.6996	19	0.3004	9.9514	4	63.4°
26.7°	9.6526	15	9.7015	19	0.2985	9.9510	4	63.3°
26.8°	9.6541	15	9.7034	19	0.2966	9.9506	3	63.2°
26.9°	9.6556	14	9.7053	19	0.2947	9.9503	4	63.1°
27.0°	9.6570	15	9.7072	18	0.2928	9.9499	4	**63.0°**
27.1°	9.6585	15	9.7090	19	0.2910	9.9495	4	62.9°
27.2°	9.6600	15	9.7109	19	0.2891	9.9491	4	62.8°
27.3°	9.6615	14	9.7128	18	0.2872	9.9487	4	62.7°
27.4°	9.6629	15	9.7146	19	0.2854	9.9483	4	62.6°
27.5°	9.6644	15	9.7165	18	0.2835	9.9479	4	62.5°
27.6°	9.6659	14	9.7183	19	0.2817	9.9475	4	62.4°
27.7°	9.6673	14	9.7202	18	0.2798	9.9471	4	62.3°
27.8°	9.6687	15	9.7220	18	0.2780	9.9467	4	62.2°
27.9°	9.6702	14	9.7238	19	0.2762	9.9463	4	62.1°
28.0°	9.6716	14	9.7257	18	0.2743	9.9459	4	**62.0°**
28.1°	9.6730	14	9.7275	18	0.2725	9.9455	4	61.9°
28.2°	9.6744	15	9.7293	18	0.2707	9.9451	4	61.8°
28.3°	9.6759	14	9.7311	19	0.2689	9.9447	4	61.7°
28.4°	9.6773	14	9.7330	18	0.2670	9.9443	4	61.6°
28.5°	9.6787	14	9.7348	18	0.2652	9.9439	4	61.5°
28.6°	9.6801	13	9.7366	18	0.2634	9.9435	4	61.4°
28.7°	9.6814	14	9.7384	18	0.2616	9.9431	4	61.3°
28.8°	9.6828	14	9.7402	18	0.2598	9.9427	5	61.2°
28.9°	9.6842	14	9.7420	18	0.2580	9.9422	4	61.1°
29.0°	9.6856	13	9.7438	17	0.2562	9.9418	4	**61.0°**
29.1°	9.6869	14	9.7455	18	0.2545	9.9414	4	60.9°
29.2°	9.6883	13	9.7473	18	0.2527	9.9410	4	60.8°
29.3°	9.6896	14	9.7491	18	0.2509	9.9406	5	60.7°
29.4°	9.6910	13	9.7509	17	0.2491	9.9401	4	60.6°
29.5°	9.6923	14	9.7526	18	0.2474	9.9397	4	60.5°
29.6°	9.6937	13	9.7544	18	0.2456	9.9393	5	60.4°
29.7°	9.6950	13	9.7562	17	0.2438	9.9388	4	60.3°
29.8°	9.6963	14	9.7579	18	0.2421	9.9384	4	60.2°
29.9°	9.6977	13	9.7597	17	0.2403	9.9380	5	60.1°
30.0°	9.6990		9.7614		0.2386	9.9375		**60.0°**
	log cos	diff.	log cot	com. diff.	log tan	log sin	diff.	Angle

Prop. Parts

Extra digit / Difference:

Extra digit	20	19
1	2.0	1.9
2	4.0	3.8
3	6.0	5.7
4	8.0	7.6
5	10.0	9.5
6	12.0	11.4
7	14.0	13.3
8	16.0	15.2
9	18.0	17.1

Extra digit	18	17
1	1.8	1.7
2	3.6	3.4
3	5.4	5.1
4	7.2	6.8
5	9.0	8.5
6	10.8	10.2
7	12.6	11.9
8	14.4	13.6
9	16.2	15.3

Extra digit	16	15
1	1.6	1.5
2	3.2	3.0
3	4.8	4.5
4	6.4	6.0
5	8.0	7.5
6	9.6	9.0
7	11.2	10.5
8	12.8	12.0
9	14.4	13.5

Extra digit	14	13
1	1.4	1.3
2	2.8	2.6
3	4.2	3.9
4	5.6	5.2
5	7.0	6.5
6	8.4	7.8
7	9.8	9.1
8	11.2	10.4
9	12.6	11.7

Extra digit	3	4
1	0.3	0.4
2	0.6	0.8
3	0.9	1.2
4	1.2	1.6
5	1.5	2.0
6	1.8	2.4
7	2.1	2.8
8	2.4	3.2
9	2.7	3.6

60°–65°

30°–35°

Angle	log sin	diff.	log tan	com. diff.	log cot	log cos	diff.		Prop. Parts
30.0°	9.6990	13	9.7614	18	0.2386	9.9375	4	**60.0°**	
30.1°	9.7003	13	9.7632	17	0.2368	9.9371	4	59.9°	
30.2°	9.7016	13	9.7649	18	0.2351	9.9367	5	59.8°	
30.3°	9.7029	13	9.7667	17	0.2333	9.9362	4	59.7°	
30.4°	9.7042	13	9.7684	17	0.2316	9.9358	5	59.6°	
30.5°	9.7055	13	9.7701	18	0.2299	9.9353	4	59.5°	
30.6°	9.7068	12	9.7719	17	0.2281	9.9349	5	59.4°	**18** **17**
30.7°	9.7080	13	9.7736	17	0.2264	9.9344	4	59.3°	1 1.8 1.7
30.8°	9.7093	13	9.7753	18	0.2247	9.9340	5	59.2°	2 3.6 3.4
30.9°	9.7106	12	9.7771	17	0.2229	9.9335	4	59.1°	3 5.4 5.1
31.0°	9.7118	13	9.7788	17	0.2212	9.9331	5	**59.0°**	4 7.2 6.8
31.1°	9.7131	13	9.7805	17	0.2195	9.9326	4	58.9°	5 9.0 8.5
31.2°	9.7144	12	9.7822	17	0.2178	9.9322	5	58.8°	6 10.8 10.2
31.3°	9.7156	12	9.7839	17	0.2161	9.9317	5	58.7°	7 12.6 11.9
31.4°	9.7168	13	9.7856	17	0.2144	9.9312	4	58.6°	8 14.4 13.6
31.5°	9.7181	12	9.7873	17	0.2127	9.9308	5	58.5°	9 16.2 15.3
31.6°	9.7193	12	9.7890	17	0.2110	9.9303	5	58.4°	**16**
31.7°	9.7205	13	9.7907	17	0.2093	9.9298	4	58.3°	1 1.6
31.8°	9.7218	12	9.7924	17	0.2076	9.9294	5	58.2°	2 3.2
31.9°	9.7230	12	9.7941	17	0.2059	9.9289	5	58.1°	3 4.8
32.0°	9.7242	12	9.7958	17	0.2042	9.9284	5	**58.0°**	4 6.4
32.1°	9.7254	12	9.7975	17	0.2025	9.9279	4	57.9°	5 8.0
32.2°	9.7266	12	9.7992	16	0.2008	9.9275	5	57.8°	6 9.6
32.3°	9.7278	12	9.8008	17	0.1992	9.9270	5	57.7°	7 11.2
32.4°	9.7290	12	9.8025	17	0.1975	9.9265	5	57.6°	8 12.8
32.5°	9.7302	12	9.8042	17	0.1958	9.9260	5	57.5°	9 14.4
32.6°	9.7314	12	9.8059	16	0.1941	9.9255	4	57.4°	**13** **12**
32.7°	9.7326	12	9.8075	17	0.1925	9.9251	5	57.3°	1 1.3 1.2
32.8°	9.7338	11	9.8092	17	0.1908	9.9246	5	57.2°	2 2.6 2.4
32.9°	9.7349	12	9.8109	16	0.1891	9.9241	5	57.1°	3 3.9 3.6
33.0°	9.7361	12	9.8125	17	0.1875	9.9236	5	**57.0°**	4 5.2 4.8
33.1°	9.7373	11	9.8142	16	0.1858	9.9231	5	56.9°	5 6.5 6.0
33.2°	9.7384	12	9.8158	17	0.1842	9.9226	5	56.8°	6 7.8 7.2
33.3°	9.7396	11	9.8175	16	0.1825	9.9221	5	56.7°	7 9.1 8.4
33.4°	9.7407	12	9.8191	17	0.1809	9.9216	5	56.6°	8 10.4 9.6
33.5°	9.7419	11	9.8208	16	0.1792	9.9211	5	56.5°	9 11.7 10.8
33.6°	9.7430	12	9.8224	17	0.1776	9.9206	5	56.4°	**11**
33.7°	9.7442	11	9.8241	16	0.1759	9.9201	5	56.3°	1 1.1
33.8°	9.7453	11	9.8257	17	0.1743	9.9196	5	56.2°	2 2.2
33.9°	9.7464	12	9.8274	16	0.1726	9.9191	5	56.1°	3 3.3
34.0°	9.7476	11	9.8290	16	0.1710	9.9186	5	**56.0°**	4 4.4
34.1°	9.7487	11	9.8306	17	0.1694	9.9181	6	55.9°	5 5.5
34.2°	9.7498	11	9.8323	16	0.1677	9.9175	5	55.8°	6 6.6
34.3°	9.7509	11	9.8339	16	0.1661	9.9170	5	55.7°	7 7.7
34.4°	9.7520	11	9.8355	16	0.1645	9.9165	5	55.6°	8 8.8
34.5°	9.7531	11	9.8371	17	0.1629	9.9160	5	55.5°	9 9.9
34.6°	9.7542	11	9.8388	16	0.1612	9.9155	6	55.4°	**5** **6**
34.7°	9.7553	11	9.8404	16	0.1596	9.9149	5	55.3°	1 0.5 0.6
34.8°	9.7564	11	9.8420	16	0.1580	9.9144	5	55.2°	2 1.0 1.2
34.9°	9.7575	11	9.8436	16	0.1564	9.9139	5	55.1°	3 1.5 1.8
35.0°	9.7586		9.8452	16	0.1548	9.9134		**55.0°**	4 2.0 2.4
									5 2.5 3.0
									6 3.0 3.6
									7 3.5 4.2
									8 4.0 4.8
									9 4.5 5.4
	log cos	diff.	log cot	com. diff.	log tan	log sin	diff.	Angle	

55°–60°